Rice Crop: Techniques to Increase Yield

Rice Crop: Techniques to Increase Yield

Editor: Radley Harwood

STATES
ACADEMIC PRESS
www.statesacademicpress.com

States Academic Press,
109 South 5th Street,
Brooklyn, NY 11249, USA

Visit us on the World Wide Web at:
www.statesacademicpress.com

ISBN: 978-1-63989-472-7 (Hardback)

Cataloging-in-Publication Data

Rice crop : techniques to increase yield / edited by Radley Harwood.
p. cm.
Includes bibliographical references and index.
ISBN 978-1-63989-472-7
1. Rice. 2. Rice--Yields. 3. Agricultural innovations. I. Harwood, Radley.

SB191.R5 R53 2022
633.18--dc23

Table of Contents

Preface VII

Chapter 1 **Genetic Analysis of Biofortification of Micronutrient Breeding in Rice (*Oryza sativa* L.)** 1
Savitha Palanisamy

Chapter 2 **Abiotic Stress Tolerance in Rice (*Oryza sativa* L.): A Genomics Perspective of Salinity Tolerance** 9
Muhammad Saeed

Chapter 3 **Anther Culture as a Supplementary Tool for Rice Breeding** 21
D.M. Ruwani G. Mayakaduwa and Tara D. Silva

Chapter 4 **Iron Biofortification of Rice: Progress and Prospects** 36
Andrew De-Xian Kok, Low Lee Yoon, Rogayah Sekeli, Wee Chien Yeong, Zetty Norhana Balia Yusof and Lai Kok Song

Chapter 5 **Methanogens Harboring in Rice Rhizosphere Reduce Labile Organic Carbon Compounds to Produce Methane Gas** 56
Prabhat Pramanik and Pil Joo Kim

Chapter 6 **Farmers' Willingness to Cultivate Traditional Rice in Sri Lanka: A Case Study in Anuradhapura District** 73
Ginigaddara Appuhamilage Sanjeewanie Ginigaddara and Sampath Priya Disanayake

Chapter 7 **Phosphorus Efficient Phenotype of Rice** 85
Josefine Kant, Takuma Ishizaki, Juan Pariasca-Tanaka, Terry Rose, Matthias Wissuwa and Michelle Watt

Chapter 8 **Heavy Metal and Mineral Element-Induced Abiotic Stress in Rice Plant** 105
Anitha Mani and Kavitha Sankaranarayanan

Chapter 9 **Improving Rice Grain Quality by Enhancing Accumulation of Iron and Zinc While Minimizing Cadmium and Lead** 136
Lei Gao and Jie Xiong

Chapter 10 **Natural Resistance of Sri Lankan Rice (*Oryza sativa* L.) Varieties to Broad-Spectrum Herbicides (Glyphosate and Glufosinate)** 160
Shyama R. Weerakoon, Seneviratnage Somaratne, E. M. Sachini I. Ekanayaka and Sachithri Munasighe

Chapter 11 **Rice Crop Rotation: A Solution for Weed Management** 178
Ananda Scherner, Fábio Schreiber, André Andres, Germani Concenço, Matheus Bastos Martins and Andressa Pitol

Chapter 12 **Microwave Weed and Soil Treatment in Rice Production** **194**
Muhammad Jamal Khan and Graham Ian Brodie

Permissions

List of Contributors

Index

Preface

Every book is initially just a concept; it takes months of research and hard work to give it the final shape in which the readers receive it. In its early stages, this book also went through rigorous reviewing. The notable contributions made by experts from across the globe were first molded into patterned chapters and then arranged in a sensibly sequential manner to bring out the best results.

Rice is a monocot which is grown as an annual plant in most parts of the world. Rice is the seed of oryza sativa grass species. It can be grown in a variety of soils which have low permeability and are free of water logging. The average rainfall required for its cultivation lies between 100 and 200 cm. Rice is a rich source of thiamine, niacin, iron, riboflavin and calcium. Rice growing ecosystems can be broadly categorized into upland rice ecosystem, rainfed lowland rice ecosystem, irrigated rice ecosystem and flood prone rice ecosystem. This book is a compilation of chapters that discuss the most vital concepts and emerging trends in the field of rice crop. From theories to research to practical applications, case studies related to all contemporary topics of relevance to this field have been included herein. This book is a vital tool for all researching or studying rice crop as it gives incredible insights into the different methods to increase yield.

It has been my immense pleasure to be a part of this project and to contribute my years of learning in such a meaningful form. I would like to take this opportunity to thank all the people who have been associated with the completion of this book at any step.

Editor

1

Genetic Analysis of Biofortification of Micronutrient Breeding in Rice (*Oryza sativa* L.)

Savitha Palanisamy

Abstract

Rice is a staple food for millions of people and has great importance in food and nutritional security. Rice is the second most widely consumed in the world next to wheat. The poorest to the richest person in the world consumes rice in one or other form. New research on the importance of micronutrients, vitamins and proteins aims at biological and genetic enrichment. Vital nutrients that the farmer can grow indefinitely without any additional input to produce nutrient-packed rice grains in a sustainable way is the only feasible way of reaching the malnourished population in India. In the present study, an attempt has been made to improve the nutritional quality of rice.

Keywords: rice, biofortification, iron, zinc, malnutrition

1. Introduction

Rice is the very life and the main staple food for more than 50% of the world's population. It provides more carbohydrate and protein in the average daily diet to supplement with essential micronutrients. The per capita consumption of rice is very high, ranging from 62 to 190 kg/year. Therefore, it is one of the most important crop plants on Earth. It provides 35–75% of the calories consumed by more than three billion Asians. Major advances have occurred in rice production during the last four decades due to the adoption of green revolution technologies. Rice production increased 136%, from 257 million tonnes in 1966 to 600 million tonnes in 2000. Rice production grew at the rate of 2% during 1970–1980 and 1.1% during the 1990s. There has been no substantial increase in rice production during the last four years. Worldwide, rice is cultivated in an area of about 152.99 million hectares with a production and productivity of 418 million metric tonnes and 4.07 tonnes per hectare, respectively. In India, rice is cultivated in 43.7 million hectares with production of 93.35 million metric tonnes and productivity of 3.18

metric tonnes per hectare (USDA, World Agriculture production, 2008). In terms of area under cultivation, India ranks first in the world, and with respect to production, China ranks first. To be healthy, human beings require more than 20 mineral elements and more than 40 nutrients, particularly vitamins and essential amino acids, all of which can be supplied by an appropriate diet. Malnutrition deficiencies common in rice consuming countries are iron (Fe) and zinc (Zn) deficiencies that occur mostly in developing countries.

2. Micronutrient deficiencies

Iron deficiency anemia is by far the most common micronutrient deficiency in the world, affecting more than two billion people. It is estimated that about 79% of the children aged between 6 and 35 months and 56% of women between the age of 15 and 49 years are anemic in India. Iron deficiency during childhood and adolescence impairs physical growth, mental development and learning capacity. In adults, iron deficiency anemia reduces the capacity to do physical labor and increases the risk of women dying during childbirth or in the postpartum period.

Zinc is required as a co-factor in over 300 enzymes and plays critical structural roles in many protein and transcriptional factors. Zinc deficiency is more extensive in developing countries, where more than 60% of the population is at risk. Zn deficiency in older children and adolescent males causes retarded growth and dwarfism, retarded sexual development, impaired sense of taste, poor appetite and mental lethargy.

To address micronutrient deficiencies, nutritionists focus on supplementation, fortification and dietary diversification. Fortified food and food supplements do not reach all those affected in the developing countries, because of weak market infrastructure and high recurring cost. Sustainable solutions to the micronutrient problem in these countries can be developed through agricultural approaches. One such approach is crop diversification and the other is to enhance the level of micronutrients in major staple food crops through plant breeding strategies, that is, biofortification of staple in rice production for both rural and urban people.

"Harvest Plus" is one of the programs to motivate the breed crop varieties to enriching in iron and zinc. The ultimate goal of the biofortification strategy is to reduce mortality and morbidity rates related to micronutrient malnutrition and to increase food security, productivity and the quality of life for poor populations of developing countries.

Exploiting genetic variation in crop plants for micronutrient content is one of the most powerful tools to change the nutrient level of a given diet on a large scale. Genetic studies are required to obtain information on the mode of inheritance of the targeted trait, variability and heritability, which are the true key phenomena determining the efficiency of the breeding programs. Based on the information generated through genetic analysis studies, desirable plants with improved levels of Fe and Zn could be fixed from advanced generation breeding materials. A human being needs 49 essential nutrients for normal metabolic activities. Inadequate consumption of even one of these nutrients will result in adverse metabolic disturbances leading to malnutrition. To consider the micronutrient malnutrition to arising world health consequences, to concentrate poor people to balance the daily diet with micronutrient enriched diet. The major micronutrient deficiencies common in rice-consuming countries are iron and zinc deficiencies.

According to the FAO (2010), lack of micronutrients in food, leading to hidden hunger, is seriously damaging the health of billions of people all over the world. Over 9 billion people are affected by undernourishment and the numbers are on the rise. Nearly 5 billion people from Asia and Pacific region and billions of people in developing and underdeveloped countries suffer due to lack of micronutrients, of which 40% of people are affected by lack of iron and zinc.

The nutritional disorder anemia due to iron deficiency is widespread in rice-consuming countries. South East Asia shows the highest prevalence of anemia in women with over 50% of pregnant women being affected. Anemia lowers work performance. It has been linked to reduced resistance to infection. Severe anemia is a significant cause of maternal deaths, while mild anemia may also affect cognitive functions. The magnitude of micronutrient (Fe and Zn) deficiency is alarming, particularly among children, women of reproductive age and pregnant and lactating women. Zinc is a critical micronutrient needed for structural and functional integrity of biological membrane and for diversification of highly aggressive radicals.

Decrease in zinc concentration in the human body results in a number of cellular disturbances and impairments such as immune dysfunctions and high susceptibility to infection of diseases, retardation of mental development and stunted growth in children.

The main strategies addressing micronutrient disorders are dietary diversification, food fortification, supplementation and biofortification. A sustainable solution for micronutrient deficiency is to enhance the level of micronutrients in major staple food crop through plant breeding strategies, that is, biofortification. According to high yielding cultivars combine effort of enriched micro nutrient rice possible through rice improvement strategies.

Successful biofortification strategies must be initiated with screening of diverse germplasm for desired micronutrient content, followed by suitable breeding methods. Genetic engineering through modification of functional pathways to improve the functional novel genes is another major approach. Genetic studies generally reveal the information on the mode of inheritance of the targeted trait. Variability and heritability are the true key phenomena determining the efficiency of the breeding program. Estimates of heritability would be helped in predicting the parents to advance generation for further selection improvement. It facilitates the breeder in finding out the heritable portion of phenotypic variance for effective selection. Genetic advance is yet another parameter for knowing the quantum of desired genes transferred to the progenies. Information on inter-relationships existing among these traits and their association with yield is also important for suggesting suitable guide for selection. Crop improvement for specific trait has been achieved through effective use of F_2 and F_3 segregating population and fixing desirable combinations [1].

Major micronutrient deficiency affected people to lose the valuable life. Since rice is the principal cereal crop in most rice-growing countries and is the staple food of the world's population, it bears significant impact on human health. Biofortification has emerged as a new way to eradicate micronutrient deficiencies. Biofortification is likely to reach rural households, especially subsistence farmers who grow and consume the harvested cereal grains, which in turn is expected to have impact in an affordable and sustainable manner.

Rice (*Oryza sativa* L.) is basically a starchy crop and it has low nutritional elements compared to millets. Though rice provides 50–80% of the energy intake of the poor, it does not provide enough essential micronutrients to eliminate hidden hunger, iron deficiency anemia (IDA) and zinc deficiency. Sufficient micronutrient in the daily diet is one of the prerequisites for human health. The main strategies addressing micronutrient disorders are dietary diversification, food fortification, supplementation and biofortification. Most of the micronutrient deficiencies can be addressed, to some extent, through biofortification. One of the ways to enhance the micronutrient level in the staple crops is biofortification breeding.

To develop new varieties with high amount of micronutrients in the rice kernel, a population with high variability serves as prime source for effective selection. Particularly, the role played by F_2 segregants, contributing much variability, is highly recognized. The F_2 and F_3 generation is the correct stage for selection in any hybridization program and fixing desirable traits in the early segregants of rice by selecting and evaluating them for desirable characters.

A scrutiny of available literature is invaluable in gaining an insight into the research problem under study. This review helps to acquire broad and general background in the given field or discipline. Comparative views of past approaches and findings can also be had through this compiled information. This could orient researchers in the desired lines of thinking, which is supposed to be a prerequisite for a scientific study.

A sincere attempt has been made to review the available literature relevant to the study and is presented under the following sub-heads:

1. Studies on variability
2. Studies on heritability and genetic advance
3. Studies on association of characters
4. Studies on path coefficient analysis
5. Studies on parent-progeny regression

2.1. Variability

The development of an effective plant breeding program is dependent upon the existence of genetic variability and it is a prerequisite for a plant breeder to work with any crop species. Genetic improvement for quantitative traits can be achieved through a clear understanding of the nature and extent of variability present in the material. The efficiency of selection in any crop largely depends on the magnitude of genetic variability available in the population.

The simple measures of variability partitions the variation into phenotypic, genotypic and environmental components. Phenotypic coefficient of variation (PCV) is the measure of total variability resulting from the genotype, environment and interaction of both. Phenotypic and genotypic coefficients of variation give the real picture of variability concealed in a population.

2.2. Heritability and genetic advance

Knowledge of heritability serves as an effective tool to the plant breeder to estimate the relative importance of the inheritance and environment on the variation observed for a character. The concept of heritability helps to discern whether phenotypic differences among individuals are due to genetic differences or due to environmental causes. Heritability in a narrow sense is defined as that fraction of the observed variance which is caused by additive genetic effect.

Similarly, the estimate of genetic advance or genetic gain for a particular character is an important parameter to evaluate the effectiveness of selection [2].

i. The amount of genetic variability

ii. The intensity of selection

Knowledge of variability, heritability and genetic advance is of great value in both stages. Estimation of heritability along with genetic gain is more useful in predicting the resultant effect through selection of the best individuals.

Estimates of heritability for different traits of economic importance are available on a variety of materials.

2.3. Association of characters

Inter-relationship of yield with other traits is considered as the most valuable while formulating selection program for yield improvement in any crop. Correlation studies pave the way to know the association between highly heritable characters with the most economic, namely the grain yield. Several authors have worked in this aspect to bring about the relationship of different characters with yield and also within the yield contributing character.

A number of independent components will influence yield, since it is a complex quantitative trait. Simultaneous selection for more characters can easily be done with the knowledge of association between yield and yield components. This association between highly heritable characters with the most economic character, the yield, can be obtained by correlation studies. Various authors have brought out the relationship between yield and yield attributing economic traits by computing genotypic and phenotypic correlation coefficients.

2.4. Path coefficient analysis

Path coefficient analysis provides an efficient means of partitioning of correlation coefficients into direct and indirect effects of the component character. Selection on the basis of direct and indirect effect is much more useful than selection for yield *per se* alone. Several authors have reported the extent of direct and indirect influence of characters on yield in rice.

2.5. Studies on parent-progeny regression analysis

Parent-offspring regression analysis in advancing generations makes it possible to study how far the genetic potentials from one generation is transferred to the next; the higher the values of regression, the higher will be the genetic effect with less environmental influence. The analysis will be helpful to select the early fixing characters in a segregating population. Parent-progeny regression analysis method helps us to ascertain the influence of environment on different characters in progenies obtained from individual selection from early generation and to study the real genetic potentiality of the progenies, which was inherited from their parents [3]. Parent-progeny regression analysis assumes no environment association between generations, so genotype × environment interaction and co-variances between parents and offspring will be zero [4]. Multiple regression analysis to fix yield attributing characters found that percentage of filled grains per panicle, biological yield and harvest index were major selection criteria for yield improvement [5]. A comparative analysis to estimate heritability for grain yield and plant height using parent-offspring regression and variance components in maize revealed broad sense heritability estimated from variance components of the progeny greater than those based on the parent–offspring regression. It was found that selection is not effective for other traits like plant height.

High narrow sense and realized heritability for the characters, 1000 grain weight, and number of grains per panicle in early generation showed the prospects of selecting for these traits in early generation itself. Multiple regression analysis in pearl millet to fix yield-attributing characters found that number of grains per spike, 1000 grain weight, totally contributed a 60% variation in grain yield, suggesting that selection should be based on the above characters for improvement in grain yield.

With regard to heritability and environmental effects of yield and yield-related traits using parent-offspring regression in F_1 progenies, environmental effects and heritability estimates were high for culm length, tillers per plant, panicles per plant and 1000 grain weight. Based on the variance estimates of the parent-offspring regression model, it was suggested that these traits with high heritability, considerable phenotypic correlation and low seasonal variability could be used for further improvement of the F_1 progenies [6].

The most widely used breeding method in rice is the hybridization of homozygous diverse genotypes followed by pedigree method of handling the segregating population in order to isolate genotypes possessing desirable characteristics of parents. The usefulness of these methods is limited because of limited parental participation, low genetic diversity, reduced recombination and rapid fixation of genes. The selection and choice of breeding method for the improvement of quantitative or qualitative character largely depends on the nature and magnitude of additive and dominance variance. The success in the improvement of cultivated variety for yield, grain quality and resistance to biotic and abiotic resistance largely depends upon the natural variability present in the population. Through hybridization among the selected genotypes, it is possible to reshuffle desired characteristics, provided the segregating generations contain large variability. Selection for quantitative characters is generally taken up in the early segregating generations. For characters like grain yield, the selection is continued till the material becomes homozygous, because such characters are controlled by large number of genes and huge number of population has to be raised for making the selection effective. This is not always true, because the effective selection was known to be restricted by close linkages between desirable and undesirable component characters and these undesirable linkages delay the utilization of full recombination potential.

Among the methods of increasing the frequency of desirable recombinants, the biparental mating or disruptive mating gets importance in the improvement of self-pollinated crops, because it increases the possibility of obtaining desirable and valuable recombinants [7]. The system of biparental mating is reported to alter the phase of linkage through forced recombination. As a result, greater amount of concealed genetic variation is released, particularly of the additive type. Since the additive genetic variance is the only variance which responds to selection, a perceptible genetic gain can be expected after biparental mating.

Quality breeding in rice has assumed greater significance due to varied consumer preference in recent years. Developing a better variety with respect to higher yield with good grain quality and multiple resistances to biotic and abiotic tolerance than the already existing ones is the prime goal of plant breeders. Thus, high yield is the foremost goal of any crop improvement program but consumer preference varies from region to region. Next to grain yield, grain quality is the important criterion considered by the plant breeders. If the newly developed variety is not accepted by the consumers due to its poor taste, texture, aroma or appearance, its usefulness is greatly impaired. In developed countries, as well as in the rice-exporting countries, physical appearance of the grain or kernel is often more important than the yield [8]. In developing countries, like India, grain quality has become greatly important as the country has become more prosperous and self-sufficient in food production.

Quality of rice is determined by a combination of many physio-chemical properties. Rice kernel size and shape are important quality components that directly influence the market value. Gelatinization temperature and amylose content are the principal determinants for cooking and eating quality of milled rice. In addition to this trait, kernel length after cooking and linear elongation ratio is desirable quality traits. Properties of cooked rice are controlled by amylose content and gelatinization temperature. Varieties with intermediate values for both parameters remain nonsticky and tender after cooking and low values for both make the rice sticky on cooking.

Author details

Savitha Palanisamy

Address all correspondence to: saviagri@gmail.com

Plant Breeding and Genetics, PGP College of Agricultural Sciences, Namakkal, Tamil Nadu, India

References

[1] Sangeetha LNE. Genetic variability studies for grain yield and grain quality traits in F_2 and F_3 population of rice (*Oryza sativa* L.) [M.Sc. (Ag.) Thesis (Unpubl.)]. TNAU, Madurai; 2013

[2] Comstock RE, Robinson HF. Estimation of the Average Dominance of Genes. Heterosis, Ames, Iowa: Iowa State College Press; 1952. pp. 494-516

[3] Lush JL. Correlation and regression of offspring on dams as a method of estimating heritability of characters. Proceedings of American Society of Animal Production. 1940; **33**:293-301

[4] Sala M. Rice breeding for biofortification with high iron and zinc content in segregating population [M.Sc. (Ag.) Thesis (Unpubl.)]. TNAU, Madurai; 2012

[5] Rao NK, Reddy KB, Naik K. Studies on genetic variability, correlation and path coefficient analysis in rice under saline soil conditions. The Andhra Agricultural Journal. 2010; **57**(4):335-338

[6] Kalaimaghal R. Studies on genetic variability of grain iron and zinc content in $F_{2,}$ F_3 generation of rice (*Oryza sativa* L.) [M.Sc. (Ag.) Thesis (Unpubl.)]. TNAU, Coimbatore; 2011

[7] Johnson HW, Robinson HF, Comstock RE. Estimates of genetic and environmental variability in soybean. Agronomy Journal. 1955;**47**:314-318

[8] Savitha P. Genetic analysis of different generation involving traditional landraces of rice (*Oryza sativa* L.) [P.hD. (Ag.) Thesis (Unpubl.)]. TNAU, Coimbatore; 2014

2

Abiotic Stress Tolerance in Rice (*Oryza sativa* L.): A Genomics Perspective of Salinity Tolerance

Muhammad Saeed

Abstract

Rice (*Oryza sativa* L.) is the main source of staple food for human population. Salinity is the major problem for agricultural production and it affects rice production globally. Different approaches have been developed and exploited to ameliorate the harmful effects of salinity on crop production. Development of salt-tolerant cultivars is the best option which ensures sustainable crop production. Genomics approaches have the potential to accelerate breeding process for the development of salt tolerant crop cultivars. Molecular mapping techniques are the most promising component of genomics. Molecular mapping approaches have greatly helped in the identification of genomic regions involved in salinity tolerance in different crop plants, including rice. Identified genomic regions associated with salinity tolerance accelerated molecular breeding efforts to develop salt-tolerant rice cultivars. Molecular mapping techniques (both linkage and association mapping) are the main components of genomics and these helped in the identification of genomic regions associated with salt-tolerance in rice. In this chapter, a detailed description of molecular mapping techniques, and major findings made by these techniques is presented. Future prospects of these techniques are also discussed.

Keywords: genome-wide association studies, genomics, quantitative trait locus mapping, rice, salinity

1. Introduction

Rice (*Oryza sativa* L.) belongs to family Poaceae and genus *Oryza*. Its genome size is approximately 430 Mb contained in 12 chromosomes. Large part of human population depends on it for staple food. Rice is a salt-susceptible crop. One third of world agricultural land is salt affected [1]. Salinity, both soil and water, has negative effect on rice production [2]. Elevated

Na^+ levels in agricultural lands are increasingly becoming a serious threat to the world agriculture. Plants suffer osmotic and ionic stress under high salinity due to the salts accumulated at the outside of roots and those accumulated at the inside of the plant cells, respectively.

Projected increase in human population demands a proportional increase in the food supply. This demand of increased food supply can be fulfilled only if we utilize all available land resources to their full potential. An associated phenomenon with the increase in human population is the decrease in world agricultural land area due to its use for human settlements. Due to these constraints, even marginal cultivable lands cannot be neglected. This urges that saline soils should be exploited to their full production potential. For good crop production on saline areas, different practices such as reclamation, agronomic adjustments, and biological amendments are used in combination. Considering sustainable crop production on these areas, the use of salt-tolerant crop cultivars seems to be most suitable option [3–5]. For development of salt-tolerant cultivars, genetic diversity with respect to salt tolerance in crops has to be evaluated. For genetic diversity assessment and identification of genomic regions associated with salt tolerance, molecular mapping approaches have made considerable contribution in different crop plants [6–14]. With the use of molecular mapping approaches, it has become possible to identify the chromosomal regions (quantitative trait loci, QTLs) associated with traits related to salt tolerance in rice. This chapter tries to cover effects of salinity on rice plant's growth and development, types of molecular mapping approaches, methodology involved in these approaches, and the achievements made through these approaches in salinity tolerance in rice to-date. It also highlights the future prospects of molecular mapping approaches. Thus, it will be a valuable resource for designing future research endeavors to genetically characterize salt tolerance mechanisms and develop salt-tolerant rice cultivars. It will also facilitate molecular breeding efforts for screening rice germplasm for salinity tolerance.

2. Effects of salinity on rice plant growth and development

Salinity affects different morphological, biochemical, and physiological attributes of rice. Salinity has negative effect on percent relative-plant height, total tillers, root dry weight, shoot dry weight, and total dry matter [15]. Biochemical attributes of rice, affected by salinity, include chlorophyll content, proline content, hydrogen peroxide content, peroxidase (POX) activity, anthocyanins, Na^+ content, K^+ content, Ca^{++} content, total cations content [11, 16]. Physiological attributes of rice, which are affected by salinity, include relative growth rate, osmotic potential, transpiration use efficiency, senescence, Na^+ uptake, K^+ uptake, Ca^{++} uptake, total cations uptake, Na^+/K^+ uptake, Na^+ uptake ratio, K^+ uptake ratio, Ca^{++} uptake ratio, Na^+/K^+ uptake ratio, and total cations uptake ratio [11, 16, 17].

Rice shows different levels of salt tolerance at leaf and whole plant level [18, 19]. Similarly, behavior of rice plants towards salt stress may be different at vegetative and reproductive phases and this may not correlate with their mean level of relative resistance [20]. It is important to know the specific salt susceptible phase of a rice variety to have a better comparison of performance among varieties under salinity stress.

Vegetative and reproductive growth potential of plant depends upon the process of photosynthesis. Increased sodium concentration in the leaf tissue negatively affects net photosynthesis and essential cellular metabolism [18, 21]. Chlorophyll content is important in photosynthesis. Reports suggest that there is no correlation between the chlorophyll content and photosynthesis under salinity stress. Net photosynthesis was reduced by a sodium concentration which did not affect chlorophyll content [18]. It implicates the disturbance by salinity stress of other cellular processes involved in photosynthesis. Sodium accumulation in the leaf also affects stomatal aperture and carbon dioxide fixation simultaneously [18] and thus it may be one of the reasons for reduced photosynthesis due to sodium accumulation. The most salt-susceptible cultivars had lowest K^+/Na^+ ratio in the leaves and exhibited strongest yield reductions [22]. Rice plant evolved different mechanisms to cope salinity stress conditions. One of these mechanisms is compartmenting salts within the plant body [23].

3. Molecular mapping approaches: types and methodology

Molecular mapping approaches are of two types, linkage mapping and association mapping, on the basis of mapping population used.

3.1. Linkage mapping

In linkage mapping, bi-parental segregating populations are used. These populations include backcross populations, doubled haploid (DH) lines, F_2 populations, introgression lines (ILs), near isogenic lines (NILs) and recombinant inbred lines (RILs). JoinMap [24], MapMaker [25] or QTL IciMapping [26] soft-wares are used for the construction of genetic linkage maps. WinQTLCartographer [27], QTL IciMapping [26], and QTLNetwork [28] programs are used for the identification of QTLs. Detailed information about input file requirements, statistical parameters thresholds, and the procedure to run the software are provided in the user manuals of these softwares.

3.2. Association mapping

Association mapping uses natural populations for mapping purposes. In this technique, commercial crop cultivars can be employed for the assessment of QTLs. First reported in humans, association mapping is now widely used in plant sciences. Assessment of marker-trait associations is facilitated by controlling underlying population structure in the used plant material for mapping purposes [29]. STRUCTURE software is used for identifying sub-populations in the used plant germplasm [30]. TASSEL software is used for the identification of QTLs in this case [31].

3.3. DNA markers used in molecular mapping approaches

In molecular mapping approaches, different types of DNA markers are used to identify QTLs. Amplified fragment length polymorphism (AFLP), restriction fragment length polymorphism (RFLP), simple sequence repeats (SSRs), sequence tagged sites (STS), simple sequence length polymorphism (SSLP), and single nucleotide polymorphism (SNP) [12, 32–35] are different types of DNA markers which are employed for genotyping in molecular mapping studies.

4. Achievements made through molecular mapping approaches with respect to salinity tolerance in rice

By using linkage and association mapping approaches, a number of QTLs linked to salinity tolerance in rice, have been identified. Detail of identified QTLs is given below.

4.1. Linkage mapping

Linkage mapping has been very successful in the identification of QTLs linked to salinity tolerance in rice. A number of significant QTLs associated with salinity tolerance in rice were identified through linkage mapping approach (**Table 1**). In these studies, the mapping populations used were F_2 population, F_3 population, $F_{2:4}$ population, near-isogenic lines, recombinant inbred lines, doubled haploid population, backcross-inbred lines, BC_3F_5 lines, BC_2F_8 advanced backcross introgression lines, and reciprocal introgression lines. Marker systems used in these studies included SSR, RFLP, SSLP, SSR AFLP, and SNPs.

Morphological parameters are supposed to be indicators of salt tolerance. There were various reports in which QTLs related to morphological traits under salt stress were identified [12, 32, 34–43]. In these mapping studies, the plant material was phenotyped at the seedling, tillering, or the maturity stage. Data for different morphological traits were recorded in these studies. These traits included seed germination (%), seedling survival days, seedling vigor, seedling root length, shoot length, fresh shoot weight, dry shoot weight, dry root weight, reduction rate of dry weight, reduction rate of fresh weight, reduction rate of leaf area, reduction rate of seedling height, tiller number, salt tolerance rating, score of salt toxicity of leaves, plant height, and grain yield-related traits. A number of significant QTLs were identified in these studies. These identified QTLs included a QTL for seedling survival days [32]; a QTL for root length flanked by restriction fragment length polymorphism (RFLP) markers RG162-RG653 [36]; QTLs with heritability values up to 53.3% [34]; two significant QTLs, *qST1* and *qST3*, for salt tolerance at seedling stage with 35.5–36.9% phenotypic variance explained values, respectively [38]; same QTLs conferring salt tolerance at both seedling and tillering stages [40], SSR marker RM223 associated with salt tolerance in rice [39], and a major QTL for straw yield, *qSY-3* [12]. These studies also suggested that it is possible to combine favorable alleles associated with salt tolerance in a single cultivar through marker-assisted selection (MAS) of main effect QTLs (M-QTLs) [42]. Similarly, pleiotropic effects were found for some QTLs which were found associated with both drought and salt tolerance [43].

There are also a number of reports of QTLs identified for different physio-biochemical traits through linkage mapping [11, 33, 44–52]. Traits which were studied in these reports were shoot Na^+ concentration; shoot K^+ concentration; leaf Na^+ concentration; leaf K^+ concentration; Na^+ uptake; K^+ uptake; Na^+ absorption; K^+ adsorption; Na^+/K^+ absorption ratio; K^+/Na^+ ratio;

Trait	Plant material used	Marker system used	Reference
Seedling survival days	RILs population	RFLP	[32]
Seed germination (%); seedling root length; seedling dry matter; seedling vigor	Doubled haploid (DH) population	RFLP	[36]
Shoot length; tiller number; shoot fresh weight	Backcross inbred lines	RFLP	[37]
Salt tolerance rating; Na^+/K^+ ratio in roots; dry matter weight of shoots	F_2 population	SSR	[34]
Survival days of seedlings; score of salt toxicity of leaves; shoot K^+ concentration; shoot Na^+ concentration; fresh weight of shoots; tiller number per plant; plant height at the tillering stage	BC_2F_8 introgression lines (IL)	SSR	[40]
Plant height; panicle length; tillers per hill; spikelets per panicle; grain yield	RILs population	SSR	[39]
Reduction rate of dry weight; reduction rate of fresh weight; reduction rate of leaf area; reduction rate of seedling height	Introgression lines	SSR	[41]
Seedling height; dry shoot weight; dry root weight; Na/K ratios in roots	RILs, $F_{2:9}$	SSR	[42]
Days to seedlings survival; score on salt toxicity symptoms on leaves; shoot K^+ concentration; shoot Na^+ concentration at seedling stage	BC_2F_8 advanced backcross introgression lines (ILs)	SSR	[43]
Plant height; root length; shoot dry weight; shoot fresh weight	RILs	SNP	[35]
Morphological and yield-related traits	F_2 population	SSR	[12]
Sodium and potassium uptake	RILs	AFLP, RFLP, SSR	[33]
Salt tolerance traits	RILs	RFLP, SSLP	[44]
Salt tolerance traits	F_2 and F_3 populations	RFLP	[45]
–	140 RILs	SSR	[47]
Leaf Na^+ concentration; K^+/Na^+ ratio; K^+ concentrations; ratio of leaf Na^+ to sheath Na^+ concentrations	RILs	RFLP, SSR	[46]
Sodium (Na^+) and potassium (K^+) in roots and shoots; Na^+/K^+ ratio in roots and shoots	Advanced backcross-inbred lines (BILs)	SSR	[48]
Na^+ and K^+ concentrations in the roots and shoots	RILs, $F_{2:9}$	SSR	[50]
Physiological traits	$F_{2:4}$ population	SSR, AFLP	[51]
Pollen fertility; Na^+ concentration and Na/K ratio in the flag leaf	F_2 population	SSR	[52]
Sodium (Na^+), potassium (K^+), and calcium (Ca^{++}) accumulation traits	F_2 population	SSR	[11]

Table 1. QTLs identified through linkage mapping studies.

ratio of leaf Na^+ to sheath Na^+ concentrations; sodium (Na^+) and potassium (K^+) in roots; Na^+ concentration and Na/K ratio in the flag leaf; and sodium (Na^+), potassium (K^+), and calcium (Ca^{++}) accumulation traits. Major discoveries in these studies included a major QTL (*QKr1.2*) identified for K^+ content in the root on chromosome 1 explaining 30% of the total variation [48]; pollen fertility, Na^+ concentration and Na/K ratio in the flag leaf were found as the most important attributes for salt tolerance at the reproductive stage in rice [52], QTLs for sodium and potassium uptake were identified on different linkage groups (chromosomes) [33] suggesting that different pathways are involved in Na^+ and K^+ uptake; and a major locus controlling Na^+ uptake (*QTLsur-7*) was identified on chromosome 7, with R^2 value of 72.57% [11].

4.2. Association mapping

In recent years, association mapping is widely used to identify QTLs in plants. Association mapping approach is relatively new arrival in plant genetics. There are some reports of association mapping for salt tolerance in rice [13, 53–58]. Main findings of these association studies are presented in **Table 2**. In these studies, rice mapping populations used consisted of European Rice Core collection (ERCC) containing 180 japonica accessions [53], 96 rice germplasm accessions including Nona Bokra [55], 220 rice accessions [56], 341 japonica rice accessions [57], 94 rice genotypes [58], and 24 indica rice genotypes [13]. Traits for which data were recorded in these studies included Na^+/K^+ ratio, survival days of seedlings, shoot K^+/Na^+ ratio, Na^+ uptake, Ca^{++} uptake, total cations uptake, Ca^{++} uptake ratio, K^+ uptake ratio, Na^+/K^+ uptake and salinity tolerance scoring. Major findings made in these studies included an observation that distribution of favorable alleles associated with salt tolerance was random in ERCC [53]; 40 new allelic variants found in coding sequences of five salt-related genes [54]; STS marker, RM22418, for *SKC1*, on Chr. 8 was found associated with salinity tolerance [55]; region containing *Saltol* was found associated with Na^+/K^+ ratio [56]; marker RM3412 was found associated to salinity tolerance at seedling stage due to its close linkage to *SKC* gene [58]; and the report that other QTLs, in addition to *Saltol*, might be involved in salinity tolerance [58]. These reports highlighted that in rice germplasm there might be other genomic regions involved in salt tolerance. These genomic regions need to be characterized in future to add a wealth

Trait	Plant material used	Marker system used	Reference
Salinity tolerance	180 japonica accessions	SNPs, SSR	[53]
Na^+/K^+ ratio equilibrium; signaling cascade; stress protection	392 rice accessions	SNPs	[54]
Salinity tolerance	96 germplasm accessions	SSR	[55]
Stress-responsive genes	220 rice accessions	SNPs	[56]
Survival days of seedlings and shoot K^+/Na^+ ratio	341 japonica rice accessions	SSR	[57]
Seedling stage salt tolerance	94 rice genotypes	SSR	[58]

Table 2. QTLs identified through association mapping studies.

of information in the present rice genetics knowledge pool. Random distribution in the rice germplasm of favorable alleles associated with salt tolerance is a worthwhile finding which should be considered while exploring and selecting crossing parents in breeding programmes.

5. Future prospects and conclusions

Climate change has affected world agriculture a lot. The most pronounced effects of climate change are the heat stress and periodic drought conditions in major rice producing countries of the world. Due to periodic drought conditions, the already existing problem of high amounts of salts in the upper surface soil has intensified. So, there is a dire need to opt for a coordinated approach to address the problem of salinity stress for rice production. Genomics has great potential to assist in this coordinated programme. With the help of molecular mapping approaches, a number of major and minor QTLs associated with salinity tolerance in rice have been identified in recent years and there are further accelerated research efforts underway in this direction. The identified QTLs are valuable resources for marker-assisted selection (MAS) to develop elite salt tolerant rice cultivars. Great task is needed to be done in this regard so that marker-assisted breeding (MAB) approach can be implemented successfully in routine breeding programmes. In future, efforts should be directed to develop climate-smart rice cultivars which can perform stably under diverse environmental conditions. Identified QTLs and rice germplasm found tolerant to salinity stress can be exploited in three major ways: (a) to understand the molecular genetics of salt tolerance in rice; (b) salinity stress tolerant rice germplasm might be incorporated into salt-tolerant rice cultivars development molecular breeding programmes; (c) identified QTLs incorporated into MAS for screening rice germplasm against salinity stress. New genes involved in salt tolerance will be identified by this approach. Genome sequence of rice, both indica and japonica subspecies, is available now. In the next phase of annotation of the rice genome, molecular mapping results can be of help in combination with the comparative genomics approach.

Lot of work related to molecular mapping for salinity tolerance in rice is to be performed yet. The main cautious point is the plant phenotyping for salt stress tolerance. Accuracy in the phenotyping work is the key in the authentic identification of QTLs related to salt tolerance. Hydroponics should be tried for this purpose. Under salinity stress conditions, phenotyping at germination, seedling, tillering, and reproductive phases require different strategies and care. In case of quantitative traits, such as salinity stress tolerance, there is pronounced effect of environment. Efforts should be made to design a judicious phenotyping plan which can minimize effect of environment. In case of plant genotyping work, robust marker systems with high resolution power such as SNPs should be preferred over other marker systems. Genotyping-by-sequencing (GBS) is another option.

Previous research efforts have pointed out that the distribution of favorable alleles, associated with salt-tolerance, is random among the rice germplasm [53]. So, it is possible to pyramid favorable alleles of salt-tolerance in an elite rice genotype through well-planned crossing programme. This elite rice cultivar will have great potential with regard to salt tolerance. In view of inland intrusion of the seas, we have to concentrate on the coastal areas to fully exploit their agricultural production potential. This is also imperative in view of alarming increase in human population and to feed this population we have to exploit every available land for agricultural production. It is hoped that genomics approaches will play a greater part in this exploitation of land by providing salt tolerant crop cultivars.

Author details

Muhammad Saeed

Address all correspondence to: saeed_pbg@gcuf.edu.pk

Department of Botany, Faculty of Life Sciences, Government College University, Faisalabad, Pakistan

References

[1] Rengasamy P. World salinization with emphasis on Australia. Journal of Experimental Botany. 2006;**57**:1017-1023

[2] Ren Z-H, Gao J-P, Li L-G, Cai X-L, Huang W, Chao D-Y, Zhu M-Z, Wang Z-Y, Luan S, Lin H-X. A rice quantitative trait locus for salt tolerance encodes a sodium transporter. Nature Genetics. 2005;**37**:1141-1146

[3] Shannon M, Rhoades JD, Draper JH, Scardaci SC, Spyres MD. Assessment of salt tolerance in rice cultivars in response to salinity problems in California. Crop Science. 1998;**38**:394-398

[4] Saeed M, Dahab AHA, Guo WZ, Zhang TZ. A cascade of recently discovered molecular mechanisms involved in abiotic stress tolerance of plants. OMICS: A Journal of Integrative Biology. 2012;**16**:188-199

[5] Singh R, Singh Y, Xalaxo S, Verulkar S, Yadav N, Singh S, Singh N, Prasad KSN, Kondayya K, Rao PVR, et al. From QTL to variety-harnessing the benefits of QTLs for drought, flood and salt tolerance in mega rice varieties of India through a multi-institutional network. Plant Science. 2016;**242**:278-287

[6] Veldboom LR, Lee M, Woodman WL. Molecular marker-facilitated studies in an elite maize population. 1. Linkage analysis and determination of QTL for morphological traits. Theoretical and Applied Genetics. 1994;**88**:7-16

[7] Yano M, Sasaki T. Genetic and molecular dissection of quantitative traits in rice. Plant Molecular Biology. 1997;**35**:145-153

[8] Barakat MN, Wahba LE, Milad SI. Molecular mapping of QTLs for wheat flag leaf senescence under water-stress. Biologia Plantarum. 2013;**57**:79-84

[9] Lohwasser U, Rehman Arif MA, Borner A. Discovery of loci determining pre-harvest sprouting and dormancy in wheat and barley applying segregation and association mapping. Biologia Plantarum. 2013;**57**:663-674

[10] Chai L, Zhang J, Pan XB, Zhang F, Zheng TQ, Zhao XQ, Wang WS, Jauhar A, Xu JL, Li ZK. Advanced backcross QTL analysis for the whole plant growth duration salt tolerance in rice (*Oryza sativa* L.). Journal of Integrative Agriculture. 2014;**13**:1609-1620

[11] Khan MSK, Saeed M, Iqbal J. Identification of quantitative trait loci for Na^+, K^+ and Ca^{++} accumulation traits in rice grown under saline conditions using F_2 mapping population. Brazilian Journal of Botany. 2015;**38**:555-565

[12] Khan MSK, Saeed M, Iqbal J. Quantitative trait locus mapping for salt tolerance at maturity stage in indica rice using replicated F_2 population. Brazilian Journal of Botany. 2016;**39**:641-650

[13] Khan MSK, Saeed M, Iqbal J. Association mapping validates previously identified quantitative trait loci for salt tolerance in rice (*Oryza sativa* L.). Molecular Breeding. 2016;**36**:172

[14] Kaur S, Iquebal MA, Jaiswal S, Tandon G, Sundaram RM, Gautam RK, Suresh KP, Rai A, Kumar D. A meta-analysis of potential candidate genes associated with salinity stress tolerance in rice. Agri Gene. 2016;**1**:126-134

[15] Razzaque MA, Talukder NM, Dutta RK, Zamil SS. Efficacy of supplemental calcium on the growth of three rice genotypes differing in salt tolerance. Journal of Plant Nutrition. 2010;**33**:571-586

[16] Chunthaburee S, Dongsansuk A, Sanitchon J, Pattanagul W, Theerakulpisut P. Physiological and biochemical parameters for evaluation and clustering of rice cultivars differing in salt tolerance at seedling stage. Saudi Journal of Biological Sciences. 2016;**23**:467-477

[17] Negrão S, Schmöckel SM, Tester M. Evaluating physiological responses of plants to salinity stress. Annals of Botany. 2017;**119**:1-11

[18] Yeo AR, Caporn SJM, Flowers TJ. The effect of salinity upon photosynthesis in rice (*Oryza sativa* L.): Gas exchange by individual leaves in relation to their salt content. Journal of Experimental Botany. 1985;**36**:1240-1248

[19] Sytar O, Brestic M, Zivcak M, Olsovska K, Kovar M, Shao H, He X. Applying hyperspectral imaging to explore natural plant diversity towards improving salt stress tolerance. Science of the Total Environment. 2017;**578**:90-99

[20] Lutts S, Kinet JM, Bouharmont J. Changes in plant response to NaCl during development of rice (*Oryza sativa* L.) varieties differing in salinity resistance. Journal of Experimental Botany. 1995;**46**:1843-1852

[21] Horie T, Karahara I, Katsuhara M. Salinity tolerance mechanisms in glycophytes: An overview with the central focus on rice plants. Rice. 2012;**5**:11

[22] Asch F, Dingkuhn M, Dörffling K, Miezan K. Leaf K/Na ratio predicts salinity induced yield loss in irrigated rice. Euphytica. 2000;**113**:109-118

[23] Yeo AR, Flowers TJ. Varietal differences in the toxicity of sodium ions in rice leaves. Physiologia Plantarum. 1983;**59**:189-195

[24] Van Ooijen JW, Voorrips RE. JoinMap ® Version 3.0: Software for the Calculation of Genetic Linkage Maps. Wageningen, The Netherlands: CPRO-DLO; 2001

[25] Lander ES, Green P, Abrahamson J, Barlow A, Daly MJ, Lincoln SE, et al. MAPMAKER: An interactive computer package for constructing primary genetic linkage maps of experimental and natural populations. Genomics. 1987;**1**(2):174-181

[26] Meng L, Li H, Zhang L, Wang J. QTL IciMapping: Integrated software for genetic linkage map construction and quantitative trait locus mapping in biparental populations. Crop Journal. 2015;**3**:269-283

[27] Basten CJ, Weir BS, Zeng ZB. QTL Cartographer, Version 1.15. Raleigh, NC: Department of Statistics, North Carolina State University; 2001

[28] Yang J, Hu CC, Ye XZ, Zhu J. QTLNetwork 2.0. Hangzhou, China: Institute of Bioinformatics, Zhejiang University; 2005

[29] Zhao K, Aranzana MJ, Kim S, Lister C, Shindo C, Tang C, Toomajian C, Zheng H, Dean C, Marjoram P, Nordborg M. An *Arabidopsis* example of association mapping in structured samples. PLoS Genetics. 2007;**3**:e4

[30] Pritchard JK, Wen W. Documentation for Structure Software. Chicago: The University of Chicago Press; 2004

[31] Bradbury PJ, Zhang Z, Kroon DE, Casstevens TM, Ramdoss Y, Buckler ES. TASSEL: Software for association mapping of complex traits in diverse samples. Bioinformatics. 2007;**23**:2633-2635

[32] Hongxuan L, Yanagihara S, Jieyun Z, Senboku T, Kangle Z, Yashima S. Identification of QTL for salt tolerance in rice via molecular markers. Chinese Journal of Rice Science. 1998;**12**:72-78

[33] Koyama ML, Levesley A, Koebner RMD, Flowers TJ, Yeo AR. Quantitative trait loci for component physiological traits determining salt tolerance in rice. Plant Physiology. 2001; **125**:406-422

[34] Yao M, Wang J, Chen H, Zhai H, Zhang H. Inheritance and QTL mapping of salt tolerance in rice. Rice Science. 2005;**12**:25-32

[35] Bimpong IK, Manneh B, El-Namaky R, Diaw F, Amoah NKA, Sanneh B, Ghislain K, Sow A, Singh RK, Gregorio G, Bizimana JB, Wopereis M. Mapping QTLs related to salt tolerance in rice at the young seedling stage using 384-plex single nucleotide polymorphism SNP, marker sets. Molecular Plant Breeding. 2013;**5**:47-63

[36] Prasad SR, Bagali P, Hittalmani S, Shashidhar H. Molecular mapping of quantitative trait loci associated with seedling tolerance to salt stress in rice (*Oryza sativa* L.). Current Science. 2000;**78**:162-164

[37] Takehisa H, Shimodate T, Fukuta Y, Ueda T, Yano M, Yamaya T, Kameya T, Sato T. Identification of quantitative trait loci for plant growth of rice in paddy field flooded with salt water. Field Crops Research. 2004;**89**:85-95

[38] Lee SY, Ahn JH, Cha YS, Yun DW, Lee MC, Ko JC, Lee KS, Eun MY. Mapping QTLs related to salinity tolerance of rice at the young seedling stage. Plant Breeding. 2007;**126**:43-46

[39] Lang N, Buu BC, Ismail A. Molecular mapping and marker-assisted selection for salt tolerance in rice (*Oryza sativa* L.). Omonrice. 2008;**16**:50-56

[40] Zang J, Sun Y, Wang Y, Yang J, Li F, Zhou Y, Zhu L, Jessica R, Mohammadhosein F, Xu J, Li Z. Dissection of genetic overlap of salt tolerance QTLs at the seedling and tillering stages using backcross introgression lines in rice. Science in China Series C: Life Sciences. 2008;**51**:583-591

[41] Kim D-M, Ju H-G, Kwon T-R, Oh C-S, Ahn S-N. Mapping QTLs for salt tolerance in an introgression line population between japonica cultivars in rice. Journal of Crop Science and Biotechnology. 2009;**12**:121-128

[42] Wang Z, Cheng J, Chen Z, Huang J, Bao Y, Wang J, Zhang H. Identification of QTLs with main, epistatic and QTL × environment interaction effects for salt tolerance in rice seedlings under different salinity conditions. Theoretical and Applied Genetics. 2012;**125**:807-815

[43] Wang Y, Zang J, Sun Y, Ali J, Xu J, Li Z. Identification of genetic overlaps for salt and drought tolerance using simple sequence repeat markers on an advanced backcross population in rice. Crop Science. 2012;**52**:1583-1592

[44] Bonilla P, Dvorak J, Mackell D, Deal K, Gregorio G. RFLP and SSLP mapping of salinity tolerance genes in chromosome 1 of rice (*Oryza sativa* L.) using recombinant inbred lines. Philippine Agriculturist. 2003;**85**:68-76

[45] Lin HX, Zhu MZ, Yano M, Gao JP, Liang ZW, Su WA, Hu XH, Ren ZH, Chao DY. QTLs for Na^+ and K^+ uptake of the shoots and roots controlling rice salt tolerance. Theoretical and Applied Genetics. 2004;**108**:253-260

[46] Haq T, Gorham J, Akhtar J, Akhtar N, Steele KA. Dynamic quantitative trait loci for salt stress components on chromosome 1 of rice. Functional Plant Biology. 2010;**37**:634-645

[47] Thomson MJ, de Ocampo M, Egdane J, Rahman MA, Sajise AG, Adorada DL, Tumimbang-Raiz E, Blumwald E, Seraj ZI, Singh RK, Gregorio GB, Ismail AM. Characterizing the *Saltol* quantitative trait locus for salinity tolerance in rice. Rice. 2010;**3**:148-160

[48] Ahmadi J, Fotokian M-H. Identification and mapping of quantitative trait loci associated with salinity tolerance in rice (*Oryza sativa*) using SSR markers. Iranian Journal of Biotechnology. 2011;**9**:21-30

[49] Cheng L, Wang Y, Meng L, Hu X, Cui Y, Sun Y, Zhu L, Ali J, Xu J, Li Z. Identification of salt-tolerant QTLs with strong genetic background effect using two sets of reciprocal introgression lines in rice. Genome. 2012;**55**:45-55

[50] Wang Z, Chen Z, Cheng J, Lai Y, Wang J, Bao Y, Huang J, Zhang H. QTL analysis of Na^+ and K^+ concentrations in roots and shoots under different levels of NaCl stress in rice (*Oryza sativa* L.). PLoS One. 2012;**7**:e51202

[51] Ghomi K, Rabiei B, Sabouri H, Sabouri A. Mapping QTLs for traits related to salinity tolerance at seedling stage of rice (*Oryza sativa* L.): An agrigenomics study of an Iranian rice population. OMICS: A Journal of Integrative Biology. 2013;**17**(5):242-251. DOI: 10.1089/omi.2012.0097

[52] Hossain H, Rahman MA, Aslam MS, Singh RK. Mapping of quantitative trait loci associated with reproductive stage salt tolerance in rice. Journal of Agronomy and Crop Science. 2015;**201**:17-31

[53] Ahmadi N, Negrão S, Katsantonis D, Frouin J, Ploux J, Letourmy P, Droc G, Babo P, Trindade H, Bruschi G, Greco R, Oliveira MM, Piffanelli P, Courtois B. Targeted association analysis identified japonica rice varieties achieving Na^+/K^+ homeostasis without the allelic make-up of the salt tolerant indica variety Nona Bokra. Theoretical and Applied Genetics. 2011;**123**:881-895

[54] Negrão S, Almadanim MC, Pires IS, Abreu IA, Maroco J, Courtois B, Gregorio GB, McNally KL, Oliveira MM. New allelic variants found in key rice salt-tolerance genes: An association study. Plant Biotechnology Journal. 2013;**11**:87-100

[55] Emon RM, Islam MM, Halder J, Fan Y. Genetic diversity and association mapping for salinity tolerance in Bangladeshi rice landraces. Crop Journal. 2015;**3**:440-444

[56] Kumar V, Singh A, Mithra SVA, Krishnamurthy SL, Parida SK, Jain S, Tiwari KK, Kumar P, Rao AR, Sharma SK, Khurana JP, Singh NK, Mohapatra T. Genome-wide association mapping of salinity tolerance in rice (*Oryza sativa*). DNA Research. 2015;**22**:133-145

[57] Zheng H, Wang J, Zhao H, Liu H, Sun J, Guo L, Zou D. Genetic structure, linkage disequilibrium and association mapping of salt tolerance in japonica rice germplasm at the seedling stage. Molecular Breeding. 2015;**35**:152

[58] Krishnamurthy SL, Sharma SK, Kumar V, Tiwari S, Singh NK. Analysis of genomic region spanning *Saltol* using SSR markers in rice genotypes showing differential seedlings stage salt tolerance. Journal of Plant Biochemistry and Biotechnology. 2016;**25**:331-336

3

Anther Culture as a Supplementary Tool for Rice Breeding

D.M. Ruwani G. Mayakaduwa and Tara D. Silva

Abstract

There is a timely need to harness biotechnology and related tools to support conventional breeding strategies, overcoming the limitations in rice production and improving quantity and quality as well as climatic and disease stress tolerance of the crop. Anther culture allows immediate fixation of homozygosity through diploidization of regenerated haploid plants and therefore serves as an efficient path for inbred line development. Anther culture has been successfully used to hasten the breeding programs in several crop species including rice. However, associated constraints still prevent the realization of its full potential. Even though anther culture technique has been effective for Japonica rice breeding, applicability for Indica rice remains limited mainly due to inherent recalcitrant genetic background. Constraints associated with Indica rice can be identified as early anther necrosis, poor callus induction and proliferation, extremely low green plant regeneration and frequent albinism. Success of androgenesis is determined by factors such as genotype, physiological status of donor plant, pollen development stage at culture, composition and physical status of culture media, culture incubation conditions and anther pretreatments. This chapter has detailed out the scope for improving the applicability of anther culture technique on rice in order to develop it as a supplementary breeding tool.

Keywords: callus induction, plant regeneration, microspores, ploidy, homozygous lines

1. Introduction

From ancient times the crop rice has served the human population as a staple food. Due to the steep increase in human population, rice growers need to increase the production as well. This has become more difficult due to the limitations in available resources. Therefore, other than the conventional strategies, it is a timely need to harness biotechnology and related tools to overcome not only the productivity barrier but also the production efficiency, quality of the product, and abiotic and biotic stress tolerance of the crop. Among a number of modern biotechnological tools to improve rice crop, anther culture plays a useful role.

Anther culture can be considered as a technique for the rapid development of fully homozygous lines. Therefore, anther culture technique provides an efficient alternative to the conventional inbred line development which is usually achieved through several cycles of inbreeding. Even though anther culture has been efficiently used as a supplementary breeding tool with Japonica rice varieties, application of this technique for Indica rice varieties is limited due to their inherent recalcitrant genetic background. Limitations of androgenesis in Indica rice result from early anther necrosis, poor callus induction and proliferation, remarkably low green plant regenerability, and frequent albinism [1, 2]. Success of the technique is determined by numerous factors such as genotype and physiological status of donor plant, pollen development stage at culture, anther wall factors, composition of culture media including nutritional sources and growth regulators, physical status of culture media, and anther pretreatments [1, 3, 4].

This chapter serves as an insight to the practical aspects of anther culture technique for it to be fully exploited for improving rice breeding.

2. Technique of anther culture

For androgenesis to be successful, normal gametophyte formation from microspores should be halted, and microspores are directed toward sporophyte development. Usually, pretreatments are required to alter the normal pollen development pathway and to trigger the androgenic response. The specific pretreatments for androgenesis that are required by different species and also varieties within species are quite variable. Therefore, a single standard method cannot be generalized for androgenesis for a given species or even a variety. However, some common protocols to be followed during anther culture are well known and documented [4].

Rice anther culture is carried out in two phases in which the initial step is to induce embryogenic calluses from microspores followed by green plant regeneration from the induced calluses [5]. Protocol for rice anther culture includes pretreatment given to panicles, surface sterilization and excision of anthers from panicles, and *in vitro* culture of anthers on a specific culture medium under aseptic conditions [6]. Response of anthers in culture is usually indicated by the gradual browning of the anther wall tissues and bursting or splitting of the anther to expose the pollen callus. Pollen callus can be expected to be formed in anthers after 3–8 weeks of culture [2]. The second phase is to regenerate green plants from the calluses using appropriate regeneration media [6]. The regenerated plants are then transplanted and acclimatized under controlled environmental conditions, and they can be subjected to chromosome doubling using antimitotic agents in order to obtain doubled haploids which can serve as homozygous lines.

3. Historical trajectory of anther culture

The possibility of changing the normal gametophytic pathway of microspores to sporophytic pathway facilitating the haploid plant development through *in vitro* culture was first reported by reference [7] on culturing immature anthers of the Solanaceous species *Datura innoxia*. Reference [8] successfully obtained haploid plants from culturing isolated anthers of *Nicotiana*. Since then, haploid development using *in vitro* culture of anthers and isolated pollen has been successful with many other crop species such as rice, wheat, maize, *Brassica*, and pepper [2, 5]. Although microspore embryogenesis has been effective with model species such as barley, rapeseed, tobacco, and wheat, some other species that are scientifically or economically important, such as *Arabidopsis*, woody plants, and legume crops, continue to be less responsive for the technique. Extensive research has been performed in order to make this important technique of developing haploids and dihaploids more robust. For the technique to be practically applied in breeding programs, anther or microspore culture should be able to permit production of haploids in very large quantities from almost any species or genotype [9].

Haploid plant production in rice through anther culture was first reported by reference [10]. Since then, many studies have been conducted improving various aspects of rice anther culture. Other than utilizing anther culture technique directly for dihaploid development, recently applications have been expanded to facilitate other biotechnological approaches such as gene transformation [3]. In Japan and China, where the Japonica rice varieties are mainly in use, anther culture technique has been extensively applied for improving the rice crop due to the amenability of Japonica rice varieties to *in vitro* anther culture [5]. However, the use of this technique as a tool for Indica rice breeding has been extremely limited due to the inherent recalcitrance associated with Indica varieties. Therefore, the potential of the technique for Indica rice breeding is yet to be fully unraveled [3].

4. Limitations associated with androgenesis

Induction of haploids in rice is associated with a number of constraints. Fine tuning of anther culture process addressing the constraints is required in order to use this technique equally well for breeding of Japonica and Indica rice. Although anther culture technique has been used to produce haploids from an array of species, success of the technique cannot be proven in respect of all genotypes of a crop species [11]. Particularly when it comes to the anther culture of Indica rice, the response remains extremely variety or genotype specific [1]. The problem is further aggravated because anther culture response is affected even by the growing season [4].

Under *in vitro* conditions, many of the anthers fail to grow in culture and thus repress the pollen from forming calluses. Some reasons for failure are the early abortion of pollen and even in situations where pollen starts to divide and produce callus and necrosis or cell death occurs very early during callus proliferation. There is also a degree of uncertainty associated with the ploidy of the resulting callus tissue as it can comprise a chimera of diploid, tetraploid, and haploid cells. Another problem that seriously affects the anther culture of cereals is the formation of albino plants during regeneration, and this can be identified as the most limiting step in the anther culture process [12]. Detailed investigations of proplastids and the plastid genome of the regenerated albino plantlets revealed that albinism is mainly due to incomplete formation of the membrane structures and different blockages in the plastid development [13]. Molecular studies carried out on anther culture of cereals such as wheat, barley, and rice have attributed the associated albinism to large-scale deletions and rearrangements in the plastid genome [14].

5. Factors affecting rice anther culture

Investigations on haploid induction through anther culture have been steadily increasing due to its importance as a supplementary breeding strategy. These studies are mainly driven with close monitoring of a number of factors that influence androgenesis in rice as described in detail below.

5.1. Genotype of the donor plants

Response to anther culture by Indica rice varieties is generally poor, and even among those that respond by producing callus, the *in vitro* morphogenic responses are highly genotype-dependent. The recalcitrance associated with the Indica types can be characterized mainly by poor callus induction response, poor regenerability of green plants, and the occurrence of a large proportion of albinos [1]. By comparison, Japonica rice varieties respond much better. Anther culture ability of Japonica varieties, Indica varieties, and their hybrids can be indicated in the following order of Japonica/Japonica > Japonica > Indica/Japonica > Indica/Indica > Indica [15]. Thus, the Japonica varieties have benefitted more from this supplementary breeding approach, and extensive practical applications have been possible. For example, 67,000–159,000 anthers from F1 hybrids of 25–36 rice crosses produced 1500–15,000 (2–10%) green plants per season for selection [16]. Anther culture technique has been used with a Japonica rice variety to purify it and in the process to develop stable new lines that are distinctly different to the parent variety [17]. On the other hand, utilizing anther culture technique for Indica rice breeding has been extremely limited due to its comparatively poor androgenic response [3] with a few exceptions [18]. However, anther culture performance in F_1 hybrids and F_2 plants could be improved when high yielding commercially grown Indica rice varieties were crossed with high anther culture-responsive Japonica varieties [19].

Reference [20] has very convincingly illustrated the extreme variability in anther culture response between Japonica and Indica varieties as 41% for a Japonica variety to 0% for an Indica variety. Not only between the subspecies but also among different varieties from the same subspecies, a considerable variation for callus induction and plant regeneration has been observed. Reference [21] stated that among seven Indica rice varieties on anther culture, callus induction frequencies varied extensively from 3.6 to 51.7%, while green plant regeneration efficiency ranged from 1.6 to 82.9%. Reference [22] reported that out of 18 Indica varieties subjected to anther culture, only five varieties were responsive for pollen callusing, and only four varieties produced regeneration response. Similarly, reference [23] verified the extremely low androgenic response of Indica varieties as they found only 1 out of 35 Indica varieties exhibited pollen callusing on N_6 medium. The use of optimal media, specifically formulated for each of different genotypes, may help to improve the low response associated with some high valued varieties [13].

5.2. Physiological status of donor plants

Success of the anther culture is greatly influenced by the physiological condition of the anther donor plants. That is mainly because physiology affects the number of viable and healthy pollen grains produced within the anthers, the endogenous levels of hormones that regulate metabolic pathways, and the nutritional status of the anther tissues [4]. During maturation of the anther donor plants, environmental factors such as light intensity, photoperiod, temperature, nutrition, and CO_2 concentration critically affect the growth and development. Also, pest infestations and control measures may have a detrimental effect on microspore development [13].

An improved androgenic response of an Indica rice variety induced by growing donor plants under specific conditions of light and day/night temperature regime was observed by reference [24]. The highest anther response was shown by anthers cultured from donor plants of variety IR43 grown until panicle emergence stage under long days (>12 h), high solar radiation (>18 Mj m^{-2}), and sunshine (>8 h) and day/night temperature (34/24°C), and a declined response was observed when the plants were grown under an environment with low values of the above conditions. They also observed that the plants grown under the field conditions were significantly superior than those grown in controlled conditions such as glasshouse or in pots near the field. Similar observations have also been reported for other cereals, such as maize and wheat. Certain chemicals such as ethereal, when applied on the donor plants, have altered their physiology, thereby enhancing the androgenic response [1]. Further, the anthers collected from the primary tillers were more responsive for anther culture than the anthers from panicles on late tillers [3]. When the anther donor plants were starved of nitrogen, anthers were able to produce much better response in *in vitro* culture compared to those that are given optimum levels of nitrogen fertilizers [2].

5.3. Pollen development stage

The pollen development stage is a critical factor that strongly affects the success of anther culture. Induction of embryogenic calluses cannot be achieved by culturing pollen in any stage of development, and the potential is restricted to specific pollen maturity stages only [4, 25]. For rice, the best responsive stage for embryogenic induction has been reported to be the middle to late uninucleate stage of microspore division, and therefore anthers need to be cultured at these specific stages [26–28]. These are very early stages of microspore development. The highest response shown by these early stages is most likely due to the fact that they are cells which have not yet been committed to gametophytic development and therefore can be forced to become proliferative. The undifferentiated cells in the callus can then assume a new pathway of development leading to sporophyte formation [4].

Although it is stated that the best responsive stages are the middle to late uninucleate microspores, the precise stage of microspores that is best suited for producing a superior anther response can vary from one genotype to another. Therefore, application of the anther culture technique requires detailed examination of pollen before culture to determine its effective development stage [3]. When culturing rice anthers, determining the microspore development stage requires nuclear staining and cytological examination of microspores prior to culture. However, repeating nuclear staining of microspores with each rice panicle, before dissecting anthers for culture, greatly impedes the anther culture process. Therefore, a distinct morphological indicator trait that correlates well with the stage of microspore maturity is commonly used during *in vitro* culture. In rice anther culture, the morphological trait that has been used is the measured distance between the nodes of the last two leaves: the flag leaf, and the penultimate leaf [27, 29]. In some cases the panicle length at the time of harvest has also been used as a visually identifiable guide [30, 31]. The use of a direct cytological marker such as the degree of starch accumulation in microspores has been identified as more accurate than the internode distance and even more convenient than the laborious nuclear staining process to rapidly assess microspore maturity. The most appropriate stage is when pollen grains just begin to accumulate starch which can be simply tested with I_2/KI solution [32].

5.4. Culture media

The two main phases of anther culture in rice, callus induction and shoot regeneration, require different nutrient regimes and growth regulators. The culture medium that best supports callus induction is often not suitable for regeneration. Therefore, the transfer of callus onto a suitable regeneration medium must be done at an appropriate time. Since the callus induction potential of a given rice variety is largely determined by the genetic makeup, significant levels of improvement in anther response cannot be expected by manipulation of nongenetic factors such as the culture medium. Nevertheless, the best responsive nutrient requirements must be chosen as an initial step in order to optimize anther culture, particularly if they are low responding Indica varieties [3].

The most commonly used basal media for anther culture are N_6 medium [33], MS medium [34], and B5 medium [35]. Generally, basal N_6 medium supplemented with plant growth regulators has been used extensively in cereal anther culture to initiate callus. Macronutrients of culture media comprises mainly of carbon and nitrogen sources. Embryogenic and morphogenic responses are elicited by supplementing the basal media with appropriate plant growth regulators at effective concentrations. Physical state of the culture medium and also culture maintenance conditions are equally important for the success of rice anther culture.

5.4.1. Carbohydrate source

A carbohydrate source is essential in tissue culture media because it serves as the main source of energy to the cultured explant tissue. Carbohydrates are also important as osmotic agents. In rice anther culture, osmotic pressure in the medium is generally regulated by applying the carbohydrate source to the medium at a particular concentration. Very high concentrations when used during the latter stages of culture seem to be deleterious for cereals [4]. The type of carbon source directly influences the anther response. Although many early studies have used sucrose as the standard carbon source, different sources have also been tested and proven effective for cereals. Maltose has been identified as a superior source of carbohydrate compared to sucrose for androgenesis in cereals. Anther culture efficiency and green plant formation of highly recalcitrant Indica rice varieties could be improved significantly when sucrose was replaced by maltose [1]. Reference [23] reported an inferior anther culture response with sucrose, as only 1 out of 23 Indica rice varieties responded with pollen callusing and green plant production on N_6 medium provided with 146 mM sucrose. When sucrose was replaced by equimolar amount of maltose, callus induction response improved from 6.3% to 10.1% and green plant regeneration from 0.6–1%. Reference [36] had observed that 20% maltose used for microspore isolation and 9% maltose used for culturing produced a genotype-independent plant regeneration response. In other cereals such as wheat, maize, and barley, maltose promoted direct embryogenesis from cultured pollen. Sucrose is rapidly broken down into glucose and fructose. The toxic effects of sucrose on androgenesis have been attributed to the sensitivity of microspores to fructose. This also causes depletion of sucrose in the medium with time [1]. Comparatively, long-term availability of maltose in the culture medium has been detected due to the slow rate of hydrolysis.

5.4.2. Nitrogen source

In culture media, inorganic nitrogen is usually supplied in the form of nitrate and/or ammonium ions. The ratio of the two nitrogen sources NO_3^-:NH_4^+ has been found to be critical for the success of anther culture in rice [3].

The N_6 basal medium which is most widely used for rice anther culture has been formulated with both these sources of nitrogen at specific concentrations. However, Indica rice varieties perform much better when lower concentration of NH_4^+ ions than normal is used in the medium [1]. Reference [37], in which the response of eight Indica rice varieties were studied on different media, found He_2 medium to be more effective than N_6 medium. He_2 medium is derived from the N_6 medium by reducing NH_4^+ concentration to half strength. In Korea, N_6-Y_1 medium which is similar to N_6 except that the $(NH_4)_2SO_4$ concentration is reduced from 3.5 to 1.5 mM has been recommended for Indica-Japonica hybrids [5]. Reference [38] reported that a significant improvement in anther culture could be made in Indica x Indica F1 hybrids using a medium with high KNO_3 and NH_4^+ ions completely replaced by an organic source of nitrogen, casein hydrolysate, at 50 mgL^{-1}.

Reference [39] studied the effect of nitrogen source on androgenesis in another Indica variety IR24 using R-2 medium as the control. R-2 has been formulated with 40 mM KNO_3 and 2.5 mM $(NH_4)_2SO_4$. When 20 mM KNO_3 was combined with the amino acid 5 mM alanine, superior green plant regeneration could be achieved. In rice anther culture, amino acids such as proline and glutamine added to the culture media have been able to increase the rate of callus induction from cultured anthers while avoiding the degeneration of anther wall tissue [4].

5.4.3. Plant growth regulators

Plant growth regulators have been widely investigated in anther culture. Supplementing *in vitro* culture media with effective growth regulators (auxins, cytokinins, or a combination of these) as appropriate is crucial for the success of androgenic response particularly from recalcitrant genotypes [4].

The growth regulator 2,4-dichlorophenoxy acetic acid (2,4-D) is commonly used in the first phase of rice anther culture, and 2,4-D provided at fairly high concentrations (2 mgL^{-1}) has produced improved rates of callus induction of up to 15% in some genotypes [24]. Also, applicability of some other auxins such as naphthalene acetic acid (NAA), phenyl acetic acid, picloram, and dicamba alone or in combination with 2,4-D has been tested for improving androgenic response. Not only the growth regulator combination but also the auxin/cytokinin balance has been found critically important for effective androgenesis. Reference [23] reported that the growth regulator regime of 2,4-D (2 mgL^{-1}), picloram (0.07 mgL^{-1}), and kinetin (0.5 mgL^{-1}) was favorable for enhancing the anther response in a large number of genotypes. Further, the type of auxin and its concentration determine the microspore development pathway. For example, the use of 2,4-D favored callus formation, whereas indole-3-acetic acid and NAA promoted direct embryogenesis from cultured anthers without an intervening callus phase [4].

Although high levels of 2,4-D were useful for increasing callus production, it has proven to have a negative effect on the next phase of culture which is regeneration from callus, particularly from recalcitrant rice genotypes [22]. A lower level of 2,4-D (0.5 mgL^{-1}) in combination with the milder auxin NAA (2.5 mgL^{-1}) and kinetin (0.5 mgL^{-1}) has been used effectively during both phases [9]. This suggests that the use of 2,4-D in the callus induction medium needs to be regulated with a compromise reached between callus induction and regeneration efficiency [24].

5.4.4. Physical state of the medium

Usually, rice anthers are cultured on solid media. However, reference [23] found increased necrosis of anther tissue when they were cultured on solid media and observed a better callusing response in liquid media. Liquid culture media are able to supply the anthers with an improved access to nutrients and plant growth regulators, and also toxic and degenerated material can be readily dispersed. During the culture of anthers from Indica×Basmati rice on liquid media, severalfold increment in green plant regeneration comparable with the rates reported for Japonica rice varieties/hybrids could be obtained [29].

However, since the rice anthers tend to settle at the bottom of the liquid cultures, this would affect respiration and result in loss of viability of the explants. These have been identified as barriers for the use of liquid media for androgenesis. When the liquid culture media was added with substances such as Ficoll, it was possible to avoid sinking of anthers due to the increased buoyancy, and therefore viability could be maintained [3]. In principle, the solidifying agent should not carry any nutritional effect. Agar is in extensive use as the gelling agent of solid culture media. However, more reproducible results have been obtained with the use of Gelrite. Starch also has been used for solidification despite the nutritional effects and its dissociation into sugar [13]. Some have found improved response by embedding anthers in agarose than culturing on semisolid or liquid media [40].

5.4.5. Culture incubation conditions

Culture temperature plays an important role in plant tissue culture. Anther cultures are usually incubated at the temperature range of 24–27°C. For two Indica rice varieties, Nona

Bokra and Pokkali, callus induction frequencies and plant regeneration responses could be improved when cultures were incubated at alternating temperature regime of 30/20°C (14/10 h) instead of constant incubation at 25°C [5]. Light regulates morphogenesis of cultured pollen and specifically darkness (low intensity of light) or alternating light and dark conditions can be preferable for embryogenic induction. Reference [41] reported the effectiveness of culture conditions such as alternating periods of light with different temperatures (12–18 h; 5000–10,000 lux/m^2 at 28°C and 12–6 h; in darkness at 22°C). Regeneration phase requires even more specific incubation conditions to achieve success, particularly for green shoot formation. Shoot regeneration from scutellum-derived callus of Indica rice was stimulated by applying osmotic stress conditions. Osmotic stress was created in tissues by altering the water content of the medium with the use of agarose and mannitol or by partial desiccation of callus. It is possible to expect similar stimulatory effects in anther-derived callus also. With osmotic stress, water content in the calluses is reduced, thus converting the callus tissues into more compact structures with better embryogenic and regeneration potential [42].

The composition of the atmosphere in the culture vessel has not been thoroughly studied despite its importance shown with tobacco [4]. Explant density and explant orientation in the culture medium also have been found to be critical in anther culture [2, 4].

5.5. Pretreatments to trigger androgenesis in rice

In many crop varieties including cereals, usually a treatment applied to excised anthers, inflorescences, or anther donor plants prior to culture is important to trigger the sporophytic development deviating from normal pollen development pathway. The type of the effective pretreatment, duration, and the time of application vary with the species or even for different varieties [1, 4]. Reference [43] reviewed the different pretreatments which are in current applications for triggering the anther culture response, and they have been classified into three categories based on their utility as widely used, neglected, and novel. These pretreatments include high temperature and chilling, high humidity, water stress, anaerobic treatment, centrifugation, sucrose and nitrogen starvation, ethanol, γ-irradiation, use of microtubule disruptive agents, electrostimulation, high medium pH, and heavy metal treatment.

5.5.1. Temperature pretreatment

Most frequently used effective method of pretreatment for rice anther culture is the low-temperature application. Harvested rice panicles are subjected to cold shock prior to the culture. However, the temperature and duration vary with the variety. Cold pretreatment given to rice anthers is known to enhance the androgenesis potential by delaying the degeneration of microspores and anther wall tissue in rice [1, 3]. Reference [44] reported that a pretreatment at 10°C for 10–30 days was necessary to induce sporophytic divisions in microspores of the Japonicas. Generally, temperatures from 8 to 10°C for 8 days have been recommended to be optimal for many varieties of rice [45]. Panicle pretreatments longer than 11 days tend to increase albino production [46]. Reference [37] reported a brief exposure to high temperature (35°C for 10 min) before the cold treatment to enhance callus induction although it adversely affected green plant production.

5.5.2. Osmotic stress

Osmotic shock has been identified as another pretreatment, which can substitute or be used in combination with cold treatment for the induction of androgenesis. Reference [47] have reported the treatment of anthers in 0.4 M mannitol solution to be effective for inducing androgenesis in microspore cultures of Indica and Japonica varieties. Sole mannitol treatment without the cold pretreatment given to anthers promoted androgenesis in anther cultures of variety IR43 from 3 to 33.4% [48]. It is described that when the anthers or isolated microspores are subjected to high osmolarity by incubating in metabolizable carbohydrates for short time, they start divisions during stress treatment and tolerate the following stress conditions [43]. Further, regenerability of callus could also be improved markedly by osmotic treatment. It is supposed to regulate the endogenous levels of auxin interacting with abscisic acid affecting the carbohydrate metabolism and thereby trigger both callus initiation and shoot regeneration responses in rice [49].

5.5.3. Sugar starvation

Not only in rice but also in many other crop species such as tobacco, wheat, and barley, sugar starvation has been found effective in induction of embryogenesis [43]. Reference [39] reported that cold pretreatment could be partially substituted by subjecting microspores for sugar starvation for 3 days during androgenesis of Indica rice. Reference [47] also confirmed that sugar starvation could be applied for Indica and Japonica rice in obtaining high-frequency embryogenesis and plantlet regeneration. Many changes induced in starved microspores at cytoplasmic and nuclear levels have been described in detail by reference [43].

5.5.4. Irradiation

Penetration of irradiation varies with the species and dependent pollen morphology and the thickness of the pollen wall [50]. Reference [51] demonstrated the stimulation of green plant regeneration from rice anther culture with the application of gamma rays at the dose of 20 Gy. Enhancement of the green plant regeneration from two- to threefold could be possible by the use of irradiation of the ^{137}Cs gamma rays, and the maximum response was elicited with the dose of 15 Gy [52].

5.6. Ploidy-level determination and doubled haploid production

Ideally, the plants developed from anther culture can be considered as haploids as they are arisen from haploid microspores. However, the actual plants resulted during the regeneration could be a mixture of haploid, diploid, or mixoploid [13]. Occurrence of non-haploids can be due to different malformations. Tissues formed from somatic tissue of anther walls, the fusion of nuclei, endomitosis within the pollen grain, or irregular microspores formed during irregular meiosis lead to the development of plants other than the haploids. Also, when the vegetative and generative nuclei are not separated by cell wall formation, non-haploids could be originated [4].

Some species are associated with a great tendency for spontaneous chromosome doubling. In such cases haploid cells are directly converted to homozygous DH plants. If spontaneous doubling is completely absent or occurs at a low frequency, the haploid plants resulted in cultures need to be converted to dihaploids by some other means such as application of chromosome doubling agents [53]. Among many such chemicals, colchicine is the most widely used anti-microtubule agent *in vivo* and *in vitro*, and oryzalin and trifluralin have also been used.

The ploidy status of regenerated plants can be determined through direct and indirect measures. Direct determinations include mitotic chromosome counts or use of the flow cytometry technique [54]. Initial attempts to ploidy determination relied on karyotypical assessments which was highly time-consuming, laborious, and difficult due to the requirement of skilled operators and was most of the time a failure due to the unavailability of dividing cells. Recently, flow cytometry has hastened the ploidy status analysis. As an alternative to chromosome counting, some other correlated measurements which do not require actively dividing cells have been reported for estimating the ploidy level. These include leaf stomatal density and size. However, these measurements alone cannot be granted sufficiently reliable [54]. When the diploids arisen from anther cultures were analyzed using SSR markers to determine their source of origin, homozygosity has been detected for all in 150 DHs except for one which turned out as a heterozygote [31]. Therefore, microsatellites and other molecular tools such as isozyme analyses and RAPD markers can be utilized to ascertain the homozygosis confirming that the calluses and plantlets have been arisen from gamete itself and not from other somatic tissues.

6. Conclusion

Anther culture technique has been recognized as an efficient alternative to the conventional inbred line development which is usually achieved through a number of inbreeding cycles. A number of critical factors have been addressed for directing the anther culture process for its optimum condition since it can be potentially developed as a supplementary breeding tool for rice.

Acknowledgements

This research on rice anther culture was graciously funded by the National Science Foundation (NSF) of Sri Lanka under the research grant RG/2011/BT/10.

Conflict of interest

No other condition or relationship has a potential conflict of interest.

Author details

D.M. Ruwani G. Mayakaduwa* and Tara D. Silva

*Address all correspondence to: ruwani.mayakaduwa@pts.cmb.ac.lk

University of Colombo, Sri Lanka

References

[1] Bhojwani SS, Pande H, Raina A. Factors Affecting Androgenesis in Indica Rice. Delhi, India: Department of Botany, University of Delhi; 2001

[2] Razdan MK. Introduction to Plant Tissue Culture. 2nd ed. USA: Science Publishers; 2003. pp. 87-106

[3] Silva TD. Indica rice anther culture: Can the impasse surpassed? Plant Cell, Tissue and Organ Culture. 2010;**100**:1-11. DOI: 10.1007/s11240-009-9616-9

[4] Germana MA. Anther culture for haploid and doubled haploid production. Plant Cell, Tissue and Organ Culture. 2011;**104**:283-300. DOI: 10.1007/s11240-010-9852-z

[5] Javed MA, Ishii T, Kamijima O, Misoo S. The role of alternating culture temperatures and maltose in enhancing the anther culture efficiency of salt tolerant indica rice (*Oryza sativa* L.) cultivars, Pokkali and Nona Bokra. Plant Biotechnology. 2007;**24**:283-287

[6] Sah BP. Response of genotypes to culture media for callus induction and regeneration of plants from rice anthers. Scientific World. 2008;**6**(6):37-43

[7] Guha S, Maheshwari SC. *In vitro* production of embryos from anthers of *Datura*. Nature. 1964;**204**:497

[8] Bourgin JP, Nitsch JP. Production of haploids *nicotiana* from excised stamens. Annales De Physiologie Vegetale. 1967;**9**(4):377

[9] Shahnewaz S, Bari MA. Effect of concentration of sucrose on the frequency of callus induction and plant regeneration in anther culture of rice (*Oryza sativa* L.). Plant Tissue Culture. 2004;**14**(1):37-43

[10] Niizeki H, Oono K. Induction of haploid rice plant from anther culture. Proceedings of the Japan Academy. 1968;**44**:554-557

[11] Weigt D, Kiel A, Nawracala J, Pluta M, Lacka A. Solid-stemmed spring wheat cultivars give better androgenic response than hollow-stemmed cultivars in anther culture. *In Vitro* Cellular & Developmental Biology: Plant. 2016;**52**:619-625

[12] Torp AM, Andersen SB. Albinism in microspore culture. In: Touraev A, Forster BP, Jain SM, editors. Advances in Haploid Production in Higher Plants. The Netherlands: Springer; 2009. pp. 155-160

[13] Foroughi-Wehr B, Wenzel G. Andro- and parthenogenesis. In: Hayward MD, Bosemark NO, Romagosa I, Cerezo M, editors. Plant Breeding-Principles and Prospects. Dordrecht: Springer; 1993. pp. 261-277. DOI: 10.1007/978-94-011-1524-7_18

[14] Kumari M, Clarke HJ, Small I, Siddique KHM. Albinism in plants: A major bottleneck in wide hybridization, androgenesis and doubled haploid culture. Critical Reviews in Plant Sciences. 2009;**28**(6):393-409

[15] Yan J, Xue Q, Zhu J. Genetic studies of anther culture ability in rice (*Oryza sativa*). Plant Cell, Tissue and Organ Culture. 1996;**45**:253-258

[16] Huang CS, Tsay HS, Chern CG, Chen CC. Japonica rice breeding using anther culture. Journal of Agriculture Research China. 1988;**37**(1):1-8

[17] Serrat X, Cardona M, Gil J, Brito AM, Moysset L, Nogues S, Lalanne E. A Mediterranean Japonica rice (*Oryza sativa*) cultivar improvement through anther culture. Euphytica. 2014;**195**:31-44

[18] Naik N, Rout P, Umakanta N, Verma RL, Katara JL, Sahoo KK, Singh ON, Samantaray S. Development of doubled haploids from an elite *indica* rice hybrid (BS6444G) using anther culture. Plant Cell, Tissue and Organ Culture. 2017;**128**:679

[19] Herath HMI, Bandara DC. Anther culture performance in selected high yielding indica (of Sri Lanka) and japonica rice varieties and their inter sub-specific hybrids. Journal of the National Science Foundation of Sri Lanka. 2011;**39**(2):149-154

[20] Miah MAA, Earle ED, Kush GS. Inheritance of callus formation ability in anther callus of rice, *Oryza sativa* L. Theoretical and Applied Genetics. 1985;**70**:113-116

[21] He T, Yang Y, Tu SB, Yu MQ, Li XF. Selection of interspecific hybrids for anther culture of indica rice. Plant Cell, Tissue and Organ Culture. 2006;**86**:271-277

[22] Guha S, Mukherjee S. Genotypic differences in the *in vitro* formation of embryoids from rice pollen. Journal of Experimental Botany. 1973;**24**:139-144

[23] Lentini Z, Reyes P, Martinez CP, Roca WM. Androgenesis of highly recalcitrant rice genotypes with maltose and silver nitrate. Plant Science. 1995;**110**:127-138

[24] Raina SK, Zapata FJ. Enhanced anther culture efficiency of indica rice (*Oryza sativa* L.) through modification of the culture media. Plant Breeding. 1997;**116**:305-315. DOI: 10.1111/j.1439-0523.1997.tb01004.x

[25] Dunwell JM. Haploids in flowering plants: Origins and exploitation. Plant Biotechnology Journal. 2010;**8**:377-424. DOI: 10.1111/j.1467-7652.2009.00498.x

[26] Niroula RK, Bimb HP. Effect of genotype and callus induction medium on green plant regeneration from anther of Nepalese rice cultivars. Asian Journal of Plant Sciences. 2009;**8**:368-374

[27] Silva TD, Ratnayake WJ. Anther culture potential of indica rice varieties, kurulu thuda and BG 250. Tropical Agricultural Research and Extension. 2009;**12**:53-56. DOI: 10.4038/tare.v12i2.2789

[28] Mishra R, Rao GJN. *In vitro* androgenesis in rice: Advantages, constraints and future prospects. Rice Science. 2016;**23**:57-68. DOI: 10.1016/j.rsci.2016.02.001

[29] Bishnoi US, Jain RK, Gupta KR, Chowdhury VK, Chowdhury JB. High frequency androgenesis in indica x basmati rice hybrids using liquid culture media. Plant Cell, Tissue and Organ Culture. 2000;**6**:153-159

[30] Cha-um S, Srianan B, Pichakum A, Kirdmanee C. An efficient procedure for embryogenic callus induction and double haploid plant regeneration through anther culture of Thai aromatic rice (*Oryza sativa* L. subsp. indica). *In Vitro* Cellular & Developmental Biology: Plant. 2009;**45**:171-179. DOI: 10.1007/s11627-009-9203-0

[31] Rout P, Naik N, Ngangkham U, Verma RL, Katara JL, Singh ON, Samantaray S. Doubled haploids generated through anther culture from an elite long duration rice hybrid, CRHR32: Method optimization and molecular characterization. Plant Biotechnology. 2016;**33**:177-186. DOI: 10.5511/plantbiotechnology.16.0719a

[32] Mayakaduwa DMRG, Silva TD. A cytological indicator allows rapid assessment of microspore maturity, leading to improved *in vitro* anther response in Indica rice (*Oryza sativa* L.). *In Vitro* Cellular & Developmental Biology: Plant. 2017;**53**(6):591-597. DOI: 10.1007/s11627-017-9855-0

[33] Chu C. The N_6 medium and its applications to anther culture of cereal crops. In: Proceedings of the Symposium on Plant Tissue Culture; Beijing: Science Press; 1978. pp. 43-50

[34] Murashige T, Skoog F. A revised medium for rapid growth and bioassays with tobacco tissue cultures. Physiologia Plantarum. 1962;**15**:473-497

[35] Gamborg OL, Miller RA, Ojima K. Nutrient requirement suspension cultures of soybean root cells. Experimental Cell Research. 1968;**50**:151-158. DOI: 10.1016/0014-4827(68)90403-5

[36] Mejza SJ, Morgant V, DiBona DE, Wong JR. Plant regeneration from isolated microspores of *Triticum aestivum*. Plant Cell Reports. 1993;**12**:149-153

[37] Reddy VS, Leelavathi S, Sen SK. Influence of genotype and culture medium on microspore callus induction and green plant regeneration in anthers of *Oryza sativa*. Physiologia Plantarum. 1985;**63**:309-314

[38] Raina SK. Tissue culture in rice improvement: Status and potential. Advances in Agronomy. 1989;**42**:339-397

[39] Ogawa T, Fukuoka H, Ohkawa Y. Plant regeneration through direct culture of isolated pollen grains in rice. Breeding Science. 1995;**45**:301-307

[40] Gill R, Kaur N, Sindhu AS, Bharaj TS, Gosal SS. Improved methods for anther and pollen culture in rice. In: Kush GS, Brar DS, Hardy B, editors. Advances in Rice Genetics. Philippines: International Rice Research Institute; 2000. pp. 503-505

[41] Vasil IK. Plants: Haploid tissue cultures. In: Kruse PF, Patterson MK, editors. Tissue Cultures Methods and Application. New York: Academic Press; 1973. pp. 157-161

[42] Jain RK, Jain S, Wu R. Stimulatory effect of water stress on plant regeneration in aromatic indica rice varieties. Plant Cell Reports. 1996;**15**:449-454

[43] Shariatpanahi ME, Bal U, Heberle-Borsa E, Touraeva A. Stresses applied for the re-programming of plant microspores towards *in vitro* embryogenesis. Physiologia Plantarum. 2006;**127**:519-534. DOI: 10.1111/j.1399-3054.2006.00675.x

[44] Matsushima T, Kikuchi S, Takaiwa F, Oono K. Regeneration of plants by pollen culture in rice (*Oryza sativa* L.). Plant Tissue Culture. 1988;**5**:78-81

[45] Zapata-Arias FJ. Laboratory protocol for anther culture technique in rice. In: Maluszynski M, Kasha KJ, Forster BP, Szarejko I, editors. Doubled Haploid Production in Crop Plants, a Manual. Dordrecht: Kluwer Academic Publishers; 2003. pp. 109-116

[46] Gupta HS, Borthakur DN. Improved rate of callus induction from rice anther culture following microscopic staging of microspores in iron alum-haematoxylin. Theoretical and Applied Genetics. 1987;**74**:95-99

[47] Raina SK, Irfan ST. High frequency embryogenesis and plantlet regeneration from isolated microspores of indica rice. Plant Cell Reports. 1998;**17**:957-962

[48] Pande H. Androgenesis in anther cultures of an indica cultivar of *Oryza sativa* L. [thesis]; University of Delhi; 1997

[49] Lee ST, Huang WL. Osmotic stress stimulates shoot organogenesis in callus of rice (*Oryza sativa* L.) via auxin signaling and carbohydrate metabolism regulation. Plant Growth Regulation. 2014;**73**:193-204

[50] Savaskan C. The effects of gamma irradiation on the pollen size of *Gossypium hirsutum* L. Turkish Journal of Botany. 2002;**26**:477-480

[51] Chen QF, Wang CL, Lu YM, Shen M, Afza R, Duren MV, Brunner H. Anther culture in connection with induced mutations for rice improvement. Euphytica. 2001;**120**(3):401-408

[52] Mkuya MS, Hua-min S, Wen-zhen L, Zong-xiu S. Effect of ^{137}Cs gamma rays to panicles on rice anther culture. Rice Science. 2005;**12**(4):299-302

[53] Ferrie AMR, Caswell KL. Isolated microspore culture techniques and recent progress for haploid and doubled haploid plant production. Plant Cell, Tissue and Organ Culture. 2011;**104**:301-309

[54] Ochatt SJ, Patat-Ochatt EMP, Moessner A. Ploidy level determination within the context of *in vitro* breeding. Plant Cell, Tissue and Organ Culture. 2011;**104**:329-341. DOI: 10.1007/s11240-011-9918-6

4

Iron Biofortification of Rice: Progress and Prospects

Andrew De-Xian Kok, Low Lee Yoon,
Rogayah Sekeli, Wee Chien Yeong,
Zetty Norhana Balia Yusof and Lai Kok Song

Abstract

Biofortification is the process of improving the bioavailability of essential nutrients in food crops either through conventional breeding or modern biotechnology techniques. Rice is one of the most demanding staple foods worldwide. Most global population live on a diet based on rice as the main carbohydrate source that serve as suitable target for biofortification. In general, polished grain or white rice contains nutritionally insufficient concentration of iron (Fe) to meet the daily requirements in diets. Therefore, iron biofortification in rice offers an inexpensive and sustainable solution to mitigate iron deficiency. However, understanding on the mechanism and genes involved in iron uptake in rice is a prerequisite for successful iron biofortification. In this chapter, the overview of iron uptake strategies in plants and as well as different iron-biofortified approaches used in rice will be outlined. Then, the challenges and future prospects of rice iron biofortification to improve global human health will also be discussed.

Keywords: agronomic practices, conventional plant breeding, genetic engineering, iron biofortification, *Oryza sativa*, transgenic rice

1. Introduction

Rice is one of the most consumed staple foods worldwide. In developing countries, people often rely on rice as their sole source of nutrition [1]. However, polished grain, known as white rice, contains limited amount of essential nutrients to sustain a good health and development [2]. Hence, those who are incapable to afford other micronutrients-rich nonstaple food for their balance diet are often at the highest risk for micronutrients deficiencies [3].

Iron deficiency is a common health disorder affecting nearly 2 billion people worldwide with other mineral and vitamin deficiency [4, 5]. Common effects of iron deficiency include anemia and impaired growth development in pregnant women and preschool children [6]. It can be easily addressed through dietary diversification, micronutrient supplements, medicines, and surgery depending of the severity of the condition [5, 7]. However, such treatments may not be available to everyone due to limitations such as geographical and financial capabilities [4]. In addition, iron is the most difficult mineral to be used in food fortification because the most soluble and absorbable compounds (e.g., $FeSO_4$) alters the taste or color of fortified food making it unappetizing while the least soluble compounds (e.g., $Fe_4(P_2O_7)_3$) are poorly absorbed by human body [8, 9]. Hence, food fortification is not a sustainable solution to mitigate iron deficiency.

Government bodies and nonprofit organization could play an important role to combat micronutrient deficiency by providing adequate food, supplements, and medicine supplies to rural areas. Nevertheless, it may not be an effective long-term solution because it is highly dependent on continuous investment, appropriate infrastructures, and transportation [10]. Hence, an alternative solution through biofortification is seen to be more efficient and cost-friendly in mitigating to micronutrient deficiency.

Iron biofortification, the process of improving the bioavailability of iron in food crops can be achieved via agronomic practices, conventional breeding, and genetic engineering. Biofortification through agronomic practices can be performed through fertilizer or foliar feeding [11]. Agronomic practices need to take bioavailability of iron at different stages into account as not all of the nutrients are transferred [12]. Several crucial factors may contribute to the nutrient loss at different stages such as bioavailability of nutrient uptake from the soil, nutrient distribution in different parts of the plants, milling or dehusking during food processing, and the ability of human to absorb and utilize the nutrients [13]. These factors need to be considered carefully to ensure successful iron biofortification through agronomic practices.

Meanwhile, conventional plant breeding involves identifying and selection of parent line, which contains desirable traits found in both parent plants. Parent lines are then crossed over for a few generations until daughter plants with both desirable nutrient and agronomic traits are observed and selected [14]. For instance, iron bean is one of the successful products through conventional plant breeding with high iron content, high bioavailability, and high yield [15, 16]. In addition, the advancement of modern biotechnology techniques, such as marker-assisted selection, improves the efficiency and precision in identification of potential lines in daughter plants [17].

To date, with the recent advancement of genetic engineering technologies served as a platform, which inspires many researchers in exploring alternative solution through genetic modification. Genetic engineering involves in removing, altering, or inserting specific sequence into the plant genome, which provides a better flexibility by silencing or overexpressing desirable gene sequences for desirable traits [18, 19]. Genetic engineering is an excellent method to obtain desirable micronutrient levels in a more effective manner by targeting specific gene of interest. However, successful biofortification via genetic engineering requires extensive knowledge and understanding of iron uptake, trafficking, and homeostasis mechanisms in plants to prevent undesirable side effects.

2. Iron uptake strategies in plants

Plants acquire iron mainly from the rhizosphere. There are abundant of iron in the soil, but only minute quantities of iron are absorbed by the plants. The availability of iron is dependent on the soil pH and soil redox potential [7]. Iron becomes less soluble in higher pH and it can be found in the form of insoluble ferric oxides. In contrast, iron becomes more soluble in low pH and they can be readily absorbed by the plant roots [20].

Micronutrient uptake and distribution in plants are heavily controlled and regulated by different uptake strategies. This allows the required amount of micronutrients to be absorbed into the plant but not high enough to exhibit toxicity effect [21]. Similarly for iron uptake in plants, there are two strategies used for iron uptake from the soil, namely reduction-based strategy and chelation-based strategy [22, 23]. Graminaceous plants are able to utilize chelation-based strategy while nongraminaceous plants utilize reduction-based strategy. However, rice is able to utilize combination of both reduction-based and chelation-based strategies as shown in **Figure 1** [22, 23].

2.1. Strategy I: reduction-based strategy

Reduction-based strategy is utilized by nongraminaceous plants. This strategy involves reducing available Fe^{3+} through the reduction activity into Fe^{2+} before being absorbed into the plant system. In reduction-based strategy, nongraminaceous plant will release protons toward the rhizosphere to decrease the pH in the surrounding soil under Fe-deficient condition. Kim [20] suggested that ATPase are responsible for releasing protons into the rhizosphere and reducing the pH of surrounding rhizosphere. The decrease in pH will increase the solubility of Fe^{3+} in the rhizosphere. In addition, NADPH-dependent Fe^{3+}-chelate reductase reduces Fe^{3+} into a more soluble form of Fe^{2+} with the help of ferric reductase oxidase 2 (FRO2). Then, Fe^{2+} will be transported into the roots via ferric ion transporter controlled by iron regulated transporter 1 (IRT1) [22].

2.2. Strategy II: chelation-based strategy

Grasses families such as maize, wheat, and rice are known as graminaceous plants. In response to iron deficiency, these plants are able to increase iron uptake through chelation-based strategy. Chelation-based strategy transports Fe^{3+} from rhizosphere into the roots with the help of soluble siderophores. Mugineic acid (MA) family phytosiderophores are natural iron chelators and they have a higher affinity toward Fe^{3+} [7]. Depending on different species, different sets of MAs will be released by the plant to surrounding rhizosphere via transporter of MAs (TOM1). For instance, rice will secrete only 2′-deoxymugineic acid (DMA), while barley secretes different types of MA such as MA, 3-epihydroxymugineic acid (epi-HMA), and 3-epihydroxy-2′-deoxymugineic acid (epi-HDMA) [22]. During iron deficiency, graminaceous plants will secrete MAs into the rhizosphere to solubilize sparingly soluble iron in rhizosphere. MAs will bind Fe^{3+} efficiently forming Fe^{3+}-MA complexes. The complexes will be transported into the root via yellow stripe 1 (YS1) transporter [22, 24].

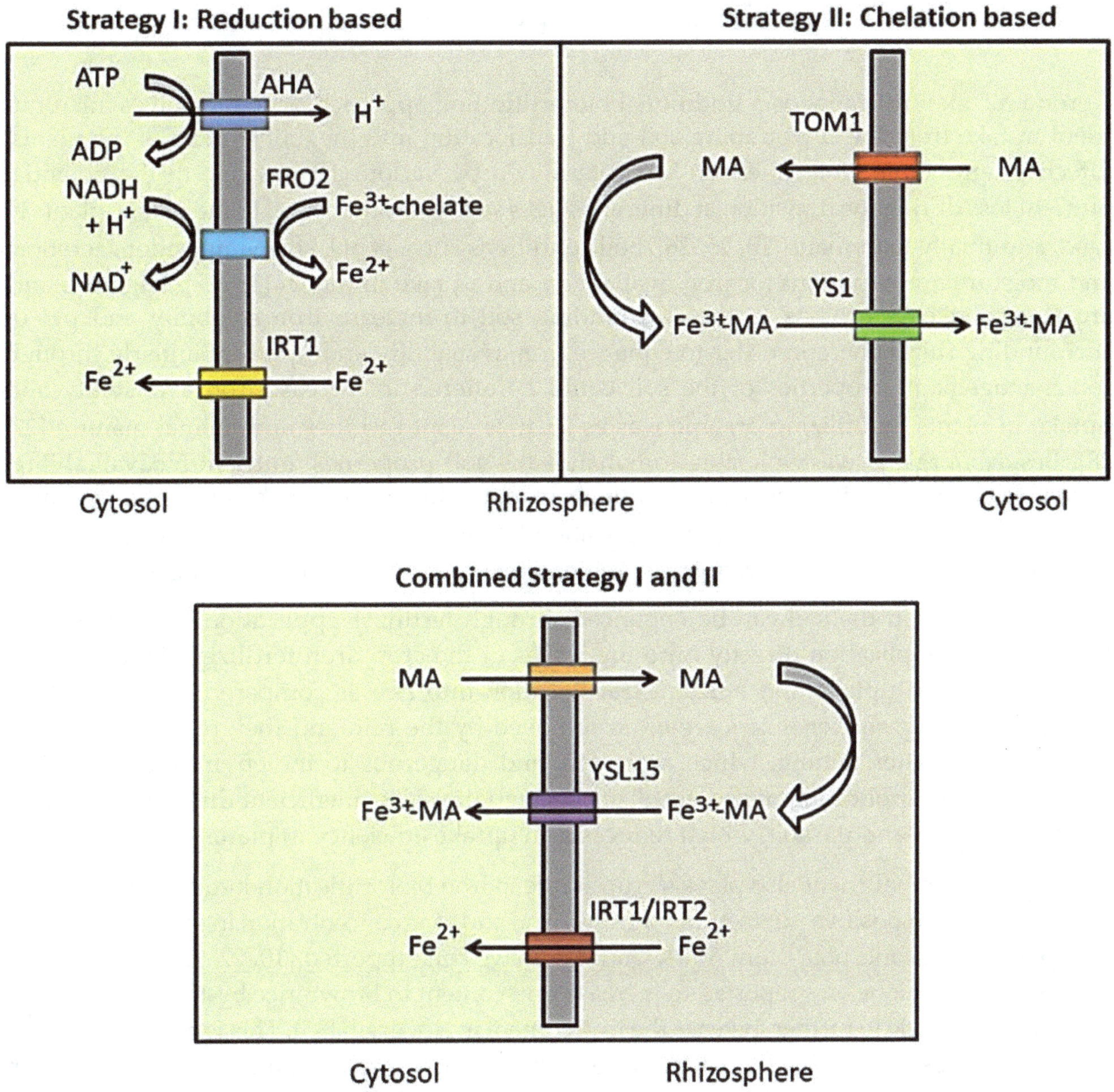

Figure 1. Iron uptake strategy by graminaceous plants and nongraminaceous plants.

2.3. Iron uptake mechanism in rice

Some graminaceous plants, in particular rice, can undergo combined strategies of reduction-based strategy and chelation-based strategy for iron uptake. Rice acquires Fe^{3+} via strategy I-like system and Fe^{2+} directly from the surroundings via IRT1 or IRT2. However, there is no increase in Fe^{3+}-chelate reductase levels detected in the roots as compared to nongraminaceous plants [20]. Possible explanation is that adaptation of rice when grown in submerged and anaerobic environment rich in Fe^{2+} compared to Fe^{3+} [10]. Similarly to strategy II, MAs will be secreted into the rhizosphere to bind with Fe^{3+} and the complexes will be transported into the root via YS-like 15 (YSL15). Between both strategies, rice is able to uptake iron from the surrounding more efficiently through Fe^{3+}-MA complexes as compared to direct Fe^{2+} uptake [22].

3. Iron biofortification via agronomic practices

Agronomic biofortification is a traditional biofortification approach, which involves micronutrient uptake from the surrounding soil and translocation into the edible parts of the plants. Effective agronomic biofortification are determined by various factors due to the potential nutrient loss during the transition at different stages such as from the soil to the plants, plants to food, and finally to humans [13, 25, 26]. Soil conditions such as pH, soil composition, aeration, and moisture are important for iron availability and uptake in plants [13, 27]. Higher plants are able to release protons to the surrounding soil to increase iron solubility and pH of surrounding rhizosphere in order to enhance iron availability and uptake. Similarly through soil management, properties of the soil could be altered to increase iron availability and uptake in plants by utilizing organic wastes such as plant residues and animal manure [27, 28]. Besides, organic wastes is able to enhance the soil properties, nutrient bioavail-ability, cation exchange capacity, and water holding capacity while providing a constant and slower nutrient release [13, 29]. However, application of organic wastes alone is insufficient to mitigate iron deficiency and it requires combination application with iron fertilizer [13].

Iron availability in the soil can be enhanced through fertilizer application onto the soil or foliar feeding application directly onto the leaves of the crop. Iron fertilizer via foliar feeding enhance iron uptake and efficient translocation into rice as compared to soil fertilizer [30–32]. However, the fertilizers are often removed by the rain and they require reapplication each time after raining, which are costly and dangerous to the environment [13, 33]. Conversely, the application of iron fertilizer through the soil is inefficient due to strong binding between iron and the soil, which reduces iron uptake efficiency in plants [13, 15].

In addition, macronutrient also plays a crucial role in iron biofortification in plants. Previously, several studies on positive interactions between iron and zinc concentration in grains with nitrogen, phosphorus, and potassium (NPK) fertilizer have been reported [10, 27, 32, 34]. The presence of nitrogen alone was reported to increase iron content in brown rice by 15% and addition of potassium is able to further increase the iron content in rice grain [35]. This is because nitrogen and phosphorus are involved in root development, shoot transport and re-localization, which improves the translocation of iron into rice grain [13, 15, 27, 33]. On the other hand, the presence of phosphorus is able to reduce toxicity in plants at the cost of reduce uptake of both iron and zinc uptake in plant due to dilution effect [13]. Hence, combined application of both NPK fertilizer and iron fertilizer could be a potential approach to increase iron bioavailability in rice [13].

4. Iron biofortification via conventional plant breeding

Conventional plant breeding has been practiced for centuries to improve the properties of food crops by identifying and developing parent plants with desired characteristics, crossing the parent plants, and selecting offspring with desired agronomic traits inherited from both parent plants [14]. An example of a product developed via plant breeding is high iron rice variety (IR68144) with high yield, disease tolerant, good tolerant to mineral deficient, and excellent seed vigor. The IR68144 rice variety was developed through crossing between semi dwarf rice cultivar, IR8 and Taichung (Native)-1. Meanwhile, IR8 is a product developed through crossing between Chinese dwarf rice variety "Dee-Geo-woo-gen" (DGWG) and Indonesia high yield rice variety "Peta" [36]. The Taichung (Native)-1 is a product of crossing between DGWG and a traditional tall variety 'Tsai-Yuan-Chung', which produces high yield and dwarf variety. Crossing between IR8 and Taichung (Native)-1 allows the development of new rice cultivar, which is semi-dwarf and contains high yield properties [37]. This rice variety is able to produce 21 μg/g (2-fold) of iron concentration in brown rice [35]. In addition, IR68144 is able to retain most of the iron content (approximate 80%) after polishing for 15 minutes compared to other varieties [10]. Furthermore, consumption of IR68144 was reported to have improvement in iron status of women [38]. This rice cultivar can serve as a stepping stone for further transgenic enhancement [10].

Even though conventional breeding is able to develop high yield and semi dwarf IR68144, this approach alone in iron biofortification is insufficient in developing a sustainable agronomic plant in terms of yield and quality [39]. This is due to the possibility of inheriting undesirable traits from the parent line as the selection process is done based on the phenotypes and new traits can only be developed after performing extensive back crossing or wide crossing [40]. For instance, low phytic acid (PA) maize mutant (*lpa241*) has demonstrated its ability to reduce PA concentration by 90% in exchange of reduced germination rate by 30% [41]. Hence, conventional breeding is best coupled with other approach such as genetic engineering and agronomic practices to enhance iron content in grains [32, 37, 42, 43].

5. Iron biofortification via genetic engineering

The advancement of genetic engineering technologies allows the advancement in molecular field including the development of transgenic plants. Characterization and analysis of gene function are performed via genetic engineering by the manipulation of gene expression. These include introducing gene of interest from other closely-related organism, RNA interference (RNAi) gene silencing, and overexpression of gene of interest [18]. Genetic engineering technologies is able to provide a more efficient and reliable method to study the relationship between genotype and the phenotype as compared to agronomic and conventional plant breeding [44, 45]. As a result, genetic engineering is preferred as an alternative for biofortification to increase the iron content in rice grains. There are five different transgenic approaches (**Table 1**) and as well as combination of different transgenic approaches (**Table 2**), which have been attempted and successfully used to enhance the iron content in rice grain.

5.1. Enhancement of iron storage in rice via *ferritin* genes

Ferritin is an iron storage protein ubiquitously present in most organisms, which is capable to store up to 4500 iron atoms in a complex and nontoxic form [65, 66]. Iron complex in soybean ferritin is readily available for human body absorption via iron uptake mechanism in the intestine [46, 62]. Thus, the first approach in iron biofortification is to enhance the expression of ferritin by introducing soybean *ferritin* (*SoyferH1* and *SoyferH2*) genes into rice.

Approach	Genes-promoter used	Rice cultivar	Fold of Fe increase	References
Improving iron storage via *ferritin* genes	*OsGluB1* pro-*SoyferH1*	*Japonica* cv. Kitaake	1.5 fold (brown grain)	[48]
	OsGluB1 pro-*SoyferH1*	*Japonica* cv. Kitaake	2 fold (polished grain)	[49]
	OsGluB1 pro-*SoyferH1*	*Japonica* cv. Taipei 309	2.2 fold (brown grain)	[51]
	OsGluB1 pro-*SoyferH1*	*Indica* cv. IR68144	3.7 fold (polished grain)	[50]
	OsGluB4 pro-*SoyferH1*	*Indica* cv. IR64	3.4 fold (polished grain)	[64]
	OsGluA2 pro-*Osfer2*	*Indica* cv. Pusa-Sugandh II	2.1 fold (polished grain)	[67]
Enhancing iron transport via *NAS* gene	*35S* pro-*OsNAS1, 2, 3*	*Japonica* cv. Nipponbare	4 fold (polished grain)	[2]
	Maize *Ubiquitin* pro-*OsNAS2*	*Japonica* cv. Kitaake	2.9 fold (polished grain)	[54]
	Maize *Ubiquitin* pro-*OsNAS3*	*Japonica* cv. Dongjin	2.9 fold (polished grain)	[55]
	35S pro-*HvNAS1*	*Japonica* cv. Tsukinohikari	2.3 fold (polished grain)	[74]
Enhancing iron influx via *OsYSL2* gene	*OsSUT1* pro-*OsYSL2*	*Japonica* cv. Tsukinohikari	4.4 fold (polished grain)	[75]
Enhancing iron uptake and translocation via *IDS3* gene	*35S* pro-barley 20-kb *IDS3* genome fragment	*Japonica* cv. Tsukinohikari	1.4 fold (polished grain)	[58]
	35S pro-barley 20-kb *IDS3* genome fragment	*Japonica* cv. Tsukinohikari	1.3 fold (brown grain)	[78]
Enhancing iron translocation via silencing *OsVITs* genes	*OsVIT1* or *OsVIT2* T-DNA insertion line	*Japonica* cv. Zhonghua11 and *Japonica* cv. Dongjin	1.4 fold (brown grain)	[59]
	OsVIT2 T-DNA insertion line	*Japonica* cv. Dongjin	1.8 fold (polished grain)	[80]

Table 1. Iron biofortification approach in rice targeting genes responsible for iron storage, iron transport, iron influx, iron uptake and translocation.

In soybean, there are two types of ferritin proteins, known as *SoyferH1* and *SoyferH2*, and both *ferritin* genes are controlled by endosperm specific promoters [47]. However, expression of multiple endosperm specific promoters (*Oryza sativa Globulin* (*OsGlb*) and *Oryza sativa Glutelin* (*OsGluB1*) promoters) did not produce a significant increase of iron concentration in

Genes-promoter used	Rice cultivar	Fold of Fe increase	References
OsGlb1 pro-*Pvferritin* *35S* pro-*AtNAS1* *OsGlb* pro-*Afphytase*	*Japonica* cv. Taipei 309	6.3 fold (polished grain)	[23]
OsGluB1 pro-*SoyferH2* *OsGlb1* pro-*SoyferH2* *HvNAS1, HvNAAT-A,-B* and *IDS3* genome fragments	*Japonica* cv. Tsukinohikari	4 fold (polished grain)	[77]
MsENOD12B pro-*AtIRT1* *OsGlb1* pro-*Pvferritin* *35S* pro-*AtNAS1* *OsGlb* pro-*Afphytase*	*Japonica* cv. Taipei 309	4.3 fold (polished grain)	[86]
Native *AtIRT1* pro-*AtIRT1* *OsGlb1* pro-*Pvferritin* *35S* pro-*AtNAS1*	*Japonica* cv. Nipponbare	4.7 fold (polished grain)	[66]
OsGluB1 pro-*SoyferH2* *OsGlb1* pro-*SoyferH2* *OsAct* pro-*HvNAS1* *OsSUT1* pro-*OsYSL2* *OsGlb1* pro-*OsYSL2*	*Japonica* cv. Tsukinohikari	6 fold (brown grain)	[47]
OsGluB1 pro-*SoyferH2* *OsGlb1* pro-*SoyferH2* *OsAct* pro-*HvNAS1* *OsSUT1* pro-*OsYSL2* *OsGlb1* pro-*OsYSL2*	Tropical *Japonica* cv. Paw San Yin	3.4 fold (polished grain)	[88]
GluA2 pro-*SoyferH1* *35S* pro-*OsNAS2*	IR64	6 fold (polished grain)	[69]

Table 2. Combinational of multiple transgene used for iron biofortification in rice.

rice grains when compared to transgenic rice with *ferritin* genes expression driven by single endosperm specific promoter [48]. On the other hand, the overexpression of soybean *ferritin* in rice has been demonstrated with at least twofold increase in iron concentration in endosperm compared to the wild-type rice [49–51, 64, 67].

Nevertheless, introducing *SoyferH2* into rice plants is preferred as *SoyferH1* is more susceptible to protease digestion causing alteration in structure in comparison to *SoyferH2*, which is more resistant to protease digestion [68, 69]. Interestingly, rice plants introduced with single soybean *ferritin* gene did not increase iron concentration in rice grain [48, 68]. This suggests that expressions of *ferritin* genes are dependent on soil composition and overexpression of *ferritin* genes as a single transgene approach may be insufficient in combating iron deficiency [48, 70].

5.2. Enhancement of iron transport in rice via *NAS* genes

The second approach involves enhancing iron transport in the plant via overexpression of genes involves in biosynthesis of MA such as *nicotianamine synthase* (*NAS*). *NAS* is able to catalyze the synthesis of nicotianamine (NA) from *S*-adenosyl methionine [23]. NA, a natural metal chelators for Fe(II) and Fe(III), are found in all higher plants and involved with metal translocation and homeostasis in plants [47, 71–73].

Rice comprises of three *NAS* genes, *OsNAS1*, *OsNAS2*, and *OsNAS3*. These genes are involved in long-distance transportation in plants and each *NAS* gene is regulated at different parts of the plants in response to iron deficiency [9, 52]. Overexpression of *NAS* gene enhances MA secretion into the rhizosphere, and thus, increasing iron uptake into the plant via chelation-based strategy [23, 53, 72]. It has been demonstrated that overexpression of rice *OsNAS1*, *OsNAS2* and *OsNAS3* [2], *OsNAS2* [54], *OsNAS3* [55], and barley *HvNAS1* [74] genes are able to increase the iron content by more than twofold in polished grain.

5.3. Enhancement of iron influx into seeds via *OsYSL2* gene

A total of 18 different *YSL* (*yellow stripe-like*) genes were identified by Koike [73] in rice. The rice *YSL2* (*OsYSL2*) is the main focus in this approach as this gene plays an important role as a metal-chelator transporter involved in translocation and accumulation of iron in endosperm [73, 75]. *OsYSL2* was found to be highly expressed in leaves of iron-deficient rice plants in contrast to other parts of the plant where no expression was detected. Therefore, Koike [73] hypothesized that this transporter is involved in long-distance transport of iron-NA complexes via phloem in response to iron deficiency in rice plant.

Consistently, it was discovered that the iron influx into the rice endosperm could be controlled through iron nicotianamine transporter *OsYSL2* [60]. Ishimaru [75] successfully demonstrated that disruption of *OsYSL2* gene in rice decreased the iron content in both brown rice and polished grain by 18 and 39%, respectively with increased iron accumulation in roots as compared to wild-type rice. Moreover, Ishimaru [75] also able to increase the iron content in rice grain up to 4-fold in polished grain through enhanced expression of *OsYSL2* using the rice sucrose transporter (*OsSUT1*) promoter. However, overexpression of *OsYSL2* may cause opposite effect similar to *OsYSL2* gene silencing in transgenic rice such that the iron concentration in roots was found higher than in both shoot and rice grain. Undoubtedly, the expression of *OsYSL2* with *OsSUT1* promoter is a promising approach in iron biofortification of rice grains.

5.4. Enhancement of iron uptake and translocation via *IDS3* gene

MAs are natural iron chelators, which are involved in transloca-tion of iron from the rhizosphere into the plant by forming complexes with iron. Different sets of *MAs* genes were found in barley, which confers the ability to synthesize different types of MAs via biosynthetic pathway of MAs [76, 77]. In addition, the presence of iron deficiency specific clone no. 2 (*IDS2*) and no. 3 (*IDS3*) in barley play an important role in combating iron deficiency [77]. The *IDS* genes enable the synthesis of different types of MAs via DMA and these genes are highly expressed in roots in response to iron deficiency [56]. On the contrary, rice lacks the ability to synthesize other types of MAs apart of DMA as rice does not contain both *IDS2* and *IDS3* genes. Having different sets of MAs enable barley become more tolerant to iron-deficient conditions as compared to rice.

Introducing *IDS3* gene from barley enables the synthesis and secretion of different types of MAs from transgenic rice into the rhizosphere [56]. In addition, formation of Fe(III)-MA complex has a better stability as compared to Fe(III)-DMA complex when grown in a slightly acidic soil [57]. This may enhance iron translocation in rice in combating iron deficiency while increasing tolerance toward iron deficiency in rice plants. Furthermore, Masuda [58] and Suzuki [78] demonstrated that *IDS3* rice lines are able to increase Fe concentrations to 1.4 and 1.3-fold for both polished and brown grains respectively compared to wild-type rice when grown in either Fe-sufficient soil or Fe-deficient soil. Thus, presence of *IDS3* gene is able to enhance iron accumulation in rice grain even when it is grown in iron-sufficient soil and as well as enhancing tolerance toward iron deficiency.

5.5. Enhancement of iron translocation via *OsVIT* gene

Zhang [59] reported on the functional characterization of rice vacuolar iron transporter genes (*OsVIT1* and *OsVIT2*). These genes were found to be expressed ubiquitously in different parts of the plants at low levels but high level expression of *OsVIT* genes were detected in the flag leaves. These genes play an important role in transportation of Zn^{2+} and Fe^{2+} into vacuoles via tonoplast [79]. In addition, knockdown of *OsVIT* genes increases Fe and Zn accumulation in the rice grains significantly while decreases Fe and Zn accumulation in the flag leaves correspondingly [80]. Knockout of *OsVIT1* and *OsVIT2* genes were able to increase the iron content in rice grain by at least 1.4-fold [59, 80]. However, this approach is only applicable when the transgenic rice is grown in unpolluted soil. This is because studies had shown that accumulation of Cd^{2+} concentration was detected in rice when it is grown in polluted soil [59]. Hence, further understanding of regulatory mechanism is required to prevent toxic metal accumulation and to ensure the crops are safe for consumption.

5.6. Combinational of multiple transgenes

Multiple gene manipulation has been successfully carried out in rice. Wirth [23] has proven the synergism of three different genes expression with the increased of iron content in rice by 6-fold through introducing *Arabidopsis thaliana NAS1* (*AtNAS1*), *Phaseolus vulgaris ferritin* (*Pvferritin*), and *Aspergillus fumigatus phytase* (*Afphytase*) genes into rice. The main purpose of introducing *phytase* genes is to reduce iron antinutrient phytate in rice. Some food may contain antinutrients like PA, which binds strongly to metal cations, such as iron and zinc, which render them insoluble [81]. Phytase is able to catalyze the hydrolysis of PA releasing the phosphate and chelated minerals [21]. Human digestive system lacks enzyme responsible for breakdown of such components [23]. Reducing antinutrients is a feasible approach to increase nutrient content in crops but it should be exercised with cautions due to many antinutrients playing important roles in both plant metabolism and human diet [21, 32]. In plants, antinutrients involve in resistance toward pests, pathogens and abiotic stress and at the same time function as anticarcinogens in human diets [61, 63, 82–85]. For instance, PA is able to reduce the risk for both colon cancer and mammary cancer through its strong metal cations binding capabilities [63, 83]. Moreover, PA display antioxidant capability by acting as inhibitor of iron-mediated hydroxyl radical (-OH) formation in food and gastrointestinal tract, which would result in lipid peroxidation and tissue damage [83, 84]. On the other hand, antinutrient lectin was found to be responsible for plant defense system by exhibiting cytotoxic activities when ingested by pests and animals [85].

Masuda [77] has demonstrated that introducing a combination of different genes responsible for MA synthesis into rice (Fer-NAS-NAAT-IDS3 lines) and result in 4-fold increase of iron accumulation in endosperm. Likewise, transgenic line expressing *AtIRT1, Pvferritin, AtNAS1,* and *Afphytase* was shown to cause a 4-fold increase of iron accumulation in polished grain [66, 86]. The *OsYSL15* or *OsIRT1* genes are predominantly expressed in roots with enhanced expression in response to iron deficiency [86]. *OsIRT1* gene encode for Fe^{2+} transporter involved in both strategy I and II. Although overexpression of *OsIRT1* alone could increase the iron content in rice grain by 1.3-fold, *OsIRT1* has the potential to further enhance the iron content when it is expressed with other genes [87].

On the other hand, combination approaches were also demonstrated to increase the iron content in rice grain by 3.4- and 6-fold when introduced into Myanmar and Japanese rice cultivar respectively [47, 88]. Both *SoyFerH2* and *OsYSL2* were strongly expressed in transgenic rice due to the vector inserted contains two gene cassettes for each gene expression driven by different promoters for each gene cassettes (*OsSUT1* promoter-*OsYSL2*, *OsGlb* promoter-*OsYSL2*, *OsGluB1* promoter-*SoyferH2*, *OsGlb* promoter-*SoyferH2*). Interestingly, Trijatmiko [69] was able to develop transgenic rice expressing *OsNAS2* and *SoyferH1* genes result in 15 μg Fe/g increased (6-fold) in polished grain. In the transgenic rice line, the transgene construct was found to be inserted with inverted repeats in a single locus. This concludes that multiple transgene insertion was able to increase the iron concentration in rice [47, 69, 88]. However, transgene cassette with duplicated or inverted repeats of transgene may not be stable and inherited after several generations due to possibility of epigenetic silencing in transgenic plants [66, 89–91]. Hence, further investigation should be conducted to elucidate the stability of transgene or different approach to maintain multiple transgene over multiple generations.

6. Challenges and future prospect

Biofortification is a promising strategy for sustainable long-term approach in combating micronutrient deficiency but successful biofortification at the cost of the environmental damage is not acceptable. In agronomic practice, leaching is one of the main concerns in application of fertilizer as it will damage the environment, but most micronutrients are not susceptible to leaching as they are able to form a strong bond with the soil [13]. However, continuous application of micronutrient fertilizer may cause accumulation of these minerals which result in toxicity. Excessive intake of iron may cause Fe^{2+} and Fe^{3+} to act as a catalyst to form noxious reactive oxygen species (ROS). ROS are strong oxidizing agents, which are able to cause detrimental effect on DNA, proteins, and lipids in plants [33]. Therefore, fertilization strategies should be devised and optimized to ensure adequate supply of iron for proper growth of agronomic plants while minimizing accumulation of iron [92]. For instance, the 4R Nutrient Stewardship principle (application of fertilizer at the right place, right rate, right time and right source) could be implemented with fertilizer application [34, 93]. Based on HarvestPlus breeding programs, the iron-biofortified rice are to meet a recommended target iron level which is approximate 30% of the estimated average requirement (EAR) or 15 μg/g (dry weight) in polished grain [69, 94]. The recommended 30% EAR could be achieved via genetic engineering approaches listed in both **Tables 1** and **2**, however, the iron concentration in rice grain decreases when evaluated under field conditions as compared to iron concentration achieved in rice grown in greenhouse [47, 69]. This demonstrates that interactions between genetic and environment play an important role in iron concentration in rice grain [69, 95]. Field experiments should be included in evaluating iron levels in iron-biofortified rice for several growing seasons as evaluating iron levels in rice grown under strictly controlled environment conditions does not simulate the conditions when grown in the field [69, 94, 96, 97].

Biofortified crops still face strict regulatory hurdles and a lack of consumer acceptance, especially in Europe, even though there have been reports on improvement in nutritional status after consuming biofortified crops [16, 38, 98, 99]. For instance, golden rice, a product developed via genetic engineering in combating vitamin A deficiency, has been announced since early 2000, but it has yet to be seen in the market [100]. Although these transgenic plants has demonstrated its high nutritional content in combating micronutrient deficiency, but as far as public safety concern, additional regulations and more stringent monitoring are implemented onto transgenic crops before being available to the public compared to conventional breeding which is more widely accepted [3]. In addition, there are possibilities of irreversibility effect on health and environment due to the effects of GM crops on health and environment are not fully understood and not sustainable in the long run [100, 101]. Furthermore, there are possibilities of the transgene in biofortified crops survived through human digestion system which allows transgenic plant DNA such as antibiotic resistance genes to be transferred into small intestine microflora [102]. Therefore, additional researches from different disciplines are required in order to elucidate the effect of biofortified crops consumption on human health. This may appease public anxiety and to gain consumer acceptance [3]. On the other hand, the recent advancement of genetic engineering technologies, such as zinc-finger nucleases, TALENs and CRISPR-Cas9, could be a potential approach in iron biofortification, which allows efficient and effective gene editing without affecting the plant as a whole [18, 44, 45]. Moreover, gene-edited crops are subjected to different regulations and monitoring from government bodies and nongovernment organizations, which are not as stringent as genetically modified crops. As a result, gene-edited crops will have a higher consumer acceptance compared to conventional genetic engineering.

While iron biofortification in rice is a promising approach in combating iron deficiency, the success of biofortification is dependent on various factors and it requires the collaboration between different parties ranging from consumer, plant breeder, multilateral organizations, national governments, and researchers from various disciplines. Without the help and adoption from plant breeders, biofortified crops are unable to be produced despite the crop has the potential to alleviate micronutrient deficiency. Hence, to gain plant breeder acceptance, biofortified crops should contain visible and favorable traits such as increased in yield, higher stress tolerant, disease resistance, and other important agronomic traits [10]. Plant breeders may be reluctant to produce the biofortified crops with the potential income risk if the consumer does not adopt with the new crop variety especially with biofortified crops having their sensory characteristics altered such as the color and taste [12]. Some biofortified crops have been introduced for production and accepted by the public in some countries despite the change in sensory characteristics [14]. These biofortified crops are orange flesh sweet potato, orange maize, yellow cassava, iron pearl millet, and iron beans. Consumer acceptance on biofortified crops is not easy and achievable in a short duration of time but it can be accomplished through thoroughly planned strategies such as spreading knowledge among the people, raising awareness of micronutrient deficiency, creating new market opportunities, and creating a demand on biofortified variety [103]. On the contrary, the success of iron biofortification would results in improved nutritional value of micronutrient-deficient affected areas in developing countries and as a first step toward improving nutritional status worldwide.

Acknowledgements

The authors thank Putra Grant (GP/2017/9572000) from Universiti Putra Malaysia and Fundamental Research Grant Scheme (FRGS/1/2014/SG05/MOSTI/1) from Ministry of Higher Education, Malaysia for the support.

Author details

Andrew De-Xian Kok[1], Low Lee Yoon[1], Rogayah Sekeli[2], Wee Chien Yeong[2], Zetty Norhana Balia Yusof[3] and Lai Kok Song[1]*

*Address all correspondence to: laikoksong@upm.edu.my

1 Department of Cell and Molecular Biology, Faculty of Biotechnology and Biomolecular Sciences, Universiti Putra Malaysia, Selangor, Malaysia

2 Malaysian Agricultural Research and Development Institute (MARDI), Kuala Lumpur, Malaysia

3 Department of Biochemistry, Faculty of Biotechnology and Biomolecular Sciences, Universiti Putra Malaysia, Selangor, Malaysia

References

[1] Yang Q, Zhang C, Chan M, Zhao D, Chen J, Wang Q, et al. Biofortification of rice with the essential amino acid lysine: Molecular characterization, nutritional evaluation, and field performance. Journal of Experimental Botany. 2016 Jul;**67**(14):4285-4296

[2] Johnson AAT, Kyriacou B, Callahan DL, Carruthers L, Stangoulis J, Lombi E, et al. Constitutive overexpression of the OsNAS gene family reveals single-gene strategies for effective iron- and zinc-biofortification of Rice endosperm. Baxter I, editor. PLoS One 2011 Sep;**6**(9):e24476

[3] Bouis HE, Saltzman A. Improving nutrition through biofortification: A review of evidence from HarvestPlus, 2003 through 2016. Vol. 12, Global Food Security. Elsevier; 2017. p. 49-58

[4] Mayer JE, Pfeiffer WH, Beyer P. Biofortified crops to alleviate micronutrient malnutrition. Vol. 11, Current Opinion in Plant Biology. Elsevier Current Trends; 2008. p. 166-170

[5] White PJ, Broadley MR. Biofortification of crops with seven mineral elements often lacking in human diets - iron, zinc, copper, calcium, magnesium, selenium and iodine. The New Phytologist. 2009 Apr;**182**(1):49-84

[6] WHO. WHO Micronutrient deficiencies [Internet]. WHO. World Health Organization; 2015 [cited 2017 Oct 9]. Available from: http://www.who.int/nutrition/topics/ida/en/

[7] Morrissey J, Lou GM. Iron uptake and transport in plants: The good, the bad, and the ionome. Chemical Reviews. 2009 Oct;**109**(10):4553-4567

[8] Hurrell R. How to ensure adequate iron absorption from iron-fortied food. Nutrition Reviews. 2002 Jul;**60**:(7):7-(7)15

[9] Singh SP, Keller B, Gruissem W, Bhullar NK. Rice nicotianamine synthase 2 expression improves dietary iron and zinc levels in wheat. Theoretical and Applied Genetics. 2017 Feb;**130**(2):283-292

[10] Sperotto RA, Ricachenevsky FK, Waldow V de A, Fett JP. Iron biofortification in rice: It's a long way to the top. Plant Science. 2012;**190**:24-39

[11] Phattarakul N, Rerkasem B, Li LJ, Wu LH, Zou CQ, Ram H, et al. Biofortification of rice grain with zinc through zinc fertilization in different countries. Plant and Soil. 2012 Dec;**361**(1-2):131-141

[12] Welch RM, Graham RD. Agriculture: The real nexus for enhancing bioavailable micronutrients in food crops. Journal of Trace Elements in Medicine and Biology. 2005 Jun;**18**(4):299-307

[13] de Valença AW, Bake A, Brouwer ID, Giller KE. Agronomic biofortification of crops to fight hidden hunger in sub-Saharan Africa. Vol. 12, Global Food Security. Elsevier; 2017. p. 8-14

[14] Saltzman A, Birol E, Oparinde A, Andersson MS, Asare-Marfo D, Diressie MT, et al. Availability, production, and consumption of crops biofortified by plant breeding: Current evidence and future potential. Annals of the New York Academy of Sciences. 2017 Feb;**1390**(1):104-114

[15] Petry N, Boy E, Wirth JP, Hurrell RF. Review: The potential of the common bean (Phaseolus vulgaris) as a vehicle for iron biofortification. Nutrients. 2015 Feb;**7**(2):1144-1173

[16] Haas JD, Luna S V, Lung'aho MG, Wenger MJ, Murray-Kolb LE, Beebe S, et al. Consuming iron biofortified beans increases iron status in Rwandan women after 128 days in a randomized controlled feeding trial. The Journal of Nutrition 2016 Aug;**146**(8):1586-1592

[17] Collard BCY, Mackill DJ. Marker-assisted selection: An approach for precision plant breeding in the twenty-first century. Philosophical Transactions of the Royal Society of London. Series B, Biological Sciences. 2008 Feb;**363**(1491):557-572

[18] Abdallah NA, Prakash CS, McHughen AG. Genome editing for crop improvement: Challenges and opportunities. GM Crops Food. 2015 Oct;**6**(4):183-205

[19] Lim YY, Lai KS. Generation of transgenic rice expressing cyclotide precursor Oldenlandia affinis kalata b1 protein. Plant Science 2017 Apr;**27**(2):671-675

[20] Kim SA, Lou GM. Mining iron: Iron uptake and transport in plants. May. 2007;**581**(12): 2273-2280

[21] Welch RM. Breeding strategies for biofortified staple plant foods to reduce micronutrient malnutrition globally. The Journal of Nutrition. 2002 Mar;**132**(3):495S-499S

[22] Ishimaru Y, Suzuki M, Tsukamoto T, Suzuki K, Nakazono M, Kobayashi T, et al. Rice plants take up iron as an Fe^{3+}-phytosiderophore and as Fe^{2+}. The Plant Journal. 2006 Feb;**45**(3):335-346

[23] Wirth J, Poletti S, Aeschlimann B, Yakandawala N, Drosse B, Osorio S, et al. Rice endosperm iron biofortification by targeted and synergistic action of nicotianamine synthase and ferritin. Plant Biotechnology Journal. 2009 Sep;**7**(7):631-644

[24] Schaaf G, Ludewig U, Erenoglu BE, Mori S, Kitahara T, von Wirén N. ZmYS1 functions as a proton-coupled symporter for phytosiderophore- and nicotianamine-chelated metals. The Journal of Biological Chemistry 2004 Mar;**279**(10):9091-9096

[25] Daud NSM, Zaidel DNA, Lai KS, Muhamad II, Mohd Jusoh YM. Antioxidant properties of rice bran oil from different varieties extracted by solvent extraction methods. Jurnal Teknologi. 2016 Jun;**78**(6-12):107-110

[26] Daud NSM, Zaidel DNA, Lai KS, Jusoh YMM, Muhamad II, Ya'akob H. Microwave-Assisted stabilisation and storage stability study of rice bran oil from different varieties. Vol. 56, Chemical Engineering Transactions. Italian Association of Chemical Engineering - AIDIC; 2017. p. 1285-1290

[27] Prasad R, Shivay YS, Kumar D. Agronomic biofortification of cereal grains with iron and zinc. Advances in Agronomy. 2014 Jan;**125**(1):55-91

[28] Haynes RJ, Naidu R. Influence of lime, fertilizer and manure applications on soil organic matter content and soil physical conditions: A review. Nutrient Cycling in Agroecosystems. 1998 Jun;**51**(2):123-137

[29] Zingore S, Delve RJ, Nyamangara J, Giller KE. Multiple benefits of manure: The key to maintenance of soil fertility and restoration of depleted sandy soils on African smallholder farms. Nutrient Cycling in Agroecosystems. 2008 Mar;**80**(3):267-282

[30] Wei Y, Shohag MJI, Yang X, Zhang Y. Effects of foliar iron application on iron concentration in polished rice grain and its bioavailability. Journal of Agricultural and Food Chemistry. 2012 Nov;**60**(45):11433-11439

[31] Yuan L, Wu L, Yang C, Lv Q. Effects of iron and zinc foliar applications on rice plants and their grain accumulation and grain nutritional quality. Journal of the Science of Food and Agriculture. 2013 Jan;**93**(2):254-261

[32] Velu G, Ortiz-Monasterio I, Cakmak I, Hao Y, Singh RP. Biofortification strategies to increase grain zinc and iron concentrations in wheat. Vol. 59, Journal of Cereal Science. Academic Press; 2014. p. 365-372

[33] García-bañuelos ML, Sida-arreola JP, Biofortification SE. Promising approach to increasing the content of iron and zinc in staple food crops. Journal of Elementology. 2014 Sep;**19**(3):865-888

[34] Bindraban PS, Dimkpa C, Nagarajan L, Roy A, Rabbinge R. Revisiting fertilisers and fertilisation strategies for improved nutrient uptake by plants. Biology and Fertility of Soils. 2015 Nov;**51**(8):897-911

[35] Gregorio GB, Senadhira D, Htut H, Graham RD. Breeding for trace mineral density in rice. Food and Nutrition Bulletin. 2000 Dec;**21**(4):382-386

[36] Peng S, Cassman KG, Virmani SS, Sheehy J, Khush GS. Yield potential trends of tropical tice since the telease of IR8 and the challenge of increasing rice yield potential. Crop Science. 1999 Nov;**39**(1):1552-1559

[37] Virmani SS, Ilyas-Ahmed M. Rice breeding for sustainable production. In: Breeding Major Food Staples. Oxford, UK: Blackwell Publishing Ltd; 2008. pp. 141-191

[38] Haas JD, Beard JL, Murray-Kolb LE, del Mundo AM, Felix A, Gregorio GB. Iron-biofortified rice improves the iron stores of nonanemic Filipino women. The Journal of Nutrition 2005 Dec;**135**(12):2823-2830

[39] Graham R, Senadhira D, Beebe S, Iglesias C, Monasterio I. Breeding for micronutrient density in edible portions of staple food crops: Conventional approaches. Field Crops Research. 1999 Jan;**60**(1-2):57-80

[40] Bhullar NK, Gruissem W. Nutritional enhancement of rice for human health: The contribution of biotechnology. Vol. 31, Biotechnology Advances. Elsevier; 2013. p. 50-57

[41] Pilu R, Panzeri D, Gavazzi G, Rasmussen SK, Consonni G, Nielsen E. Phenotypic, genetic and molecular characterization of a maize low phytic acid mutant (lpa241). TAG Theoretical and Applied Genetics. 2003 Oct;**107**(6):980-987

[42] Welch RM, Graham RD. Breeding for micronutrients in staple food crops from a human nutrition perspective. Journal of Experimental Botany. 2004 Feb;**55**(396):353-364

[43] Jeng TL, Lin YW, Wang CS, Sung JM. Comparisons and selection of rice mutants with high iron and zinc contents in their polished grains that were mutated from the indica type cultivar IR64. Journal of Food Composition and Analysis. 2012 Dec;**28**(2):149-154

[44] Gaj T, Gersbach CA, Barbas CF, III. ZFN, TALEN, and CRISPR/Cas-based methods for genome engineering. Trends in Biotechnology 2013 Jul;**31**(7):397-405

[45] Yin H, Kauffman KJ, Anderson DG. Delivery technologies for genome editing. Nature Reviews. Drug Discovery. 2017 Mar;**16**(6):387-399

[46] Theil EC. Iron homeostasis and nutritional iron deficiency. The Journal of Nutrition. 2011 Apr;**141**(4):724S-728S

[47] Masuda H, Ishimaru Y, Aung MS, Kobayashi T, Kakei Y, Takahashi M, et al. Iron biofortification in rice by the introduction of multiple genes involved in iron nutrition. Scientific Reports. 2012 Jul;**2**(1):543

[48] Qu LQ, Yoshihara T, Ooyama A, Goto F, Takaiwa F. Iron accumulation does not parallel the high expression level of ferritin in transgenic rice seeds. Planta. 2005 Oct;**222**(2):225-233

[49] Goto F, Yoshihara T, Shigemoto N, Toki S, Takaiwa F. Iron fortification of rice seed by the soybean ferritin gene. Nature Biotechnology. 1999 Mar;**17**(3):282-286

[50] Vasconcelos M, Datta K, Oliva N, Khalekuzzaman M, Torrizo L, Krishnan S, et al. Enhanced iron and zinc accumulation in transgenic rice with the ferritin gene. Plant Science. 2003 Mar;**164**(3):371-378

[51] Lucca P, Hurrell R, Potrykus I. Fighting iron deficiency anemia with iron-rich rice. Journal of the American College of Nutrition. 2002 Jun;**21**(Suppl 3):184S-190S

[52] Inoue H, Higuchi K, Takahashi M, Nakanishi H, Mori S, Nishizawa NK. Three rice nicotianamine synthase genes, OsNAS1, OsNAS2, and OsNAS3 are expressed in cells involved in long-distance transport of iron and differentially regulated by iron. The Plant Journal. 2003 Nov;**36**(3):366-381

[53] Bashir K, Ishimaru Y, Nishizawa NK. Iron uptake and loading into rice grains. Rice. 2010 Sep;**3**(2-3):122-130

[54] Lee S, Kim Y-S, Jeon US, Kim Y-K, Schjoerring JK, An G. Activation of rice nicotianamine synthase 2 (OsNAS2) enhances iron availability for biofortification. Molecules and Cells. 2012 Mar;**33**(3):269-275

[55] Lee S, Jeon US, Lee SJ, Kim Y-K, Persson DP, Husted S, et al. Iron fortification of rice seeds through activation of the nicotianamine synthase gene. Proceedings of the National Academy of Sciences of the United States of America. 2009 Dec;**106**(51):22014-22019

[56] Kobayashi T, Nakanishi H, Takahashi M, Kawasaki S, Nishizawa NK, Mori S. Vivo evidence that Ids3 from Hordeum Vulgare encodes a dioxygenase that converts 2′-deoxymugineic acid to mugineic acid in transgenic rice. Planta. 2001 Apr;**212**(5-6):864-871

[57] von Wirén N, Khodr H, Hider RC. Hydroxylated phytosiderophore species possess an enhanced chelate stability and affinity for iron(III). Plant Physiology 2000 Nov;**124**(3):1149-1158

[58] Masuda H, Suzuki M, Morikawa KC, Kobayashi T, Nakanishi H, Takahashi M, et al. Increase in iron and zinc concentrations in rice grains via the introduction of barley genes involved in phytosiderophore synthesis. Rice. 2008 Sep;**1**(1):100-108

[59] Zhang Y, Xu YH, Yi HY, Gong JM. Vacuolar membrane transporters OsVIT1 and OsVIT2 modulate iron translocation between flag leaves and seeds in rice. The Plant Journal. 2012 Nov;**72**(3):400-410

[60] Schroeder JI, Delhaize E, Frommer WB, Guerinot M, Lou HMJ, Herrera-Estrella L, et al. Using membrane transporters to improve crops for sustainable food production. Nature. 2013 May;**497**(1):60-66

[61] Metabolism SAM. Cellular functions of IP6: A review. In: Anticancer Research. 1999. pp. 3733-3736

[62] Lönnerdal B. Soybean ferritin: Implications for iron status of vegetarians. The American Journal of Clinical Nutrition 2009 May;**89**(5):1680S-1685S

[63] Thompson LU, Zhang L. Phytic acid and minerals: Effect on early markers of risk for mammary and colon carcinogenesis. Carcinogenesis. 1991 Nov;**12**(11):2041-2045

[64] Oliva N, Chadha-Mohanty P, Poletti S, Abrigo E, Atienza G, Torrizo L, et al. Large-scale production and evaluation of marker-free indica rice IR64 expressing phytoferritin genes. Molecular Breeding. 2014 Jan;**33**(1):23-37

[65] Theil EC. Ferritin: At the crossroads of iron and oxygen metabolism. The Journal of Nutrition. 2003 May;**133**(5 Suppl 1):1549S-1553S

[66] Boonyaves K. Wu T-Y, Gruissem W, Bhullar NK. Enhanced grain iron levels in Rice expressing an iron-regulated metal transporter, nicotianamine synthase, and ferritin gene cassette. Frontiers in Plant Science. 2017 Feb;**8**(1):130

[67] Paul S, Ali N, Gayen D, Datta SK, Datta K. Molecular breeding of Osfer2 gene to increase iron nutrition in rice grain. GM Crops Food. 2012 Oct;**3**(4):310-316

[68] Masuda H, Aung M, Nishizawa NK. Iron biofortification of rice using different transgenic approaches. Rice. 2013 Dec;**6**(1):40

[69] Trijatmiko KR, Dueñas C, Tsakirpaloglou N, Torrizo L, Arines FM, Adeva C, et al. Biofortified indica rice attains iron and zinc nutrition dietary targets in the field. Scientific Reports 2016 Jan 25;**6**(1):19792

[70] Lephuthing MC, Baloyi TA, Sosibo NZ, Progress TTJ. Challenges in improving nutritional quality in wheat. In: Wheat Improvement, Management and Utilization. InTech. 2017. pp. 345-364

[71] Hell R, Iron SUW. Uptake, trafficking and homeostasis in plants. Planta. 2003 Feb;**216**(4): 541-551

[72] Takahashi M, Terada Y, Nakai I, Nakanishi H, Yoshimura E, Mori S, et al. Role of nicotianamine in the intracellular delivery of metals and plant reproductive development. The Plant Cell. 2003 Jun;**15**(6):1263-1280

[73] Koike S, Inoue H, Mizuno D, Takahashi M, Nakanishi H, Mori S, et al. OsYSL2 is a rice metal-nicotianamine transporter that is regulated by iron and expressed in the phloem. The Plant Journal. 2004 Aug;**39**(3):415-424

[74] Masuda H, Usuda K, Kobayashi T, Ishimaru Y, Kakei Y, Takahashi M, et al. Overexpression of the barley nicotianamine synthase gene HvNAS1 increases iron and zinc concentrations in rice grains. Rice. 2009 Dec;**2**(4):155-166

[75] Ishimaru Y, Masuda H, Bashir K, Inoue H, Tsukamoto T, Takahashi M, et al. Rice metal-nicotianamine transporter, OsYSL2, is required for the long-distance transport of iron and manganese. The Plant Journal. 2010 Feb;**62**(3):379-390

[76] Bashir K, Inoue H, Nagasaka S, Takahashi M, Nakanishi H, Mori S, et al. Cloning and characterization of deoxymugineic acid synthase genes from graminaceous plants. The Journal of Biological Chemistry. 2006 Oct;**281**(43):32395-32402

[77] Masuda H, Kobayashi T, Ishimaru Y, Takahashi M, Aung MS, Nakanishi H, et al. Iron-biofortification in rice by the introduction of three barley genes participated in mugineic acid biosynthesis with soybean ferritin gene. Frontiers in Plant Science. 2013 May;**4**(1):132

[78] Suzuki M, Morikawa KC, Nakanishi H, Takahashi M, Saigusa M, Mori S, et al. Transgenic rice lines that include barley genes have increased tolerance to low iron availability in a calcareous paddy soil. Soil Science & Plant Nutrition. 2008 Feb;**54**(1):77-85

[79] Kim SA, Punshon T, Lanzirotti A, Li L, Alonso JM, Ecker JR, et al. Localization of iron in Arabidopsis seed requires the vacuolar membrane transporter VIT1. Science. 2006 Nov;**314**(5803):1295-1298

[80] Bashir K, Takahashi R, Akhtar S, Ishimaru Y, Nakanishi H, Nishizawa NK. The knock-down of OsVIT2 and MIT affects iron localization in rice seed. Rice. 2013 Nov;**6**(1):31

[81] White PJ, Broadley MR. Biofortifying crops with essential mineral elements. Vol. 10, Trends in Plant Science. Elsevier Current Trends; 2005. p. 586-593

[82] Saied IT, Up-regulation SAKM. Of the tumor suppressor gene p53 and WAF1 gene expression by IP6 in HT-29 human colon carcinoma cell line. Anticancer Research. 1998 May;**18**(3):1479-1484

[83] Zhou JR, Erdman Jr. JW. Phytic acid in health and disease. Vol. 35, Critical Reviews in Food Science and Nutrition. Taylor & Francis Group; 2009. p. 495-508

[84] Chrispeels MJ, Raikhel NV. Lectins, lectins genes, and their role in plant defense. The Plant Cell. 1991 Jan;**3**(1):1-9

[85] Zajdel A, Wilczok A, Węglarz L, Dzierżewicz Z. Phytic acid inhibits lipid peroxidation in vitro. BioMed Research International. 2013 Oct;**2013**(1):1-6

[86] Boonyaves K, Gruissem W, Bhullar NK. NOD promoter-controlled AtIRT1 expression functions synergistically with NAS and ferritin genes to increase iron in rice grains. Plant Molecular Biology. 2016 Feb;**90**(3):207-215

[87] Lee S, Chiecko JC, Kim SA, Walker EL, Lee Y, Guerinot ML, et al. Disruption of OsYSL15 leads to iron inefficiency in rice plants. Plant Physiology. 2009;**150**(2):786-800

[88] Aung MS, Masuda H, Kobayashi T, Nakanishi H, Yamakawa T, Nishizawa NK. Iron biofortification of Myanmar rice. Frontiers in Plant Science. 2013 May;**4**(1):158

[89] Kumpatla SP, Hall TC. Recurrent onset of epigenetic silencing in rice harboring a multi-copy transgene. The Plant Journal. 1998 Apr;**14**(1):129-135

[90] Tang W, Newton RJ, Weidner DA. Genetic transformation and gene silencing mediated by multiple copies of a transgene in eastern white pine. Journal of Experimental Botany. 2006 Dec;**58**(3):545-554

[91] Rajeevkumar S, Anunanthini P, Epigenetic SR. Silencing in transgenic plants. Frontiers in Plant Science. 2015 Sep;**6**(1):693

[92] Wang YD, Wang X, Wong YS. Generation of selenium-enriched rice with enhanced grain yield, selenium content and bioavailability through fertilisation with selenite. Food Chemistry. 2013 Dec;**141**(3):2385-2393

[93] IPNI Issues 4R Plant Nutrition Manual I Nutrient Stewardship [Internet] [Cited 2017 Oct 9]. Available from: http://www.nutrientstewardship.com/4r-news/ipni-issues-4r-plant-nutrition-manual/

[94] Vasconcelos MW, Gruissem W, Iron BNK. Biofortification in the 21st century: Setting realistic targets, overcoming obstacles, and new strategies for healthy nutrition. Current Opinion in Biotechnology. 2017 Apr;**44(1)**:8-15

[95] Yap WS, Lai KS. Biochemical properties of twelve malaysia rice cultivars in relation to yield potential. 2017;**11**(4):1-7

[96] Phenotyping TR. For drought tolerance of crops in the genomics era. Frontiers in Physiology. 2012 Sep;**3**(1):347

[97] Gaudin ACM, Henry A, Sparks AH, Slamet-Loedin IH. Taking transgenic rice drought screening to the field. Journal of Experimental Botany. 2013 Jan;**64**(1):109-117

[98] Lindenmayer GW, Stoltzfus RJ, Prendergast AJ. Interactions between zinc deficiency and environmental enteropathy in developing countries. Archive of Advances in Nutrition. 2014 Jan;**5**(1):1-6

[99] Talsma EF, Brouwer ID, Verhoef H, Mbera GNK, Mwangi AM, Demir AY, et al. Biofortified yellow cassava and vitamin a status of Kenyan children: A randomized controlled trial. The American Journal of Clinical Nutrition. 2016 Jan;**103**(1):258-267

[100] Wesseler J, Zilberman D. The economic power of the golden rice opposition. Environment and Development Economics. 2014 Dec;**19**(6):724-742

[101] Mulualem T. Application of bio-fortification through plant breeding to improve the value of staple crops. Biomedicine and Biotechnology 2015 Jan;**3**(1):11-19

[102] Netherwood T, Martín-Orúe SM, O'Donnell AG, Gockling S, Graham J, Mathers JC, et al. Assessing the survival of transgenic plant DNA in the human gastrointestinal tract. Nature Biotechnology. 2004 Feb;**22**(2):204-209

[103] Nestel P, Bouis HE, Meenakshi J V, Pfeiffer W. Biofortification of staple food crops. The Journal of Nutrition 2006 Apr;**136**(4):1064-1067

5

Methanogens Harboring in Rice Rhizosphere Reduce Labile Organic Carbon Compounds to Produce Methane Gas

Prabhat Pramanik and Pil Joo Kim

Abstract

Submerged rice paddy soils are one of the major anthropogenic sources of methane (CH_4) emission to the atmosphere. Methane is the second most important greenhouse gas after carbon dioxide. Methanogens are strictly anaerobic microorganisms and CH_4 is the metabolic end product of those methanogens. Methane is produced by methanogens through multi-step enzyme-mediated process. Methanogens convert labile organic carbon compounds in CH_4 and application of organic matter in submerged rice field significantly increased CH_4 emission from soil to the atmosphere. The rate of methanogenesis may be determined by quantifying biomarkers namely methyl coenzyme M reductase A (mcrA) gene and coenzyme M (2-mercaptoethane sulphonate) in soil. Nickel ions are present as cofactor in enzymes involved in methanogenesis. Methane emission can be mitigated by application of EDTA at suitable rate in the soil of submerged rice field.

Keywords: methane emission, methanogens, biomarkers, EDTA application, rice paddy soil

1. Introduction

In the era of development and globalization, emissions of greenhouse gases (GHGs) are unavoidable consequences, and that increases atmospheric temperature causing global warming. A greenhouse gas is a substrate in the atmosphere that absorbs and emits radiation within the thermal range. This process is the fundamental cause of the greenhouse effect and global warming [1]. Without GHGs, the average temperature of earth's surface would be about −18°C (0°F) [2], rather than present average of 15°C (59°F) [3]. The primary GHGs in the earth's atmosphere are water vapor, carbon dioxide (CO_2), methane (CH_4), nitrous oxide (N_2O) and chlorofluorocarbon (CFC). Human activities since the beginning of the industrial revolution (Taken as year 1750) have resulted 40% increased in the atmospheric carbon dioxide concentration from 280 ppm in

1970 to 400 ppm in 2015 [4]. Carbon dioxide (CO_2) is the most important GHG in atmosphere in terms of its emitted volume. The other GHGs are CH_4, N_2O, CFC compounds etc. Methane is the second most important GHG emitted to the atmosphere on volume basis and it has 25 times higher global warming potential (GWP) as compared to equivalent amount of CO_2 [5]. The half-life of CH_4 in the atmosphere is about 25 years, which is also much higher than that of CO_2. Due to these characteristics, CH_4 is considered as one of the most notorious GHGs having potential of causing global warming to the atmosphere. The CH_4 concentration in the earth's atmosphere has been increased by 150% since 1750. Methane accounts for 20% of the total radiative forcing from the entire long-lived and globally mixed GHGs, excluding water vapor.

2. Chemistry of submerged rice paddy soil

Agriculture is one of the most important anthropogenic sources of CH_4 emission to the atmosphere and about 11% of the total CH_4 is emitted from submerged rice paddy soils. Rice is the staple food for more than half of world population and about 90% of that rice is cultivated under submerged condition [6]. During rice cultivation, rice seedlings are transplanted after flooding the soil, and water is removed (drained out) few days before crop harvesting. Therefore, in case of transplanted rice, soil remains submerged for at least 85–90% of the total cultivation duration. Such submerged rice paddy soil is the most important anthropogenic source of CH_4 emission to the atmosphere.

Submerged condition for such a prolonged duration makes rice paddy soils different from soils of upland crops. Submerged condition cuts off air transportation between soil and atmosphere. Flooding of rice paddy soil disconnects gas exchange between soil and air. Under this situation, molecular diffusion is the main mechanism to enter oxygen and other gases from atmosphere to the interstitial water. However, this process is 10,000 times slower than the diffusion through gas-filled pores in soil [7]. Thus the oxygen diffusion rate suddenly decreases when a soil reaches saturation by water [8]. Evans and Scott [9] noted that the concentration of oxygen in the water used for saturating a soil decreased to one - hundredth of its initial value in 75 minutes. The major characteristics of submerged soils are:

a. Absence of molecular oxygen

Flooding of land disconnected gas exchange between soil and air. Under submerged condition, oxygen along with other atmospheric gases enters into the soil only by molecular diffusion in the interstitial water. It was observed that gas diffusion under submerged soil condition is 10,000 times slower than diffusion through gas-filled pores [7]. Hence, soil of submerged rice paddy soil losses all its molecular oxygen as soil microorganisms use-up the oxygen present in soil within a few hours.

b. Oxidized mud – water interface

A submerged soil; however, is not completely devoid of oxygen. The top-most layer of few-millimeter thick soil, saturated with water (in mud form) remains oxygenated. The chemical properties of this oxidized interface are completely different from underneath top-soil.

c. Exchanges between mud and water

The presence of molecular oxygen in the soil-water interface makes it a sink of several redox reactions in soil and controls availability of phosphate and other nutrients in submerged soil. The presence of oxygen in the soil-water interface profoundly affects the N economy of submerged rice paddy soils. Ammonium-N released from broadcasted chemical fertilizer or from applied organic matter is converted to nitrate in the oxygenated interface on soil.

d. Soil reduction

An acute reduced state makes the major difference between chemical reactions of a submerged soil and aerated soil. Excluding the thin oxygenated layer in the soil-water interface, submerged soils have a negative oxidation-reduction potential (Eh value) due to anaerobic condition. Under such condition, dominant form of elements are NH_4^+, H_2S, Mn^{2+}, Fe^{2+} and CH_4 instead of NO_3^-, SO_4^{2-}, Mn^{4+}, Fe^{3+} and CO_2.

2.1. Oxidation – reduction (redox) potential

Under submerged condition, aerobic soil microorganisms consume oxygen during their metabolism and that in turn gradually depletes oxygen pool making the soil anaerobic in reaction [10]. The redox potential (E_h) value in submerged soil starts decreasing after 3–4 days of flooding and sharply decreases with time. The Eh values in submerged anaerobic soils vary around −200 eV values throughout the rice cultivation duration [11]. Such anaerobic reducing environment is one of the prime factors for determining the rate and quantity of CH_4 production in rice paddy soil.

3. Methanogens and CH_4 production

The average global CH_4 emission from rice fields is approximately 20–40 Tg CH_4 year^{-1}, which accounts for 11% of the total anthropogenic CH_4 emissions [12]. It had already been reported that rice production will be increased from 473 million tons of 1990 to approximately 781 million tons by 2020 to fulfill the food demand of the world population and that proportionately increase CH_4 emission from rice paddy soils by 40–50% [13].

Methane is mainly produced during decomposition of organic matter by strictly anaerobic methanogens under intense reduced condition [14]. At the initial state of rice cultivation, the rate of CH_4 emission is generally low; however, the flux gradually increases with plant development and with enhanced anaerobic condition [15, 16]. Anaerobic conditions of submerged rice paddy soil favors CH_4 production and the highest CH_4 emission is generally observed after the soil E_h value dropped below −200 eV [17].

Both cold- and hot-water extractable organic carbon (C) compounds are labile fraction of soil organic C. Low molecular weight organic compounds namely low molecular weight organic acids, carbohydrates are considered as labile organic C compounds in soil [18]. Labile organic C compounds rather than total organic C pool acts as the energy source for heterotrophic microorganisms like methanogens in soil [19]. Methane is the metabolic end product of methanogens [20] and methanogens reduce simple carbonaceous compounds namely CO_2, carbohydrates and/ or simple carboxylic acids like formate, acetate through multi-step enzyme-mediated methanogenesis to generate ATP and to produce CH_4 as the end product.

Methanogens are generally specific to their substrate requirement. Based on the ability to utilize carbonaceous compounds as energy source, methanogens may be classified as acetophilic methanogens and hydrophilic methanogens [21]. The acetophilic methanogens transform acetate ions into CH_4, while the hydrophilic methanogens utilize hydrogen and CO_2 as their energy source.

4. Factors affecting CH_4 production in soil

Methanogenesis or the process of CH_4 production is an enzyme-mediated multi-step biochemical process and kinetics of this process depends on several factors. The most important and prominent factor of CH_4 production is the availability of initial carbonaceous compounds in soil. Addition of organic materials in flooded rice field promotes CH_4 emission (**Figure 1**) by providing readily available C source to methanogens [22, 23]. Improvement in crop production by organic amendment conflicts with mitigation strategies of CH_4 emission [24]. Hence, it may be believed that methanogens degrade applied organic matter to produce CH_4 under strictly anaerobic conditions [14]. However, this is the exaggeration of the truth.

Application of organic matter increases total population and activity of microorganisms including methanogens in rice paddy soil [25]. Therefore application of organic substrates significantly increases CH_4 emission from rice field [26]. However, methanogens does not have potential to degrade carbonaceous polymers like cellulose due their inability to produce cellulase enzyme. In fact, there is a synergistic effect of cellulolytic microorganisms on methanogens and CH_4 formation in submerged rice paddy soil. Cellulose, a 1,4-β-linked glucan, contributes 20–30% of the organic biomass [27] and application of organic matter provides a significant C source in the form of cellulose to soil microbial community in soil.

The hydrolysis of carbonaceous polymers (mainly cellulose) is an important pathway to convert added organic C into CH_4 and anaerobic cellulolytic microorganisms play a significant role in that process [28]. Incubation of rice paddy soil with different amounts of carboxymethyl cellulose (CMC) under anaerobic condition in a close-vessel produced variable amount of CH_4 after 3 days (**Figure 2**). The amount of generated CH_4 within that period was proportional to the quantity of added CMC in soil. Therefore, cellulolytic materials of applied organic substrates were initially degraded by cellulolytic microorganisms into low molecular weight organic acids and/ or carbohydrates, which are then utilized by methanogens to produce CH_4 under anaerobic condition of submerged rice paddy soils.

In submerged rice paddy soil, the flux of CH_4 emission depends on the amount as well as nature of applied organic matter [29]. The rate of CH_4 emission is dependent on the nature i.e. degree of stabilization of applied organic matter. During decomposition, carbonaceous compounds like cellulose and hemicelluloses are readily stabilized through mineralization and converted into humified substrates. Therefore, composts contain lesser amount of easily

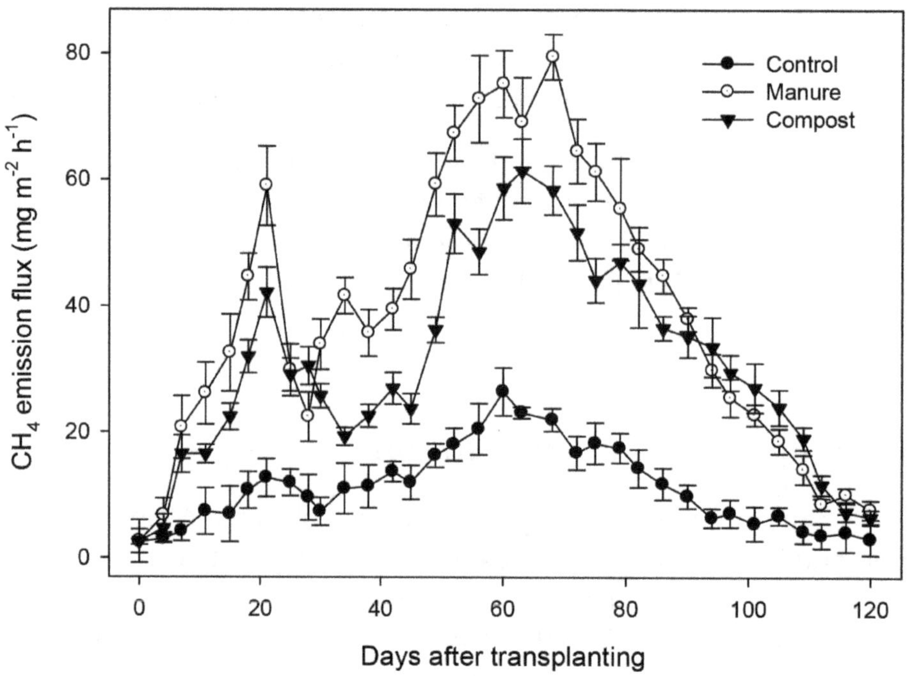

Figure 1. Changes in methane emission flux from rice paddy soils as affected by air-dried and composted dairy cow manures.

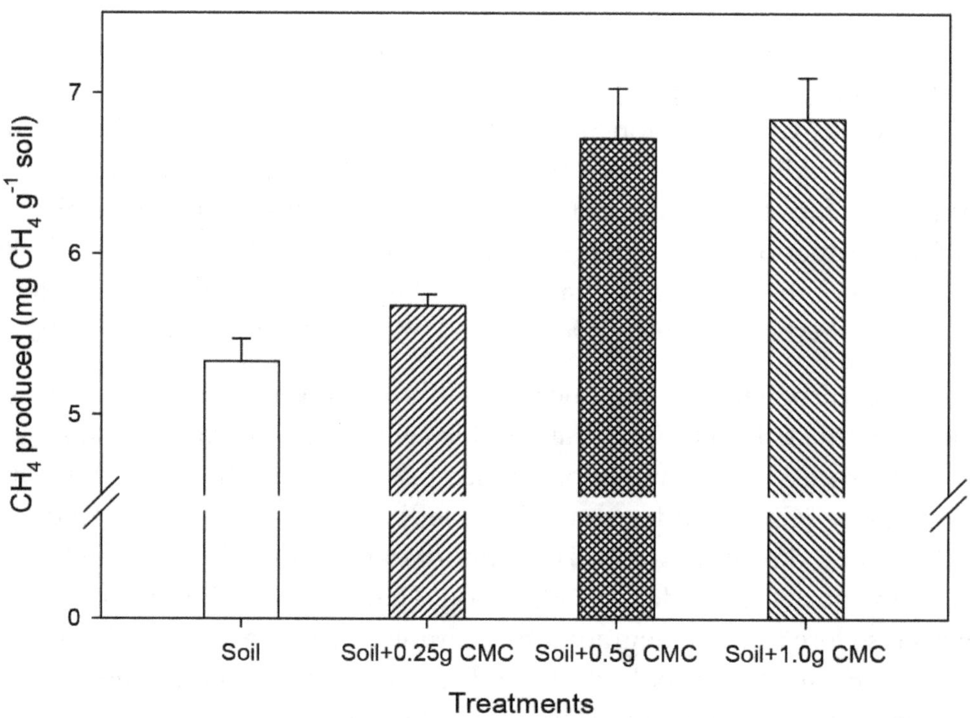

Figure 2. Changes in CH_4 production as affected by CMC application.

decomposable compounds like cellulose and hemicelluloses compounds as compared to their initial substrates. Lower cellulose contains leads to fewer cellulolytic microbial populations in soil and that generates lower amount of labile organic C compounds in soil. Therefore, reduced availability of precursor carbonaceous compounds is responsible for lower CH_4

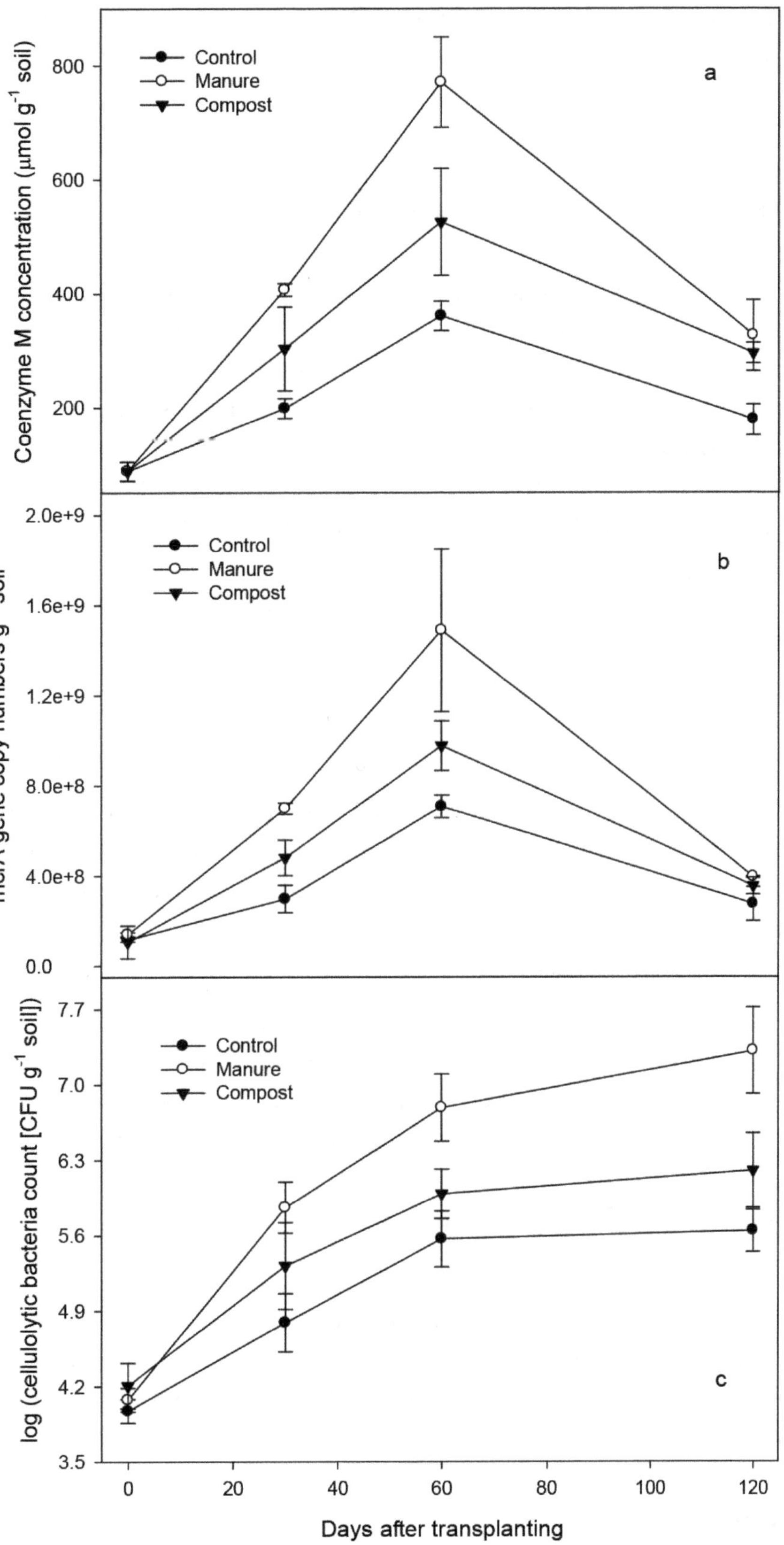

Figure 3. Coenzyme M concentrations (a), mcrA gene copy numbers, (b) and cellulolytic bacterial populations, (c) as affected by manure and compost applications in rice paddy soils.

emission from compost treated rice paddy soil. Therefore, degree of stabilization or humification of applied organic substrates is inversely related to the flux of CH_4 emission, which may be reduced by ~20% by application of compost instead of air-dried cattle manure (**Figure 3**).

5. Biomarkers of methanogens: quantification

In many anoxic environments, methanogenesis is the predominant terminal electron accepting process involved in organic matter mineralization and that process is catalyzed by methanogenic *Archaea*. This group of microorganisms represents a unique but phylogenetically diverse group of prokaryotes [30]. The most widely used method for measuring the rate of methanogenesis is the quantification of CH_4 flux in a specific system within unit time interval [17]. Though this analysis is closely related to the metabolic functions of methanogenic community; it does not directly quantify methanogens [31]. One of the most convincing processes of quantifying methanogens is the direct determination of methanogens numbers on specific culture. However, isolating methanogenic *Archaea* remains a fastidious process because of the slow growth of these *Archaea* and also for their extreme intolerance to oxygen [32]. Growth of methanogens is also restricted to the availability of specific organic substrates and metal ions to complete their metabolic process [33]. Culturable microorganisms isolated by specific enriched medium can only detect a small portion (2–5%) of the total microbial community in soil.

Still researchers often prepared one specific culture medium for each one of the various methanogenic *Archaea* species to fulfill their specific requirements [34]. However, there are reports of versatile media like SAB medium, which is capable of supporting growth of wide spectrum of methanogens [33]. Hence, due to these limitations and difficulties, methanogens are preferably quantified by measuring their biomarkers.

Biomarkers are compounds that have a biological specificity in the sense that they are produced only by a limited group of organisms [35]. A variety of compounds such as fatty acids and ether lipids are used in microbial ecology and related fields like organic geochemistry to detect groups of organisms or their remains in natural or artificial ecosystems [36, 37].

5.1. Methyl coenzyme M reductase A (mcrA) gene

The mcrA gene is responsible for synthesizing methyl coenzyme M reductase enzyme, which is involved in the production of CH_4 during the final stage of methanogenesis. Culture-dependent and culture-independent techniques targeting 16S rRNA and methyl coenzyme M reductase (*mcrA*) genes have been used to assess the phylogenetic diversity of methanogens assemblages [38].

5.2. DNA extraction and quantification

The DNA may be extracted from natural and/or enriched samples using any suitable kit following manufacturers' protocol. The quality of the extracted DNA is observed in an agarose gel.

The extracted DNA is amplified by PCR in a final volume of 25 μl containing 2 μl of undiluted template DNA, 1 μl each of forward and reverse primers (10 mM) and 12.5 μl of Taq polymerase enzyme [39]. For detecting the presence of methanogens, a forward primer with 32-mer and a reverse primer with 23-mer were developed after testing against 23 species of methanogen representing all five recognized orders of this group of *Archaea* [40]. The two oligonucleotide primers were a forward primer, 5′-GGTGGTGTMGGATTCACACARTAYGCWACAGC-3′ and a reverse primers, 5′-TTCATTGCRTAGTTWGGRTAGTT-3′. The methanogen diversity in a sample may be studied by analyzing amplified DNA (or PCR product of extracted DNA) through denatured gradient gel electrophoresis (DGGE) [41].

Total population of methanogens can be determined from extracted DNA by quantitative PCR (qPCR) or real-time PCR (RT-PCR) using PCR efficiency, 110.5%; slope of the standard curve, −3.093; y-intercept, 5.134 and correlation coefficient, 0.9949 [31]. The C_t for the no template control was 24.03 and >26.5 for all the no-reverse transcriptase control. The qPCR results (mcrA gene copy numbers ng^{-1} DNA) of extracted DNA show significant correlation with specific methanogenic activities against H_2 and CO_2 gases. Steinberg and Regan [42] developed the TaqMan qPCR probe assay for successfully determining the environmental abundance of different phylogenetic groups of methanogens, including several groups with few or no cultivated members.

5.3. 2-mercaptoethane sulphonate (coenzyme M)

Methane production in soil is a complex enzyme-mediated multi-step process and methanogens reduce simple carbonaceous compounds like CO_2 and H_2, formate, methanol, methykamines and/ or acetate into CH_4 gas [43]. In the penultimate step of methanogenesis, coenzyme M (Co-M), 2-mercaptoethane sulphonate, is methylated and generated methyl Co-M is reduced by methanogens to CH_4 gas involving previously-mentioned methyl Co-M reductase enzyme [44]. Therefore, irrespective of the preference towards initial carbonaceous compounds, Co-M could be considered as the precursor of CH_4 formation [45]. The whole methanogenesis is intracellular and Co-M is synthesized inside methanogen cells [46]. The conversion factor (0.39 ± 0.07 fmol $cell^{-1}$) of Co-M to methanogens could be used for quantitative estimation of methanogen abundance and methanogenic activity in soil.

5.4. Coenzyme M quantification

Pramanik and Kim [47] developed a HPLC-based technique for quantifying Co-M in soil. Pure Co-M is detected at 270 nm wavelength using UV detector and a mixture of acetonitrile and 0.05 M trichloroacetic acid (TCA) solution at flow rate of 0.5 ml min^{-1} is used as mobile phase during this analysis.

Coenzyme M is an intracellular compound of methanogens; hence, rupturing cells of methanogens is mandatory prior to extraction. The lysis buffer was prepared by mixing Tris-HCl solution (pH 8.0), ethylenediaminetetraacetic acid (EDTA) solution (pH 8.0) and NaCl solution. The Co-M was extracted from fresh soil using lysis buffer through consecutive

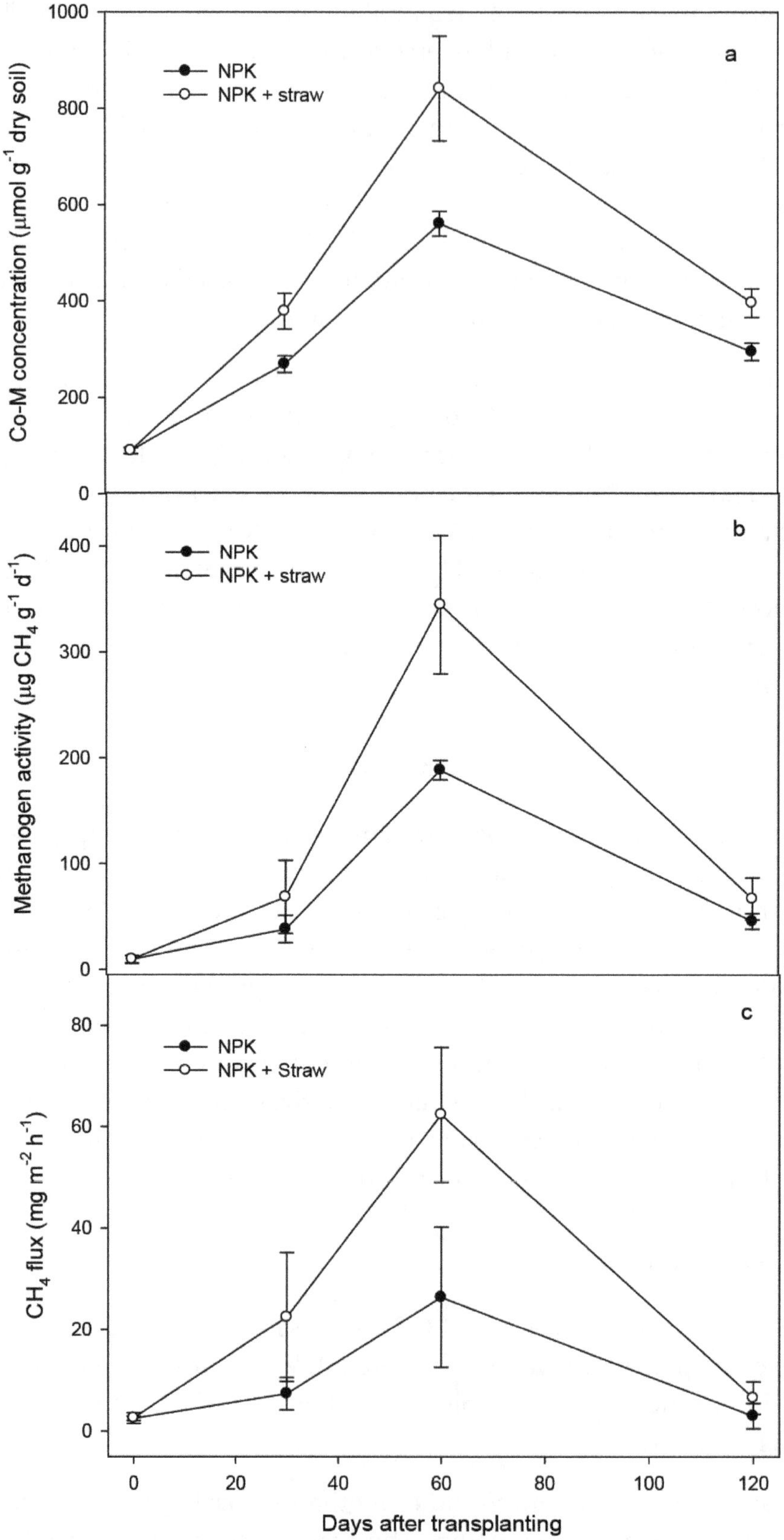

Figure 4. Changes in coenzyme M concentration (a) and methanogen activity (b) and methane flux (c) from soil during rice cultivation (bars represent standard errors).

homogenization by vortex shaker and sonication [47]. The Co-M was precipitated from ethanolic solution of the supernatant and re-dissolved in distilled water prior to the HPLC analysis. The precision of this method to quantify Co-M in the soil matrix was >97%. The high recovery rate (90.3 ± 8.1%) indicated that Co-M is not adsorbed to the ionic sites of soil colloids and the measured values are very close to the actual Co-M content in soil (**Figure 4**).

Methanogen activity in soil is linearly correlated to the Co-M content (r = 0.857*) of soil and the mean conversion factor between these two parameters is 155.03 ± 14.20 μg CH_4 produced $mmol^{-1}$ Co-M d^{-1} [47]. Therefore, it could be stated that both *mcrA* gene copy numbers and the concentration of Co-M could be quantified as biomarkers of methanogens for precise estimation of methanogenic activity in submerged rice paddy soils.

6. Mitigation techniques of CH_4 emission

Rice cultivation under flooded condition is regarded as an important source of CH_4 emission in soil. Methane is produced during decomposition of organic matter under anaerobic condition and simple carbonaceous compounds are biochemically reduced by methanogenic *Archaea* to form CH_4 [48]. During this reduction process, an electron donor is required to transfer the electron and availability of such electron donors might control the flux of CH_4 emission in rice paddy soil. Iron (Fe) is a transition metal with partially filled d-orbital in its configuration. Lee et al. [48] observed that Fe in active form (ionic form) may accept electrons required for reductive methanogenesis process and that in turn decreased CH_4 emission flux from rice paddy soil. Methane emission from submerged soil may be reduced approximately 14.5% by application of byproduct of steel industry as silicate fertilizers. Those byproducts of steel industry provide adequate silicate ions, which are necessary for higher rice productivity [49] and also for inducing resistance to biotic and abiotic stress [50]. However, Fe present in silicate fertilizers absorbs part of free electrons generated in the system and that restricts the terminal electron transfer during methanogenesis. This property enabled to reduce CH_4 emission from conventional (chemical fertilizer treated) rice paddy soils. However, Fe-enriched silicate fertilizer is not a strong mitigating agent; in fact a contrasting effect of silicate fertilizers on CH_4 emission was observed in organic matter treated rice paddy soil.

6.1. Effect of EDTA on CH_4 emission

In the last step of methanogenesis, methyl coenzyme M reductase (MCR) enzyme is involved in the reduction of methyl Co-M to CH_4 and nickel (Ni) ion is involved as the cofactor F_{430} in MCR enzyme [44]. Hence, availability of Ni determined the concentration of cofactor F_{430}, which in turn controls the activity of MCR enzyme [51]. Therefore, the rate of CH_4 production is enhanced by addition of Ni-salts to rice paddy soil [52]. Transition metals like Ni form soluble complexes with EDTA and that increases the solubility of Ni in soil. However, Pramanik and Kim [53] revealed that methanogens could not assimilate Ni^{2+} ions from Ni-EDTA complexes and suffered starvation for Ni^{2+} ions when stoichiometric amount of Ni salt and EDTA were mixed in culture broth. Finazzo et al. [54] stated that Ni is present as Ni(I) in F_{430} and conversion of Ni(I) to Ni(II) provides necessary electron for reduction of methyl coenzyme to CH_4. The presence of Ni as Ni-EDTA complex might retard this electron transfer during methanogenesis. Application of Ni as Ni-EDTA complex limited bioaccumulation of Ni by methanogens and that lower Ni content in methanogen biomass significantly ($P \leq 0.05$) reduced the rate of CH_4 production [53].

Ethylenediaminetetraacetic acid (EDTA) is a strong chelating agent and often shows adverse effect on plants when applied in higher doses. Pramanik and Kim [55] established that EDTA application at smaller doses (up to 5.0 ppm) proportionately reduces the flux of CH_4 emission from rice paddy soils without suppressing crop productivity (**Figure 5**). However, EDTA application at higher doses may increase CH_4 emission and also adversely affected maturity and productivity of rice. Therefore, application of 5.0 ppm EDTA is possibly the most rational dose to mitigate CH_4 flux from rice paddy soil. The activity of methanogens in EDTA-treated soil was significantly lower than that of control treatment. However, activity of all the microorganisms (microbial respiration) in soil was initially decreased due to EDTA application during rice cultivation. Application of EDTA enhances availability of nutrients especially nitrogen content in soil solution and that in turn gradually boosts microbial activity in soil. After 30 days of rice cultivation, microbial respiration of EDTA-treated soils did not differ significantly from that of control soil. Application of EDTA leads up to 18.1% reduction in CH_4 emission flux during submerged rice cultivation.

Unlike Fe-enriched silicate fertilizers, EDTA is also effective to mitigate CH_4 emission from organic amended submerged soils. Application of EDTA did not suppress the rate of organic

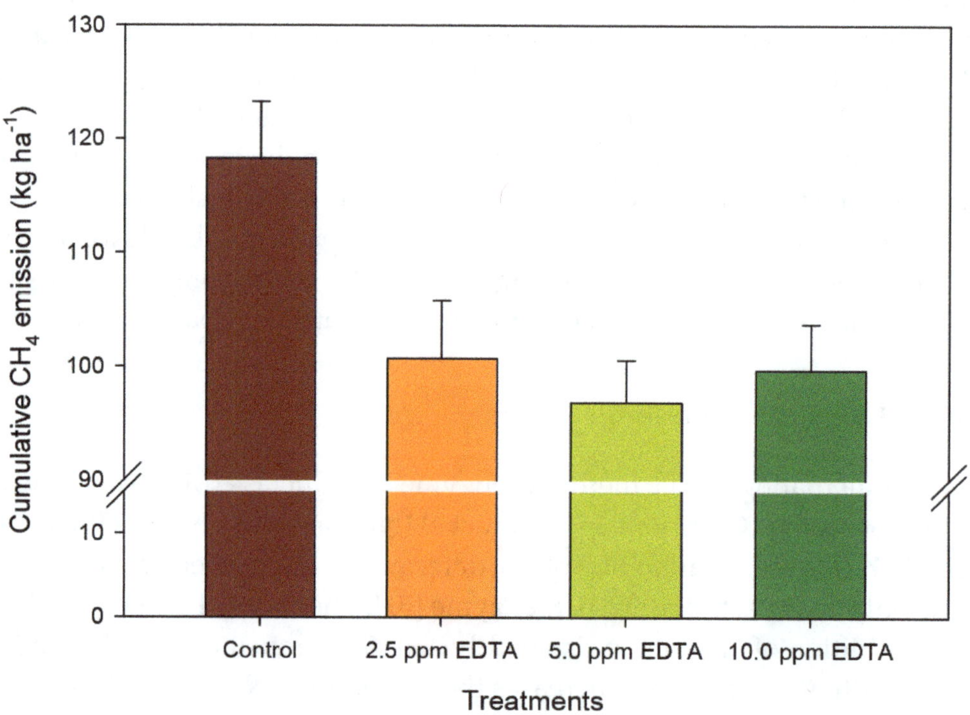

Figure 5. Cumulative CH_4 emission from different rice paddy soils.

matter mineralization; hence, the concentration of labile organic C compounds was not decreased due to EDTA application in soil. In spite of higher abundance of precursor materials (labile organic C compounds) in soil, EDTA application leads up to 22.5% reduction in CH_4 emission from straw incorporated rice paddy soils [56].

One noticeable drawback of EDTA application is that EDTA-treated soils have higher nitrate N content, which acts as the precursor for nitrous oxide formation under anaerobic reducing of submerged soil condition. Higher nitrate N content enhanced the flux of nitrous oxide, another greenhouse gas having 290 times higher global warming potential than equivalent amount of carbon dioxide. Study revealed that total global warming potential in 5.0 ppm EDTA-treated soil was 14.5% lower than that of control soil (not treated with EDTA) during rice cultivation.

Organic amendment increased C-to-N ratio, which in turn decreased the rate of mineralization and nitrate N content in soil. Therefore, the adverse effect of EDTA application could be minimized by organic amendment in soil. However, incorporation of organic matter in submerged rice paddy soil facilitated the risk of CH_4 emission to the atmosphere. Optimum combination of EDTA and compost may effectively reduce the net global warming potential due to CH_4 and N_2O emissions. It was observed that 5.0 Mg organic substrates ha^{-1} and 5.0 ppm EDTA combination had global warming potential 11186.17 ± 749.35 kg CO_2-equiv. ha^{-1}, which was 20.8% lower than that of only organic amended rice paddy soil (**Figure 6**).

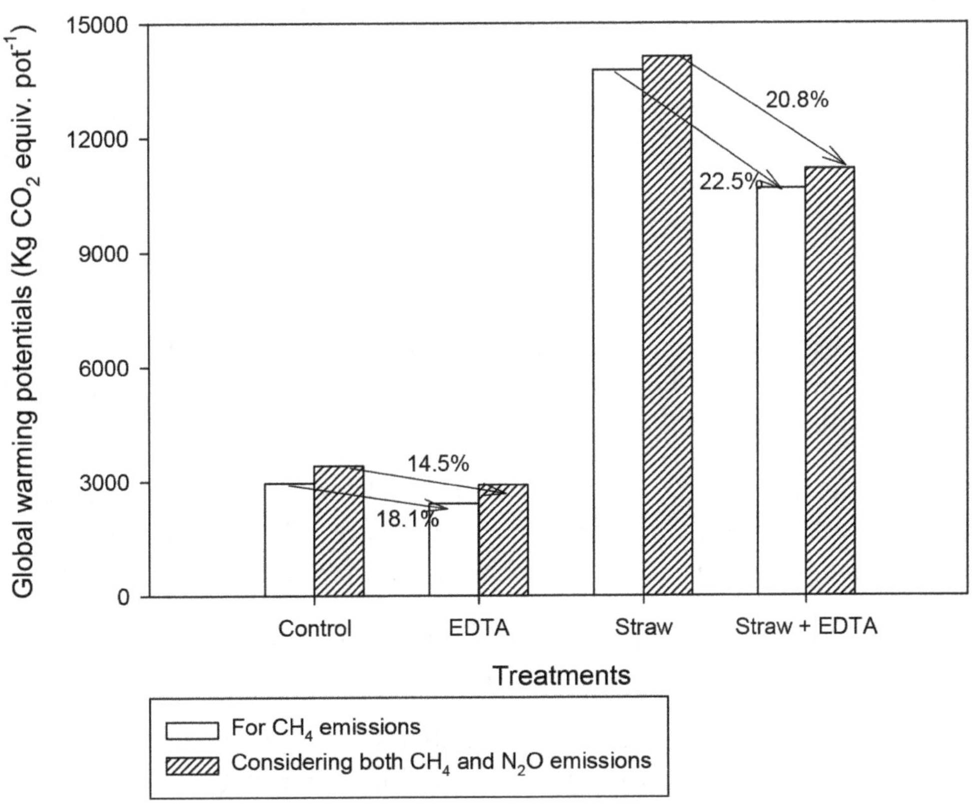

Figure 6. Global warming potentials due to methane (CH_4) and nitrous oxide (N_2O) emissions from submerged rice paddy soils.

7. Conclusion

Soil of submerged rice fields is the most important anthropogenic source of CH_4 emission to the atmosphere. Methanogens, a group of strictly anaerobic microorganisms, convert labile organic C compounds into CH_4 gas and that emits from soil to the atmosphere. Activity of methanogens may be quantitatively studied by measuring mcrA gene and/ or coenzymeM biomarkers in soil.

Acknowledgements

This work was supported by Basis Science Research Program through the National Research Foundation of Korea (NRF) funded by the Ministry of Education (NRF-2015R1A6A1A03031413).

Author details

Prabhat Pramanik[1] and Pil Joo Kim[2]*

*Address all correspondence to: pjkim@gnu.ac.kr

1 Soils Department, Tocklai Tea Research Institute, Tea Research Association, Jorhat, Assam, India

2 Division of Applied Life Science, Gyeongsang National University, Jinju, South Korea

References

[1] IPCC. In: McCarthy JJ, Canziani OF, Leary NA, Dokken DJ, White KS, editors. Climate Change 2001, Impacts, Adaptation and Vulnerability. Cambridge University Press, Cambridge; 2001

[2] Horton R, De Mel M, Peters D, Lesk C, Bartlett R, Helsingen H, Bader D, Capizzi P, Martin S, Rosenzweig C. Assessing Climate Risk in Myanmar. New York, NY, USA: Center for Climate Systems Research at Columbia University, WWF-US and WWF-Myanmar; 2016

[3] Karl TR, Melillo JM, Peterson TC. Global Climate Change Impacts in the United States. U.S. Global Change Research Program. New York: Cambridge University Press; 2009

[4] Blasing TJ. Recent Greenhouse Gas Concentrations, Carbon Dioxide Information Analysis Center (CDIAC). 2014. http://cdiac.ornl.gov/pns/current_ghg.html

[5] IPCC. Climate Change 2007:Mitigation of Climate Change — Contribution of Working Group III to the Fourth Assessment Report of the Intergovernmental Panel on Climate

Change. Cambridge, United Kingdom and New York, NY, USA: Cambridge University Press; 2007

[6] Cassman KG, Pingali PL. Extrapolating trends from long-term experiments to farmer's fields: The case of irrigated rice systems in Asia. In: Barnett V, Payne R, editors. Agricultural Sustainability: Economic Environmental and Olsen S.R. And Statistical Considerations. New York: Wiley; 1995. pp. 63-84

[7] Greenwood DJ, Goodman D. Direct measurement of the distribution of oxygen in soil aggregates and in columns of fine soil crumbs. Journal of Soil Science. 1967;**18**:182-196

[8] Saree S, Ponphang-nga P, Limtong P, Chidthaisong A. Soil carbon sequestration affected by cropping changes from upland maize to flooded rice cultivation. Journal of Sustainable Energy and Environment. 2012;**3**:147-152

[9] Evans DD, Scott AD. Polarographic method of measuring dissolved oxygen in saturated soil. Soil Science Society of America Proceedings. 1955;**19**:12-16

[10] Yu K, Patrick WH Jr. Redox window with minimum global warming potential contribution from rice soils. Soil Science Society of America Journal. 2004;**68**:2086-2091

[11] Haque MM, Kim SY, Pramanik P, Kim GY, Kim PJ. Optimum application level of winter cover crop biomass as green manure under considering methane emission and rice productivity in paddy soil. Biology and Fertility of Soils. 2013;**49**:487-493

[12] Yan X, Akiyama H, Yagi K. Global estimations of the inventory and mitigation potential of methane emissions from rice cultivation conducted using the intergovernmental panel on climate change guidelines. Global Biogeochemical Cycles. 2009;**23**:GB2002

[13] Anastasic C, Dowding M, Simpson VJ. Future CH_4 emission from rice production. Journal of Geophysical Research. 1992;**97**:7521-7525

[14] Garica JL, Patel BKC, Ollivier O. Taxonomic, phylogenetic and ecological diversity of methanogenic archaea. Anaerobe. 2000;**6**:205-226

[15] Ali MA, Lee CH, Lee YB, Kim PJ. Silicate fertilization in no-tillage rice farming for mitigation of methane emission and increasing rice productivity. Agriculture, Ecosystems and Environment. 2009;**132**:16-22

[16] Kern J, Hellebrand HJ, Gömmel M, Ammon C, Berg W. Effects of climatic factors and soil management on the methane flux in soils from annual and perennial energy crops. Biology and Fertility of Soils. 2012;**48**:1-8

[17] Ali MA, Lee CH, Kim PJ. Effects of silicate fertilizer on reducing methane emission during rice cultivation. Biology and Fertility of Soils. 2008;**44**:597-604

[18] Brinkmann T, Hörsch P, Sartorius D, Frimmel FH. Photoformation of low-molecular-weight organic acids from Brown water dissolved organic matter. Environmental Science & Technology. 2003;**37**:4190-4198

[19] Kumar A, Sharma MP. Review of methodology for estimation of labile organic carbon in reservoirs and lakes for GHG emission. Journal of Materials and Environmental Sciences. 2014;**5**:653-660

[20] Welte C, Deppenmeire U. Proton translocation in methanogens. In: Rosenzweig AC, Ragsdale SW, editors. Methods in Methane Metabolism: Methanogenesis, Chapter Thirteen. Cambridge: Academic Press, Elsevier; 2011. pp. 257-280

[21] Wraith MA, Sharma R. Technical review of methods to enhance biological degradation in sanitary landfills. Water Quality Rsearch Journal of Canada. 1998;**33**:417-437

[22] Wassmann R, Shangguan XJ, Toig M, Cheng DX, Wang MX, Papen H, Rennenberg H, Scilon W. Spatial and seasonal distribution of organic amendments affecting methane emission from Chinese rice fields. Biology and Fertility of Soils. 1996;**22**:191-195

[23] Neue HU, Wassmann R, Kludze HK, Wang B, Lantin RC. Factors and process controlling methane emissions from rice fields. Nutrient Cycling in Agroecosystems. 1997;**49**:111-117

[24] Hsu YW, Singh SK, Chiang MY, Wu YY, Chang IF. Strategies to lower greenhouse gas level by rice agriculture. African Journal of Biotechnology. 2009;**8**:126-132

[25] Weber S, Lueders T, Friedrich MW, Conrad R. Methanogenic populations involved in the degradation of rice straw in anoxic paddy soil. FEMS Microbiology Ecology. 2001;**38**:11-20

[26] Kusel K, Wagner C, Drake HL. Enumeration and metabolic product profiles of the anaerobic microflora in the mineral soil and litter of a beech forest. FEMS Microbiology Ecology. 1999;**29**:91-103

[27] Berg B, Laskowski R. Litter Decomposition: A Guide to Carbon and Nutrient Turnover. Advances in Ecological Research, 38. Amsterdam: Elsevier; 2006. pp. 4-23

[28] Pramanik P, Kim PJ. Evaluating changes in cellulolytic bacterial population to explain methane emissions from air-dried and composted manure treated rice paddy soils. The Science of the Total Environment. 2014a;**470-471**:1307-1312

[29] Pramanik P, Haque MM, Kim SY, Kim PJ. C and N accumulations in soil aggregates determine nitrous oxide emissions from cover crop treated rice paddy soils during fallow season. The Science of the Total Environment. 2014b;**490**:622-628

[30] Friedrich MW. Methyl-coenzyme M reductase genes: Unique functional markers for methanogenic and anaerobic methane-oxidizing archaea. Methods in Enzymology. 2005;**397**:428-442

[31] Morris R, Schauer-Gimenez A, Bhattad U, Kearney C, Struble CA, Zitomer D, Maki JS. Methyl coenzyme M reductase (*mcrA*) gene abundance correlates with activity measurements of methanogenic H_2/CO_2-enriched anaerobic biomass. Microbial Biotechnology. 2014;**7**:77-84

[32] Wolfe RS, Metcalf WW. A vacuum-vortex technique for preparation of anoxic solutions or liquid culture media in small volumes for culturing methanogens or other strict anaerobes. Anaerobe. 2010;**16**:216-219

[33] Khelaifia S, Raoult D, Drancourt M. A versatile medium for cultivating methanogenic archaea. PLoS One. 2013;**8**:e61563

[34] Widdel F. Growth of methanogenic bacteria in pure culture with 2-propanol and other alcohols as hydrogen donors. Applied and Environmental Microbiology. 1986;**51**: 4156-4162

[35] Boschker HTS, Middelburg JJ. Stable isotopes and biomarkers in microbial ecology. FEMS Microbiology Ecology. 2002;**40**:85-95

[36] Parker RJ. Analysis of microbial communities within sediments using biomarkers. In: Hetcher M, Gray RTG, Jones JG, editors. Ecology of Microbial Communities. Cambridge: Cambridge University Press; 1987. pp. 147-177

[37] Tunlid A, White DC. Biochemical analysis of biomass, community structure, nutritional status, and metabolic activity of microbial community in soil. In: Bollag JM, Stotzky G, editors. Soil Biochemistry. New York: Marcel Dekker; 1992. pp. 229-262

[38] Lehmacher A, Klenk HP. Characterization and phylogeny of MCRII, a gene-cluster encoding an isozyme of methyl coenzyme-M reductase from hyperthermophilic *Methanothermus fervidus*. Molecular & General Genetics. 1994;**243**:198-206

[39] Garcia-Maldonado JQ, Bebout BM, Celis LB, Lopez-Cortes A. Phylogenetic diversity of methyl-coenzymeM reductase (*mcrA*) gene and methanogenesis from trimethylamine in hypersaline environments. International Microbiology. 2012;**15**:33-41

[40] Luton PE, Wayne JM, Sharp RJ, Riley PW. The mcrA gene as an alternative to 16S rRNA in the phylogenetic analysis of methanogen populations in landfill. Microbiology. 2002;**148**:3521-3530

[41] Lopez-Cortes A, Lanz-Landazuri A, Garcia-Maldonado JQ. Screening and isolation of PHB-producing bacteria in a polluted marine microbial mat. Microbial Ecology. 2008;**56**:112-120

[42] Steinberg LM, Regan JM. mcrA-targeted real-time quantitative PCR method to examine methanogen communities. Applied and Environmental Microbiology. 2009;**75**:4435-4442

[43] Boone DR, Whitman WB, Rouviere P. Diversity and taxonomyof methanogens. In: Ferry JG, editor. Methanogenesis: Ecology, Physiology, Biochemistry and Genetics. New York: Chapman and Hall; 1993. pp. 35-80

[44] Kaster AK, Moll J, Parey K, Thauer RK. Coupling of ferredoxin and heterodisulfide reaction via electron bifurcation in hydrogenotrophic methanogensic archaea. Proceedings of the National Academy of Sciences. 2011;**108**:2981-2986

[45] Ferry JG, Kastead KA. Methanogenesis. In: Cavicchioli R, editor. Archaea: Molecular and Cellular Biology. ASM Press, Washington, DC; 2007, p. 288-314

[46] Thauer RK. Biochemistry of methanogenesisa tribute to Marjory Stephenson. Microbiology. 1998;**144**:2377-2406

[47] Pramanik P, Kim PJ. Quantitative determination of 2-mercaptoethane sulphonate as biomarker for methanogens in soil by high performance liquid chromatography using UV detector. Soil Biology and Biochemistry. 2012;**55**:140-145

[48] Lee CH, Park KD, Jung KY, Ali MA, Lee D, Gutierrez J. Effect of Chinese milk vetch (*Astragalus sinicus* L.) as a green manure on rice productivity and methane emission in paddy soil. Agriculture, Ecosystems and Environment. 2010;**138**:343-347

[49] Ma K, Qiu QF, Lu YH. Microbial mechanism for rice variety control on methane emission from rice field soil. Global Change Biology. 2010;**16**:3085-3095

[50] Ma JF, Takahashi E. Effect of silicon on the growth and phosphorus uptake of rice. Plant and Soil. 1990;**126**:115-119

[51] Thauer RK, Kaster AK, Seedorf H, Buckel W, Hedderich R. Methanogenic archaea: Ecologically relevant differences in energy conservation. Nature Reviews. 2008;**6**:579-591

[52] Williams PEV, Innes GM, Brewer A, Magadi JP. The effects of growth, food intake and rumen volume of including untreated or ammoniatreated barley straw in a complete diet for weaning calves. Animal Products. 1985;**41**:63-74

[53] Pramanik P, Kim PJ. Effect of limited nickel availability on methane emission from EDTA treated soils: Coenzyme M an alternative biomarker for methanogens. Chemosphere. 2013;**90**:873-876

[54] Finazzo C, Harmer J, Bauer C, Jaun B, Duin EC, Mahlert F, Goenrich M, Thauer RK, Doorslaer SV, Schweiger A. Coenzyme B induced coordination of coenzyme M via its thiol group to Ni (I) of F430 in active methyl-coenzyme M reductase. Journal of the American Chemical Society. 2003;**125**:4988-4989

[55] Pramanik P, Kim PJ. Mitigate CH_4 emission by suppressing methanogen activity in rice paddy soils using ethylenediaminetetraacetic acid (EDTA). Geoderma. 2014c;**219-220**: 58-62

[56] Pramanik P, Kim PJ. Contrasting effects of EDTA applications on the fluxes of methane and nitrous oxide emissions from straw-treated rice paddy soils. Journal of the Science of Food and Agriculture. 2017;**97**:278-283

6

Farmers' Willingness to Cultivate Traditional Rice in Sri Lanka: A Case Study in Anuradhapura District

Ginigaddara Appuhamilage Sanjeewanie Ginigaddara and Sampath Priya Disanayake

Abstract

Increasing health threats is a common problem among both rice growers and consumers in many parts of Sri Lanka and in the Asian region in general. Increasing trends in growing and consuming traditional rice could be observed in searching solutions for these problems. This study explored objectively the factors affecting willingness to grow traditional rice and its varietal selection in Anuradhapura district of Sri Lanka. 100 traditional and 100 non-traditional rice growers were selected using stratified sampling method for the field survey and data were analyzed descriptively, using logistic regression and factor analysis. Results revealed that 67% of the male farmers were willing to cultivate traditional rice over improved varieties and 65.6% traditional rice cultivation was observed among families with non-communicable diseases. Awareness of medicinal and nutritional values of traditional rice, land extent, farm gate price, age and education level of farmers, farming experience and farming system significantly affects ($P \leq 0.05$) the willingness to cultivate traditional rice while factors related to varietal attributes, personal, market and production, respectively affects selection of traditional rice variety. Results conclude that farmers are willing to cultivate traditional rice in Anuradhapura district of Sri Lanka and selection of traditional rice varieties is most affected by varietal attributes which are adoptable to existing environmental conditions and personal factors like presence of non-communicable diseases, age of the farmers, education level and experience for farming.

Keywords: willingness to cultivate, traditional rice, varieties, rice farmers, Sri Lanka

1. Introduction

Rice production in Sri Lanka has grown into a self-sufficiency level during the last two to three decades. Rice being the staple food of Sri Lankans and having 114 kg per capita consumption plays a major role in providing energy, protein and fat to the whole population

[1]. Approximately 1,210,140 hectares of lands was devoted for paddy cultivation in 2015 and 2,992,333 metric tons of total rice availability for human consumption from domestic sources exceeded 2,310,321 metric tons of total rice requirement in 2015 while demonstrating the 129.32% of self-sufficiency rate [2]. Many programmes have been implemented in Sri Lanka to fulfill the national demand for rice through the effectively bred high-yielding varieties which are resistant to different stresses [3]. Nonetheless, people are more health conscious and interested in purchasing nutritional good quality products as the level of human health awareness increases among Sri Lankans. Traditional rice is being considered as more healthy and nutritious among Sri Lankans; a considerable demand is generating for traditional rice varieties in both local and international markets.

Sri Lanka is one of Asian countries which had a rich treasure of over 2400 traditional rice varieties [4] with identical nutritional and medicinal values [5]. Besides, these varieties are with varied maturity periods and highly resistant to extreme climatic conditions, pest and disease attacks [1] and adoptable to various soil and geographical conditions in the country. Out of these 2400 varieties, over 400 varieties are popular among Sri Lankan farmers at present. With the introduction of high-yielding newly improved varieties around the 1960s, traditional rice varieties became vanished from the Sri Lankan farming environments [6]. Thanks to the small-scale farmers, seeds of some of these old traditional varieties are still preserved in the gene banks. Moreover, traditional rice cultivation occupies a significant place among the rural communities in dry zone from past to present and even beyond the dry zone mainly due to threatening of people by common health problems like non-communicable diseases. In that scenario, farmers have increased the extent of traditional rice cultivation by adopting indigenous cultivation methods and organic inputs. Some non-governmental organizations have taken initiatives to promote traditional rice cultivation in the country advocating farmers to produce own seed paddy, cultivating rice for own consumption and fulfilling the market demand by selling surplus. Specially, districts like Anuradhapura, Polonnaruwa, Puttalam, Vavuniya, Kurunegala, Kegalle, Matale, Kandy, Rathnapura, Gampaha, Colombo, Galle, Matara, Monaragala, Badulla, Ampara, Batticaloa and Akkaraipattu are predominantly adopted to cultivate traditional rice varieties in the last few decades [7]. Importantly, the tolerance of traditional rice cultivars for submergence and salinity conditions has popularized them among farmers from those problematic areas. Compared to the improved rice varieties, some of traditional rice varieties are capable of being raised in nurseries for 2–3 months and tolerating water scarcities and heavy rainfalls or floods. And also, the strong vigorous stem helps to withstand against the heavy rains, winds and drought conditions. Their vigorous seeds are tolerant to other adverse conditions like waterlogging and drought. Therefore, most of the traditional rice varieties have the potential to cope with the drastic climatic changes which are generally detrimental to paddy cultivation [8]. Further, medicinal values of traditional rice have been experienced by Sri Lankans over few decades [9]. The lower starch hydrolysis rate lowers in vitro digestion rate which is suitable for diabetic patients. Most of the cultivars own officinal properties of preventing diabetic conditions, fatty liver, blood pressure and muscle recovering from free radicals and controlling weight, gallstones and protection against breast cancer. Namely, *Suwandel, Madathawalu, Kaluheenati, Suduheenati, Kuruluthuda, Pachchaperumal, Ma wee, Hatadaa wee, Rathdel, Kahamala* and *Kahawanu* are some of the most popular traditional rice varieties among the rice farmers and among traditional rice consumers in present Sri Lanka.

Anuradhapura, one of the largest districts in Sri Lanka, is situated in the North Central Province. The total land area is 6664 km^2, and the total human population is 856,232. Moreover, cultivated land extent under paddy is reported as 254,296 hectares including cultivations under major and minor irrigation and rainfed systems which produced 4612 kg of average paddy yield per hectare which was more or less similar to the country's average paddy production per hectare of 4527 kg. Rice farming is practiced by 65% of the population and nearly 17% of labor force is engaged with related occupations in rice cultivation in Anuradhapura district [10].

At present, there are increasing health threats from non-communicable diseases, namely, diabetes, high blood pressure, renal failures, variety of cancers and high blood cholesterol among both farmers and consumers in dry zone of Sri Lanka where the Anuradhapura district is located specifically and all over the island in general. As both growers and consumers highly believe that these health threats are basically due to unhealthy food habits and poor quality of the food items that they are consuming every day. Therefore, an increasing interest in growing and consuming traditional rice over improved varieties in this district could be observed. Thus, there is an increasing trend of the rice farmers in Anuradhapura district to cultivate traditional rice at least in few perches in their lands to be used for their family consumption, while some farmers are cultivating them in large scale expecting higher price in the market. Different farmers prefer to grow different varieties of traditional rice in their paddy fields, and there is a lack of research and findings on which factors trigger rice farmers to cultivate traditional rice over newly improved or hybrid rice in their paddy lands in different scales and which varieties of traditional rice do they prefer to cultivate over the others. Hence, this study was conducted to identify the factors affecting the willingness to cultivate traditional rice and to select their varieties for cultivation by rice farmers in Anuradhapura district of Sri Lanka.

2. Methodology

This experimental study focused on rice farmers who are growing traditional rice varieties compared to nontraditional rice varieties in Anuradhapura district and was conducted in five divisional secretariat (DS) divisions in Anuradhapura district, namely, *Padaviya, Medawachchiya, Rambewa, Thalawa* and *Rajanganaya,* where most of the farmers are engaged in traditional rice cultivation (**Figure 1**).

Two hundred paddy farmers were interviewed with a pretested questionnaire for identifying willingness to cultivate traditional rice and factors affecting the selection of traditional rice varieties. Three-stage stratified random sampling method was utilized to select 100 traditional rice growers and 100 nontraditional rice growers. At the first stage, five DS divisions were selected purposely where the highest farmers' registration under paddy cultivation has been reported. Accordingly, the *Grama Niladhari* (GN) divisions which reported the highest farmers' registration pertaining to the above five divisions were selected at the second stage, and at the final stage, both traditional rice farmers and nontraditional rice farmers were randomly selected for the questionnaire survey proportionately to the total farmers registered in the above five DS divisions. Additionally, focus group discussion and key personal interviews were conducted during the study. Secondary data were collected from different publications of

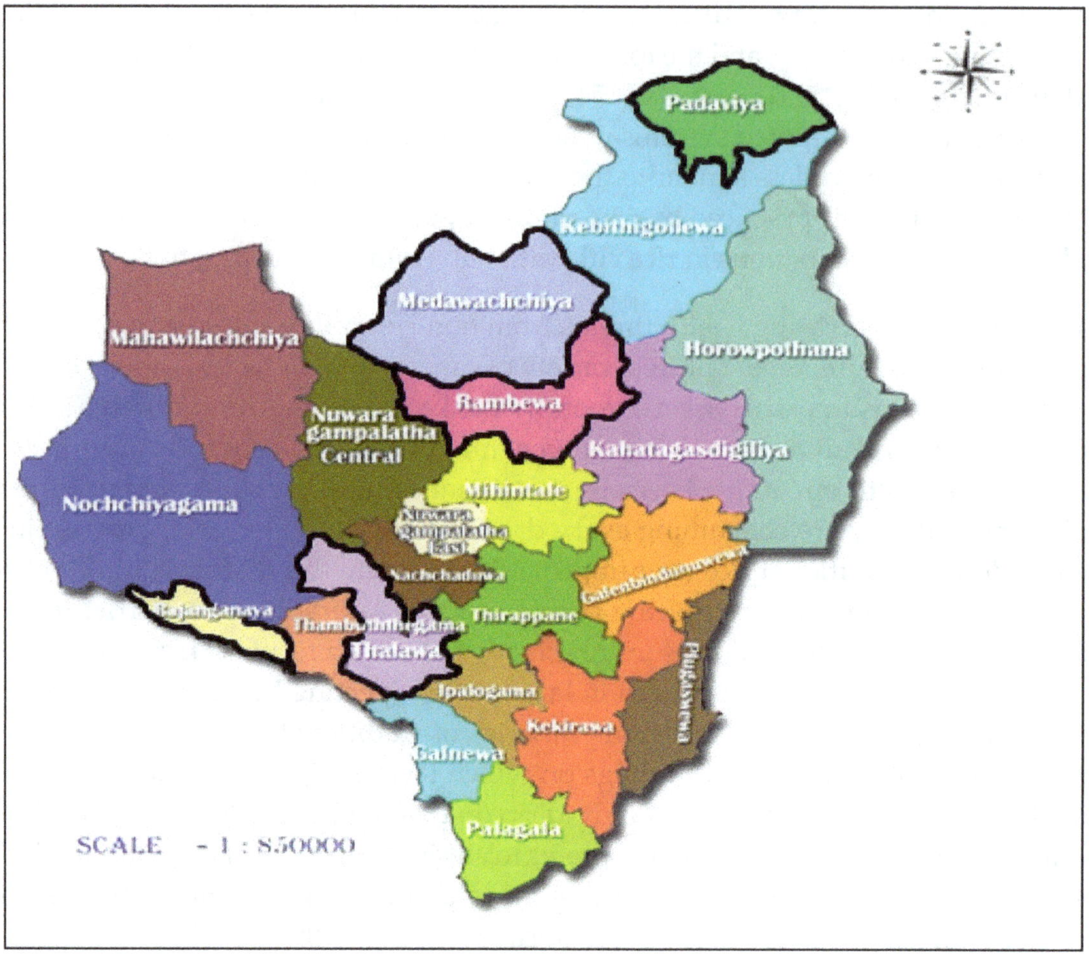

Figure 1. Map of selected Divisional Secretariat (DS) divisions.

Department of Census and Statistics, Central Bank of Sri Lanka and relevant research reports, project reports, journal articles and newspaper articles.

The collected data and information were subjected to logistic regression analysis to analyze the factors affecting willingness to grow traditional rice varieties and factor analysis to identify the factors affecting varietal selection.

3. Results and discussion

3.1. Demographic characteristics

Demographic characteristics of the study sample are summarized in **Table 1**. Mean age of the traditional rice-growing farmers was 48 years, while nontraditional rice-growing farmers have 51 years of mean age. Results revealed insignificant differences in household size (four members) and the number of years attained to a particular formal education (10 years) between these two categories of farmers. Moreover, for both farmer categories, at least two family members are available as family labor mainly consisting of the household head and his/her spouse. The land extent under traditional rice cultivation is smaller (1.34

Parameter	Traditional rice farmers		Nontraditional rice farmers	
	Mean	**±SD**	**Mean**	**±SD**
Age of the respondent (years)	47.87 (48*)	11.54	51.31 (51)	11.15
Household size (number)	4.10 (4)	1.26	3.74 (4)	1.43
Educational level (years)	10.19 (10)	2.31	9.33 (9)	2.87
Available family labor (number)	2.22 (2)	1.07	2.16 (2)	1.095
Land extent (Ac)	1.34	1.90	1.91	1.713
Yield (kg/Ac)	1199.60	521.77	1695.88	701.38
Farm gate price (Rs./kg)	50.24	14.90	31.32	5.28
Farming experience (years)	22.95	12.96	5.96	7.77
	Percentage (%)		**Percentage (%)**	
Gender of the respondent				
Male	67		92.6	
Female	33		7.4	
Presence of non-communicable diseases				
Yes	65.6		78.7	
No	31.4		21.3	
Awareness of traditional rice				
Yes	100		55.6	
No	0		44.4	
Farming system				
Organic	86.5		1.9	
Inorganic	8.1		87.0	
Mixed	5.4		11.1	

*Numbers in parenthesis are rounded numbers.

Table 1. Demographic characteristics.

acres) compared to improved rice cultivation (1.91 acres). The reason is that many of the traditional rice growers tend to cultivate traditional rice in their small paddy plots only for their family consumption as they are more health conscious, affected by non-communicable diseases (65.6%) and more aware (100%) on the nutritional and medicinal value of traditional rice varieties.

Compared to nontraditional rice varieties (Rs. 31.00), traditional rice varieties have higher farm gate price (Rs. 50.00) which shows comparatively better potential market for traditional rice. With respect to the existing farming systems, 86.5% of traditional rice farmers are practicing organic farming methods in their rice farms which ensure environmental, social and economic benefits to the society, while majority of the nontraditional rice farmers (87%) apply inorganic fertilizers and agrochemicals in their rice fields expecting a higher yield for better income.

3.2. Factors affecting willingness to grow traditional over nontraditional rice varieties

Logistic regression analysis was used to determine the factors affecting willingness to cultivate traditional rice by rice farmers. The binomial logistic analysis of measured variations in the outcome explained by predictors was significant (*Pr* < 0.001). Awareness of traditional rice, presence of non-communicable diseases in the household, land extent, yield, farm gate price, gender of respondent, age of the respondent, educational level, family labor availability, household size, farming experience and farming system were included in the logistic regression.

Table 2 explains the strength and the direction of the effect of each factor on the willingness to cultivate traditional rice. As revealed by the results, awareness of medicinal and nutritional values of traditional rice (*Pr = 0.0062*) show strong positive associations with the willingness to cultivate traditional rice compared to nontraditional rice. Land extent of rice cultivation shows a negative significant effect (*Pr = 0.0404*) on cultivation of traditional rice. It reveals that when farmers are having increased land extents, they are reluctant to cultivate traditional rice varieties. This is mainly due to the lower potential yield of traditional rice varieties over nontraditional (improved or hybrid) rice varieties. Most importantly, farm gate price shows a positive relationship with the willingness to cultivate traditional rice (*Pr = 0.0076*). Since the selling price is higher, farmers are more willing to go for traditional rice in their paddy fields over nontraditional rice varieties. Age of the respondent (*Pr = 0.0141*) and farming experience (*Pr = 0.0169*) positively affected the decisions-making regarding cultivation of traditional rice

Parameter	Estimate	Pr> chisq
Intercept	10.451	0.0946
Awareness of traditional rice	7.2517	**0.0062**
Presence of non-communicable diseases in the household	-12.372	0.5697
Land extent	-1.4726	**0.0404**
Yield	0.0015	0.2742
Farm gate price	0.3572	**0.0076**
Gender of the respondent	4.1364	0.0058
Age of the respondent	0.1586	**0.0141**
Educational level	0.2797	**0.0403**
Family labour availability	-1.8423	0.0887
Household size	0.7163	0.2626
Farming experience	0.2752	**0.0169**
Farming system	-6.1856	**0.0116**

Table 2. Maximum likelihood estimates for factors affecting willingness to cultivate traditional rice varieties.

varieties where farmers are able to ascertain more knowledge on the importance of healthy consumption and food habits, awareness of non-communicable diseases and adverse impact of synthetic fertilizer and other chemicals to human and environment through their maturity with age and cumulative farming experiences. Further, the education level of the farmers shows a positive impact toward getting into traditional rice cultivation which emphasized that when people are more educated, they tend to consider their health and nutrition compared to illiterate people. Both traditional and nontraditional rice farmers practice organic and inorganic farming systems and mixture of these two. As revealed by the results, farming system has significantly negative effect on willingness to cultivate traditional rice.

The presence of non-communicable diseases, yield, gender of the respondent, available family labor and household size were not significantly ($Pr < 0.05$) associated with willingness to cultivate traditional rice.

When farmers enhance their awareness on medicinal and nutritional values of traditional rice (OR = 4.025), nontraditional paddy farmers are also more likely to cultivate traditional paddy varieties (**Table 3**). The results further prove that when farmers increase their land extent by 1 acre (OR = −4.361), many of the traditional rice farmers tend to be relied on nontraditional rice cultivation. Moreover, the odds of being traditional rice cultivator is higher for higher farm gate price (OR = 1.704), old age of the farmers (OR = 1.17), higher educational level of the farmers (OR = 1.756) and more farming experience of the farmers (OR = 1.317).

3.3. Factor analysis for selection of traditional rice varieties for cultivation

Factor analysis was executed to determine the factors affecting varietal selection by traditional rice farmers. Principal component analysis (PCA) emphasized that multiple observed variables

Parameter	Point estimates	95% confidence limits
Awareness of traditional rice	**4.205**	**1.105–9.790**
Presence of non-communicable diseases in household	2.356	0.882–6.290
Land extent	**−4.361**	**1.066–17.832**
Yield	1.002	0.999–1.004
Farm gate price	**1.704**	**0.544–1.911**
Gender of the respondent	1.238	0.746–1.882
Age of the respondent	**1.17**	**0.948–1.448**
Educational level	**1.756**	**0.394–1.851**
Family labor availability	0.158	0.019–1.322
Household size	2.047	0.585–7.146
Farming experience	**1.317**	**1.051–1.656**
Farming system	**0.001**	**0.001–0.055**

Table 3. Odd ratio of logistic regression analysis.

have similar patterns of responses because of their association with an underlying latent variable [11]. Accordingly, it examines underlying variable in a number of observed variables of factors which affect selection of varieties.

The variance of the independent variables, explained by each principal component, is given by Eigen values. Any factor with an Eigen value ≥ 1 explains more variance than a single observed variable. **Figure 2** presents the scree plot used to identify six numbers of factors affecting varietal selection of traditional rice farmers where the first five factors (Eigen value ≥ 1) are highly affected factors loading from scree plot (**Table 4**).

The factor analysis revealed that four factors are affecting selection of varieties of traditional rice by rice farmers (**Table 5**). The first factor, namely, the varietal attributes, includes tolerance to pest and diseases, tolerance to drought conditions and tolerance to salinity. Factors, namely, presence of non-communicable diseases, gender, age, educational level and farming experience, are consolidated into the second factor which is named as personal factors. The third factor comprises of market-related attributes such as farm gate price and availability of buyers, while the fourth factor includes production-related resources, namely, cultivated land extent, yield, availability of family labor and farming system.

With the increase of world population and hence the galloping food demand, high-yielding rice varieties were highly popular among the cultivators. This practice hitherto has led to serious "genetic erosion"—the loss of traditional varieties from agroecosystems [12, 13] in the rice production sector. Due to their incredible health benefits, it has made them a pleasing choice for consumers who are suffering from diabetes, overweight or regulating their sugar intake. Cultivation and consumption of traditional rice varieties are not restricted to certain places in Sri Lanka, because consumption of these varieties in both national and international has been very consistent.

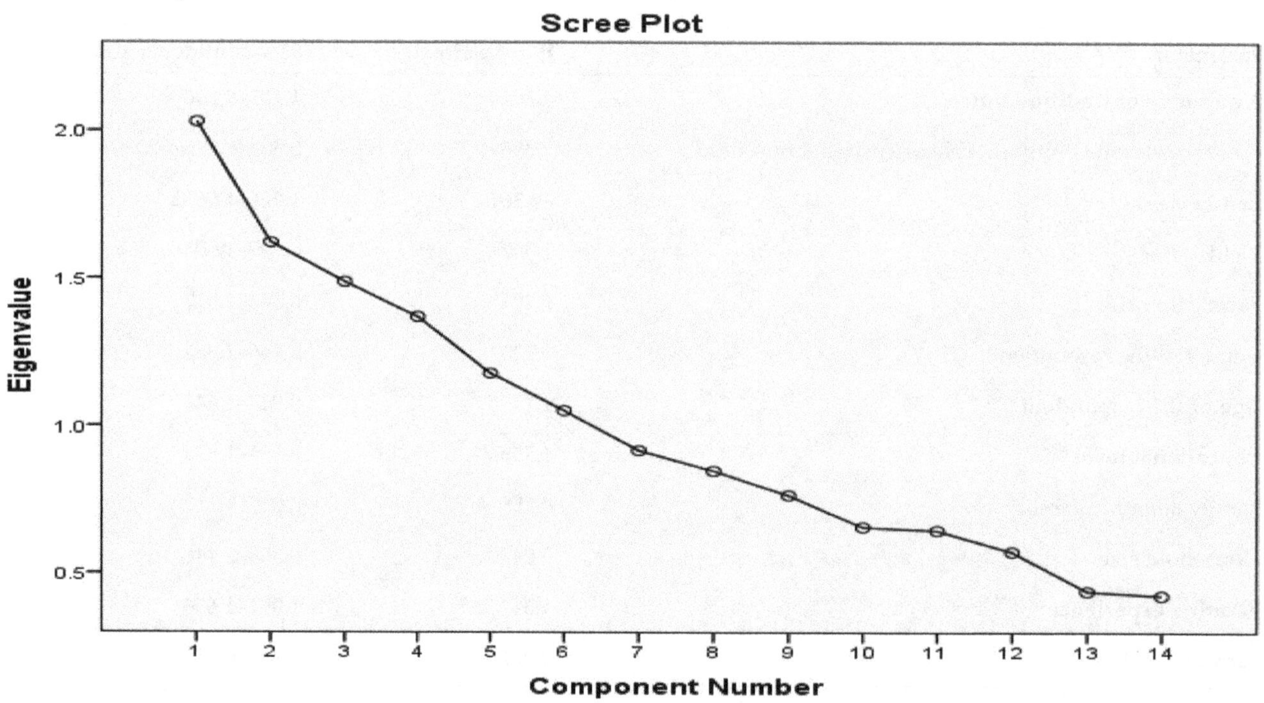

Figure 2. Scree plot for components.

Component	Initial Eigenvalues			Extraction Sums of Squared Loadings			Rotation Sums of Squared Loadings		
	Total	% of Variance	Cumulative %	Total	% of Variance	Cumulative %	Total	% of Variance	Cumulative %
1	2.030	14.498	14.498	2.030	14.498	14.498	1.586	11.328	11.328
2	1.620	11.574	26.072	1.620	11.574	26.072	1.514	10.811	22.139
3	1.487	10.621	36.693	1.487	10.621	36.693	1.494	10.674	32.813
4	1.369	9.779	46.472	1.369	9.779	46.472	1.493	10.664	43.477
5	1.177	8.410	54.881	1.177	8.410	54.881	1.419	10.137	53.614
6	1.051	7.506	62.387	1.051	7.506	62.387	1.228	8.774	62.387
7	.916	6.543	68.930						
8	.847	6.051	74.981						
9	.765	5.464	80.445						
10	.656	4.689	85.134						
11	.645	4.607	89.741						
12	.573	4.090	93.831						
13	.439	3.136	96.967						
14	.425	3.033	100.000						

Table 4. Total variance explained by factor loadings.

People also credit traditional varieties with other health benefits, such as giving sensations of cooling in the body; improving vocal clarity, eyesight and fertility; maintaining body sugar levels; and mitigating rashes. Among the local communities, many of traditional rice varieties are popular due to their inheriting characteristics. For example, *Suwandel* variety which has a milky taste upon cooking is highly recommended to be eaten by hard-working people. According to Ayurvedic medicine, this variety is known to promote fair and glowing skin, improves the functioning of the excretory system, enhances vocal clarity, increases the male sexual potency and helps to control diabetes and constipation. Likewise, variety *Pachchaperumal* is a highly nutritious red rice cultivar which helps to cool the body, is preferred by patients who are suffering from diseases like diabetes and cardiovascular complications and is also good for patients with high blood pressure. Another example is that of variety *Madathawalu*, which is able to remove toxic components especially some cancer causative agents from the human body. This variety can clean the blood circulation system and promote the activity of sweating glands. It strengthens the immune system and adds to the nutritive value of the cooked rice for lactating mothers and infants.

Besides, it is known to all that the paddy production system is extremely vulnerable to climate change impacts. Traditional agricultural practices coupled with indigenous rice varieties have proven to be more successful in facing climate change and its related threats such as droughts, floods, attack of pests and disease outbreaks. These traditional rice varieties have strong characteristics that help them survive climate change impacts compared to newer varieties used in conventional paddy cultivation. Hence, the traditional rice cultivation is not only a solution for health concerns but also a way of achieving sustainability via conservation of agricultural

Parameter	Component				
	1	2	3	4	5
Presence of non-communicable diseases in the household	−0.330	**0.718**	−0.574		
Land extent		−0.291	−0.260	**0.489**	0.622
Yield		0.284		**0.586**	
Farm gate price		0.368	**0.482**	−0.360	
Gender of the respondent	−0.299	**0.433**	0.254		
Age of the respondent	−0.265	**0.710**			0.392
Educational level	0.447	**−0.553**			
Family labor availability	0.263			**0.499**	
Farming experience	−0.329	**0.282**	0.561		
Farming system	0.434			**0.603**	
Availability of buyers			**0.318**		−0.651
Tolerance to pest and diseases	**0.578**	0.302			
Tolerance to drought	**0.576**			−0.351	
Tolerance to salinity	**0.636**			−0.300	

Extraction method: principal component analysis.
[a]Five components extracted.

Table 5. Component matrix of factor analysis.

practices which promise more congenial environment for future generation. According to [14], Indian farmers in a district in Uttar Pradesh rediscovered the advantages of traditional rice cultivation which were resistant to drought condition and have not been susceptible to diseases and fetched better market prices. Therefore, traditional rice cultivation does not restrict to a particular region or area. Therefore, findings of this study could be generalized to other areas in Sri Lanka and Asian region as a whole.

4. Conclusion

It seems logical to conclude from this study that awareness of medicinal and nutritional value of traditional rice varieties, land extent, farm gate price, age of the respondent, education level, farming experience and farming system have significant influence on the willingness of farmers to opt traditional rice cultivation in Anuradhapura district of Sri Lanka. The trend in willingness to grow traditional rice in Anuradhapura district showed that 59% of farmers attached to families with non-communicable disease are willing to grow traditional rice compared to farmers from healthier families. The study further revealed that two of the most vital factors responsible for selection of traditional rice variety are personal choices and varietal attributes. Hence, the study brings few recommendations to enhance the structured organization of traditional rice-growing farmers to make crop agronomic and management information and potential marketing information available through government policy interventions in order to empower the traditional rice-cultivating farmers in Anuradhapura district and throughout the country as a whole.

Author details

Ginigaddara Appuhamilage Sanjeewanie Ginigaddara* and Sampath Priya Disanayake

*Address all correspondence to: sanjeewanieg@gmail.com

Faculty of Agriculture, Rajarata University of Sri Lanka, Anuradhapura, Sri Lanka

References

[1] Rebeira SP, Wickramasinghe HAM, Samarasinghe WLG, Prashantha BDR. Diversity of grain quality characteristics of traditional rice (*Oryza sativa* L.) varieties in Sri Lanka. Tropical Agricultural Research. 2014;**4**:570-578

[2] Department of Census and Statistics, Annual Report. 2015. Available from: http://www.statistics.gov.lk/agriculture/Paddy%20Statistics/PaddyStats.htm [Accessed: Feb 15, 2017]

[3] Wickramasinghe HAM, Noda T. Physicochemical properties of starches from Sri Lankan rice varieties. Food Science and Technology Research. 2008;**14**(1):49-54

[4] Dharmasena PB. Traditional Rice Farming in Sri Lanka Economic S3 Review April/May 2010. Available from: http://dl.nsf.ac.lk/bitstream/handle/1/14119/ER-36-(1-2)_48.pdf?sequence=2 [Accessed: Feb 15, 2017]

[5] Abeysekera WKSM, Premakumara GAS, Ratnasooriya WD. Antioxidant Properties of Some Sri Lankan Traditional Red Rice (*Oryza sativa L.*). 2011. Available from: http://www.cmb.ac.lk/wp-content/uploads/2014/02/Antioxidant-Properties-Of-Some-Sri-Lankan-Traditional-Red-Rice.pdf [Accessed: Jun 12, 2017]

[6] Morishima H, Oka HI. Genetic erosion in wild and cultivated species. Rice Genetic Newsletter. 1995;**12**:168-171

[7] Anonymous. SRI Progress in Sri Lanka through Oxfam Australia. 2010. Available from: http://sri.ciifad.cornell.edu/countries/srilanka/SL_OxfamAusAACRP_report_2010_11.pdf [Accessed: Jun 15, 2017]

[8] Rodrigo C. Use of traditional paddy cultivation as a mean of climate change adaptation in Sri Lanka. The Island Newspaper. Sep 19, 2013

[9] Anonymous. Traditional Rice for a Healthy Life. 2014. Available from: http://laska.asia/buzzzz-247/traditional-rice-for-a-healthy-life/ [Accessed: Jun 20, 2017]

[10] District Statistics Handbook. Anuradhapura District. 2015. Available from: http://www.statistics.gov.lk/DistrictStatHBook.asp?District=Anuradhapura&Year=2015 [Accessed: Sep 6, 2017]

[11] Mittal S, Mehar M. Socio-economic factors affecting adoption of modern information and communication technology by farmers in India: Analysis using multivariate probit model. The Journal of Agricultural Education and Extension. 2015:**22**(2):199-212

[12] Porceddu E, Ceoloni C, Lafiandra D, Tanzarella O, Scarascia A, Mugnoza GT. Genetic resources and plant breeding: Problems and prospects. In: Miller TE, Koebner RMD, editors. Proceedings of the 7th International Wheat Genetics Symposium. Cambridge (United Kingdom): Institute of Plant Science Research; 1988. pp. 7-21

[13] Singh RK. Genetic resource and the role of international collaboration in rice breeding. Genome. 1999;**42**:635-641

[14] Singh V, Prakash S. Indian Farmers Rediscover Advantages of Traditional Rice Varieties. Available from: http://satavic.org/indian-farmers-rediscover-advantages-of-traditional-rice-varieties/ [Accessed: Sep 5, 2017]

7

Phosphorus Efficient Phenotype of Rice

Josefine Kant, Takuma Ishizaki,
Juan Pariasca-Tanaka, Terry Rose,
Matthias Wissuwa and Michelle Watt

Abstract

The ideal phenotype to cope with P deficiency is suggested to be a larger root system, both in terms of length and foraging area, coupled with a high capacity for P solubilization from compounds exuded from roots. Greater soil exploration results in a large number of roots in the top soil, longer roots in general with more cortical aerenchyma, more and longer root hairs, and a shift in mycorrhizal and bacterial colonization. However, these assumptions often result from experiments in highly controlled, sterile and soil-free conditions using model plants or single ecotypes where results are then extrapolated to all genotypes and plant species. In recent years this generalization has been questioned. Here, we summarize recent rice research analyzing the natural diversity of rice root systems under P deficiency. Interestingly, while some of the high yielding genotypes do show the expected, large root system phenotype, some have a surprisingly small root system—as little as a quarter of that of the large root system varieties—but achieve similar yield and P uptake under P deficiency. This effect has recently been termed root efficiency, which we discuss in this chapter in conjunction with root foraging traits.

Keywords: phosphorus deficiency, root system, root hair, root type, xyloglucantransferase

1. Introduction

Rice (*Oryza sativa*) is the most important source of calories for millions of people [1] and, like all crops, its growth and yield is enabled by taking up water and nutrients from the soil through its root system. Yield can be constrained by many abiotic and biotic factors, including drought, nutrient deficiencies, and pathogen infections. Second, only to nitrogen (N), phosphorus (P) is an essential major nutrient necessary for biosynthesis of nucleotides, proteins, bio membranes, and energy metabolism such as the provision of ATP. In contrast to most N forms, P is highly sorbed to soil particles and is therefore considered to be an immobile nutrient in soils [2]. Estimates suggest that more than half of all agricultural soils are P deficient [3] which substantially affects plant growth, both below and above ground, to the extent that P is often the most yield-limiting nutrient in resource-poor farming systems in south-east Asia and Africa. In developed nations, P and N deficiencies are generally rectified by the application of mineral fertilizers to the soil, often in excess of that taken up by the crop. This leads to rapid leaching of N, more slowly also of P, into the ground water, eventually leading to eutrophication of water systems which leads to algal blooms, anaerobic water, and death of fish and other fauna [4]. In addition, mineral phosphate fertilizers are usually produced by mining rock phosphate and, although several predictions with variable numbers are present, these resources are limited and might be depleted in 50–100 years [5]. Therefore, understanding how P deficiency shapes root systems and how plants can overcome P deficiency is critical for future breeding of P efficient crops to ensure nutrition of the ever-growing world population in the future.

Due to the immobile nature of P in soil, increasing root foraging for P is theoretically one of the best means by which to improve the efficiency of P acquisition by plants. Research in other crops and model plants have provided indications as to what the ideal root phenotype to cope with P deficiency should look like (reviewed in [3, 6]). Overall, it is suggested that a larger root system, both in length as well as in area, is beneficial for better yield performance in low P conditions. This higher soil exploration includes a large number of roots in the top soil (soil surface until 25 cm depth), longer roots in general, with more cortical aerenchyma, smaller diameter, more and longer root hairs, more mycorrhiza, altered bacterial colonization, and higher exudation of P-solubilizing chemicals.

In this chapter, we will (1) give a general literature overview of strategies for P uptake and its optimization and (2) review the proposed ideotype for rice P efficiency and present an experiment for its optimization using transgenic rice.

1.1. Exudation of P-solubilizing chemicals

Mobilization of soil-bound P via the release of P-solubilizing compounds from roots is widely proposed to be a key mechanism by which many plant species enhance P acquisition in soils. These compounds include phosphatase enzymes capable of mobilizing organic P, protons that acidify the soil to solubilize calcium (Ca)-bound P and organic compounds including phytosiderophores and carboxylates that compete with P in ligand exchange reactions [7]. The release of protons and carboxylates is a strategy that is particularly important for species from the family Proteaceae, which have often evolved in extremely P deficient soils [8]. Other species that form proteiod—or cluster—roots under P deprivation such as the model legume species *Lupinus albus*, are also highly efficient at mobilizing P from aluminum (Al)- and iron (Fe)-P complexes via efflux of a targeted surge of carboxylates (predominantly citrate and malate) and protons from the cluster roots under P deficiency [9].

Beyond species that form cluster roots, however, the role of compounds exuded from roots is less clear. Experimental evidence that non-cluster root forming species can mobilize significant amounts of P through exudation of carboxylate is lacking [10] and the concentrations of carboxylates required to mobilize P in incubation studies are much higher than those thought to occur in the rhizosphere of most plants [11]. As an example, a recent study using near-isogenic wheat (*Triticum aestivum*) lines that differed in citrate efflux failed to find evidence that higher citrate efflux led to any improvements in P uptake or crop yields [12]. The capacity for enhanced root exudation in rice to mobilize soil P is discussed in detail below.

1.2. Association with mycorrhizal fungi

Arbuscular mycorrhizal (AM) fungi are recognized for enhancing nutrient availability, notably P, to most plants [13]. AM fungi are obligate biotrophs that establish mutualistic associations with the roots of over 90% of all plant species via a complex system of intraradical and extraradical hyphae, in which the external mycelium of AM fungi acts as an extension of host plant roots, thus increasing the effective surface area to absorb nutrients and water [13–15]. The transfer of nutrients from AM fungi to plant roots is typically facilitated by highly branched fungal structures within the root cortex, known as arbuscules [15]. The transfer of nutrients from AM fungi to roots occurs in exchange for sugars generated from photosynthesis and is typically facilitated by specific transporters expressed at the interface of plant root and arbuscule [15, 16].

To what extent the AM symbiosis is beneficial to crop species depends on environmental factors and varies between different crops and among varieties of the same crop. It is generally assumed that high rates of P fertilization diminish the potential benefits from AM symbiosis [17]. Further, one may assume that crops with a rather fine and dense root system would benefit less from the increase in the effective surface area provided by external hyphae compared to crops with less fine roots. Within crop species, varietal differences may be of importance and should ideally be explored in crop improvement programs. However, the multiple interactions between AM fungi, other soil microorganisms and plant roots, each potentially affected by soil properties, climatic factors, and crop management, have made a selection for a stable increase of AM-variety symbiotic efficiency a rather challenging task.

1.3. Association with beneficial bacteria and fungi other than AM

Application of free-living soil bacteria and fungi to plants can increase plant growth and nutrition status. As early as 1948 it was reported that isolated soil bacteria could enhance plant P solubilized from calcium phosphate [10]. Today efforts continue toward identifying microbiota-root interactions beyond AM that enhance crop phosphorus acquisition. These are driven by commercial and regulatory pressures to supplement mined P reserves and optimize the recycling of P from soil and biomass pools [18], and by the increasingly comprehensive genome-based methods to characterize microbiomes [19] to explore to increase agricultural fertility [20].

Beneficial microbiota includes those contributing directly to plant or soil P fertility processes, and those indirectly contributing via control of plant host diseases, soil toxicities, and weeds that compete with crop resource uptake. P fertility mechanisms include (1) promoting greater root surface area, (2) increasing inorganic P availability in the soil solution, and (3) altering organic P pools in soil (e.g. rates of turnover). The evidence is mounting that microbiota can change molecular events within the plant (e.g. induce proton release for rhizosphere acidification, gene regulation involved in ion uptake) [21].

Management of beneficial microbiota is mainly done through inoculation onto the plant (seed or shoots) of microbes that are isolated from a given environment. However, management of the naturally-associating rhizosphere microbiota through the plant (e.g. plant genetics and breeding) or soil (e.g. rotations, tillage) is also performed. Molecular characterization of the whole root and soil microbiomes will likely lead to a merging of inoculant approaches with the direct engineering of rhizosphere microbiota [22], as plant hosts are intimately connected with microbiota whose genes can be transferred with that of the host genome [23]. Although many examples report beneficial responses to bacteria or fungal strains on plants in pots in glass-houses, there are few translations to farmers' fields with formulated, scaled production [18].

1.4. Aerenchyma formation

Aerenchyma refers to plant tissue containing enlarged gas spaces, formed in roots and shoots of wetland and dryland species through cell death "lysigenous" or cell separation "schizogenous" [24, 25]. Formation of aerenchyma can be constitutive or induced by abiotic stresses such as waterlogging [26], drought, and nutrient deficiency including phosphorus, nitrogen and sulfur deficiency [27–32]. Therefore, the presence of aerenchyma can differ even within root segments when heterogeneous abiotic conditions, such as those in field soils, are present [30, 33, 34].

Several reports from maize (*Zea mays*), bean (*Phaseolus vulgaris*), and barley (*Hordeum vulgare*) results suggested that aerenchyma formation reduces the metabolic costs of soil exploration [35–37] through the decrease of root nutrient content and respiration [31]. Nutrients such as P, released during the aerenchyma formation (cell death), could be reutilized by the plant, for example, in the continued growth of apical cells. Increased soil exploration for more P resources, coupled with less P required for root functional maintenance, would be an advantageous trait under P-limitation.

1.5. Root hair production

Root hairs are unicellular extensions of epidermal cells that constitute most of the root's surface area [38]. Root hairs have been under investigation for more than a century [39], and it was shown decades ago that root hairs are involved in P uptake from soil. P uptake by root hairs was indirectly demonstrated for wheat and barley [40], and directly for rye (*Secale cereale*) root hairs [41]. Enhanced root hair development is consistently listed as one of the cheapest and most general adaptations to P deficiency [3]. And yet, how and if their presence is beneficial for P uptake remains under discussion. Within the molecular model plant *Arabidopsis thaliana*, a study investigating phenotypic reactions of >150 ecotypes to local P levels revealed unexpected results. Half of the tested genotypes did not show any root hair reaction to local low P, while one quarter responded with a production of shorter and less, and the other quarter with longer and more root hairs [42]. Another study showed that maize root hairs were more responsive to soil moisture than to soil P level [43]. While lower water content correlated with the production of more and longer root hairs, no correlation was found with soil P concentration. A recent study on barley seedlings developing normal or very short root hairs came to the conclusion that root hairs are instrumental for water uptake from drying soil when plant transpiration rates are high [44].

2. P deficiency responses found in rice

Rice roots face highly dynamic soil conditions; possibly the most complex of all cereal grains. Soils have repeated flooding, irrigation and drying events, and tilling and compacting results in very soft soils, hardpans, and furrows, with dramatic variation in aeration, pH and nutrient conditions from the surface to depth [45]. These transitions are greatest in lowland systems, but also occur in upland systems.

2.1. The rice root system

The rice root system consists of the first emerging embryonic seminal roots as well as post-embryonic crown roots emerging from shoot-borne nodes. In addition, all of these main root axiles can branch, forming lateral roots (LRs) of first and higher orders. A unique feature of rice is the formation of very distinct classes of these LRs—S- and L-type [46]. The most abundant LRs are the short and thin S-type LRs. L-type LRs are much longer, have a larger diameter and form branch roots, thus producing higher order LRs (**Figure 1**).

2.2. Root efficiency in rice

In order to analyze the natural variation of P deficiency tolerance, a large panel of diverse rice genotypes (>200 genotypes) was grown in upland fields, both in a sufficient fertilized and P deficient fields, and their root systems were analyzed and correlated to yield performance and P uptake. Interestingly, while some of the high yielding genotypes did show the estimated, large root system phenotype under P deficiency, some genotypes had a surprisingly small root system—as little as a quarter of the large root system varieties—but achieved the same yield and P uptake under P deficiency [47]. This effect has been observed in a number of field trials and recently been termed '**root efficiency**'. What could be the basis of this root efficiency? Which traits could enable a smaller root system to take up P as efficiently as or even more efficiently than a bigger one? The efficiency could be based on any or all of the aforementioned P fertility traits for plants, and these will be elaborated on for rice in the following sections.

2.3. Exudation of P-solubilizing chemicals by rice

Rice plants exude a range of compounds from their roots, and the amount and composition of these exudates change in response to P deficiency [48–50]. A number of lines of enquiry suggest that increased efflux of carboxylates, particularly citrate, enhances the capacity of rice genotypes to acquire P. Ref. [48] reported an increase in citrate efflux under P deficiency in seven rice genotypes, and citrate efflux rates were correlated with the tolerance of these rice genotypes to P deficiency (defined as ratio of biomass grown with sparingly soluble P source compared to a soluble P source). Also found was increased citrate efflux from rice roots under P deficiency. Subsequent modeling studies indicated the importance of citrate efflux in P solubilization and uptake by rice growing under aerobic conditions [51]. However, as pointed out in a review of root traits associated with the efficient acquisition of soil P by rice [6], there is no direct evidence that increased efflux of citrate, or any other carboxylate, is responsible for higher P acquisition efficiency for any rice genotype.

Figure 1. The rice root system structure. Top soil washed off to display the root system of a field-grown rice plant ca. 50 DAS (A). Seven DAS seedlings with many seminal roots (left) or a high proportion of lateral roots (right) (B). Stereomicroscopic image demonstrating the rice root characteristics (C). Light microscopic images of rice roots grown in nutrient solution (D) or soil (E). MR: main root (seminal or crown root), LR: lateral root, L: L-type LR, S: S-type LR, 2O: second order LR, RH: root hair.

Protons and phosphatase enzymes are also released into the rhizosphere from rice roots. The release of protons typically occurs due to an imbalance in cation/anion uptake by roots [52]. Genes encoding phosphatases are up-regulated in roots under P deficiency [53], but we are not aware of any published studies that have demonstrated that increased phosphatase efflux from the roots of rice is linked to enhanced P uptake, or that such a trait confers greater P uptake in any given rice genotype.

2.4. Rice association with mycorrhizal fungi

Rice is a host for AM fungi [15, 16] and although root colonization has been observed under irrigated lowland and upland conditions, it is generally assumed that the AM symbiosis is more important in the aerobic upland rice [54]. It has been shown that P transporters exist in rice that are specifically induced in roots colonized by AM fungi and that these P transporters (*OsPT11* and *OsPT13*) facilitate the transfer of Pi from AM fungus to plant [16]. Compared to some other crop species, however, our knowledge regarding the AM fungi community colonizing rice roots in the field remains limited, and the extent to which AM symbiosis may be exploited to benefit rice yield directly in the field is unknown. Studies comparing AM colonization in the field in a set of diverse rice genotypes indicated that considerable variation in colonization rates exists (Wissuwa, unpublished). Further, all root samples taken showed gene expression of the *OsPT11* transporter, suggesting P transfer from AM fungus to rice roots commonly occurred. Yet neither colonization rates nor gene expression levels were correlated to the large genotypic differences observed for P uptake in that set of rice genotypes (Wissuwa, unpublished), whereas P uptake correlated strongly to root size [47]. Earlier studies in one set of near-isogenic rice lines differing in root size and P uptake showed these differences remained unchanged in sterilized soil [55]. Based on the limited evidence available to date, one may tentatively conclude that the AM symbiosis contributes less to rice genotypic differences in P uptake compared to root attributes such as size, fineness or root hair length and density [56].

2.5. Rice association with beneficial bacteria and fungi other than AM

There are intense research efforts to apply rhizobacteria to rice to boost its productivity and increase environmental sustainability because of the enormous value of this crop and the resources used globally. Beneficial microbiota are tested to (1) reduce methane emissions [57]; (2) increase nitrogen uptake [58]; (3) fix atmospheric nitrogen within the root [59]; (4) reduce diseases [60], and (5) increase P nutrition, and to a lesser extent, that of Fe and other micronutrients [61]. In field experiments with farmer participation, beneficial microbiota is being combined with fertilizer applications. This mode of application and level of participation appears to be required for repeatedly validating inoculants that demonstrate high efficacy in the field [62].

The types of microorganisms previously tested for improved P uptake by rice include bacteria of the genera: *Rhizobium* [63], *Pseudomonas, Azotobacter, Azospirillum* [61, 59], and *Enterobacter* [64], as well as bacteria within a consortium of mixed genera [65]. Applied organisms may reside at the root-soil interface, or within the root (endosphere). The endosphere aerenchyma may have more stable gaseous conditions than the root-soil and outer rhizosphere zones and is a target zone for beneficial isolates, referred to as endophytes [66]. Possible functions of microbiota for P fertility are the same as those proposed for other crops (reviewed in [10, 67–69]), with the exception of much greater emphasis on endophytic diazotrophic (nitrogen fixing) bacteria, perhaps driven by the unique aerenchyma environment in rice, and/or rice-specific responsiveness to effectors associated with endophytic colonization [70]. Given the niche opportunity for endophytes within rice axile root aerenchyma, it may be beneficial to look for consortia of microorganisms or native rhizosphere (including endosphere) profiles that promote these developmental features within the roots.

2.6. Aerenchyma formation in rice

Rice roots can form lysigenous (cell death) aerenchyma in both forms: constitutive, developed from the apical parts of the roots toward the base; and inducible, promoted in all parts of the roots under anaerobic conditions such as waterlogging, drought or nutrient deficiency.

The advantage of aerenchyma formation for more soil exploration with fewer nutrients required may be greater in rice than maize, for example, due to its greater tendency for aerenchyma formation and enhanced root length under low nutrient treatments. For example, phosphorus deficiency causes a 20% increase in the percent cortical area converted to aerenchyma in rice [71]. Likewise, aerenchyma formation is enhanced by both nitrogen and oxygen-deficient conditions [72]. In addition to aerenchyma, rice possesses a barrier to radial O_2 loss (ROL) to the external environment, which promotes diffusion of O_2 toward the root apex. The O_2 increases the redox potential in the rhizosphere and causes the oxidation of Mn^{4+} and Fe^{2+} forming the plaque on the surface of rice roots [73] (which may reduce the uptake of phytotoxic elements into plant tissue).

Although the benefits of aerenchyma and ROL barrier formation are widely reported, the molecular mechanisms that regulate their formation are not completely understood. Since the rice genome has been fully sequenced and many tools are available, further insights into the molecular determinants of aerenchyma and other rice-specific tissues are expected, in order to improve rice cultivars by using modern breeding techniques.

2.7. Root hair formation in rice

An often heard assumption is that root hairs are adaptations specific to limiting conditions and are not needed for growth in optimal conditions. Consequently, elite varieties adapted to high input, optimized soil conditions should produce short and few root hairs and be less responsive to stress conditions compared to landraces.

In the next sections, we will review recent findings regarding natural variation in rice root hair formation in response to P deficiency, growth conditions, and in respect to the rice root system structure. Finally, we will also give an example of transgenic plant generation in the attempt of increasing rice P uptake under deficient conditions.

2.7.1. Root hair variation within root types

Lateral roots were recently proposed to have different, specialized functions depending on their developmental type [74]. Supporting this proposal, we showed that the thinner lateral roots of rice produced shorter and fewer root hairs [75], and found a positive linear relationship between lateral root diameter and root hair length [56]. These results were reproduced in different growth conditions, with main roots (seminal and crown axile roots) consistently producing the longest and most root hairs, followed by L-type and S-type lateral roots (**Figure 2**). On the other hand, all root types within a genotype had the same tendencies regarding root hair length and density, and differences between genotypes were stable per root type [56]. This indicates that the phenotypic potential per genotype is first determined by its genome and then by the environment.

2.7.2. *Influence of root hair longevity*

Very few studies have addressed the question of root hair longevity and how long root hairs contribute to plant water and nutrient uptake. In [56] a higher proportion of living rice root hair cells was found in the low P compared to a P-fertilized field, over five tested genotypes. It can be concluded that longer living root hairs might be an adaptation to low P conditions to prolong root hair contribution to P uptake. Nevertheless, in the top soil layer fewer living root hairs were found compared to the subsoil [56]. This can most likely be attributed to the presence of younger root segments including root tips in the deeper soil layer compared to the oldest root segments in the top soil. In future experiments, more emphasis should be laid on longevity not just of root hairs, but also of lateral and main axile roots as harnessing natural variation in this trait could lead to improvements in breeding for more P deficiency tolerant varieties.

2.7.3. *Root hairs impacted by experimental conditions*

Our recent findings highlighted the vast influence that growth medium has on root hair formation [75]. Nutrient solution led rice seedlings to produce many, long root hairs on their main roots, while the discrepancy between root types was much smaller in soil-grown plants. We were able to reproduce these results in small plastic containers (50 ml volume) already 7 days after germination (DAG) (**Figure 2**). In addition to soil and nutrient solution, we grew rice seedlings in an artificial soil (vermiculite) and an agar solution without nutrient addition. Vermiculite led to very similar root hair growth to the soil, while roots grown in agar formed root hairs of comparable length and density as those found in nutrient solution.

One possible explanation for the observed differences among growth conditions is the absence or presence of physical restriction which can be found in soil particles, but not in a nutrient solution. To test this, another experiment was conducted to investigate the effect of growth medium strength on root hair length in rice seedlings. The experiment comprised six concentrations of agar with two replicate boxes per concentration. Root hair length after 3 weeks of growth significantly decreased with increasing agar concentration (**Figure 2**), proving that soil strength/presence of physical restrictions is one-factor influencing root hair growth. Detection of rice root hairs in undisturbed soil using a synchrotron approach also substantiated this assumption as root hairs growing in pores were found to grow to three times the length of those restricted by soil particles [75].

2.7.4. *Genotype-dependent root hair formation*

The commonly-stated response to P deficiency of production of longer and more root hairs may depend on genotype. For example, a wide selection of genotypes of the sup-populations indica, aus, temperate japonica, aromatic, and tropical japonica were tested in buffered, diffusion-limited solid-phase solution, and genotype-dependent variable responses to the low P condition was found [71]. We recently showed that in the soil, under controlled conditions in a greenhouse as well as in the field, some genotypes even form shorter and/or fewer root hairs in low P compared to sufficiently P-fertilized soil [56]. This highlights that increased root hair production is not a general, inevitable event upon P deficiency, but a specialized adaptation of some genotypes. Similarly, **root efficiency** (P uptake per unit root length) mentioned above

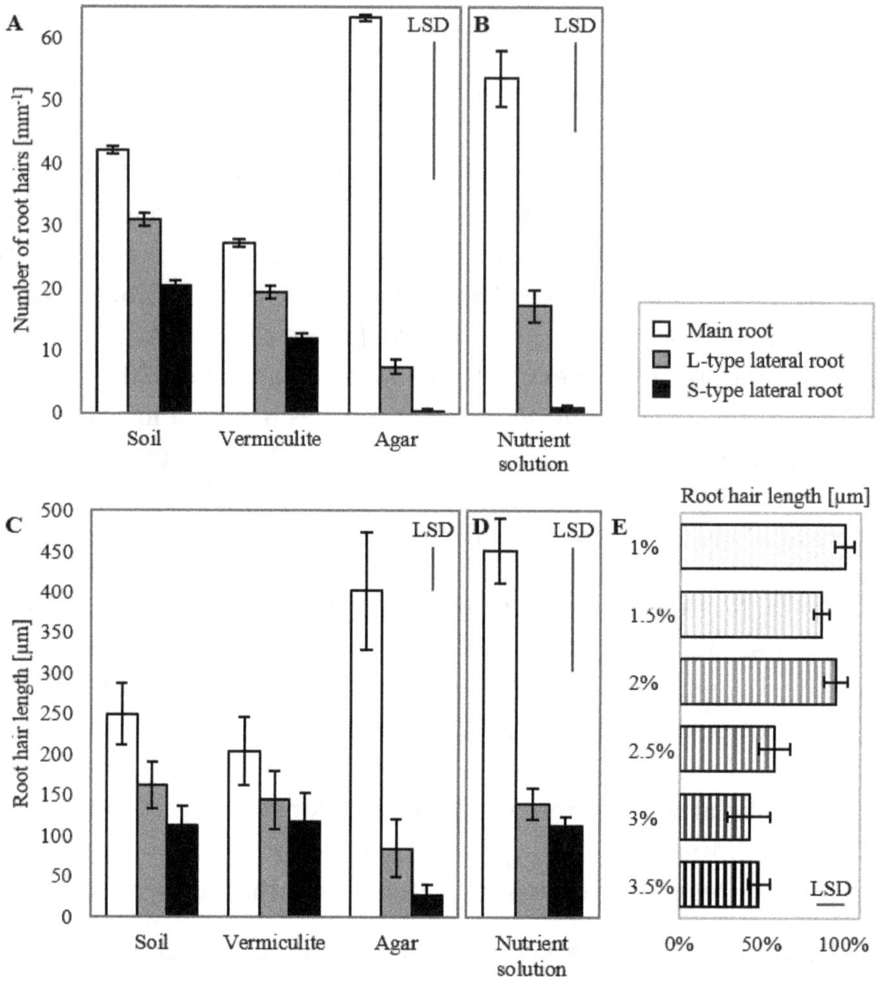

Figure 2. Dependence of root hair production on growth conditions. Determination of root hair density (A, B*) and length (C, D*) on main roots, L-type and S-type LRs. Shown are mean values over four genotypes of 4 (A, C) or 5 (B, D) replicates; +/− standard error determined at 7 DAS (A, C) or 35 DAS (B, D). Soil and nutrient solution represent low P conditions while vermiculite and agar did not receive any nutrients. Half-strength Yoshida solution with increasing amount of agar was used to measure Nipponbare root hair length (E). Least significant difference (LSD) values are indicated per graph. Please note *: data shown in (B, D) are a sub-set of data published in [56].

may not be coupled with increased root hair production, although the assumption could be that smaller root length and weight could be associated with more hairs for a given unit of P uptake. For example, one root efficient genotype had substantially increased root hair length and density upon P deficiency (DJ123). Yet another root efficient genotype (Santhi Sufaid) exhibited shorter and fewer root hairs in P deficient soil, while the inefficient genotype (Sadri Tor Misri) that took up the most P and had the greatest root weight, produced the most root hairs of all tested genotypes [56]. These results lead us to the conclusion that root hairs may respond to P deficiency for some genotypes, and may contribute to root efficiency in others, but their development is not a predictable, universal response to P.

2.8. Transgenic plants for increased P uptake: An example study

Here we present an example experiment to optimize or increase P uptake by rice roots using a transgenic approach. Our candidate genes are based on a previous study of transcriptomes of a P deficiency intolerant and a tolerant rice geno-type. Several genes putatively associated with root cell wall loosening and root hair extension were found to have higher expression in the roots of the tolerant genotype [53]. We chose one gene to study that encodes a putative cell wall modifying enzyme, a Xyloglycantransferase (Os11g33270.1), designated XTH2, and produced transgenic plants.

The ectopic overexpression of a dozen T_2 lines was tested, and three with very high, high, and moderately increased *XTH2* expression, were selected for further phenotypic characterization (**Figure 3**). Cross and longitudinal sections were prepared to compare the root anatomies of control and *XTH2* overexpressing lines and no apparent differences were found. After 4 weeks of growth either in P deficient or sufficient nutrient solution, several root parameters

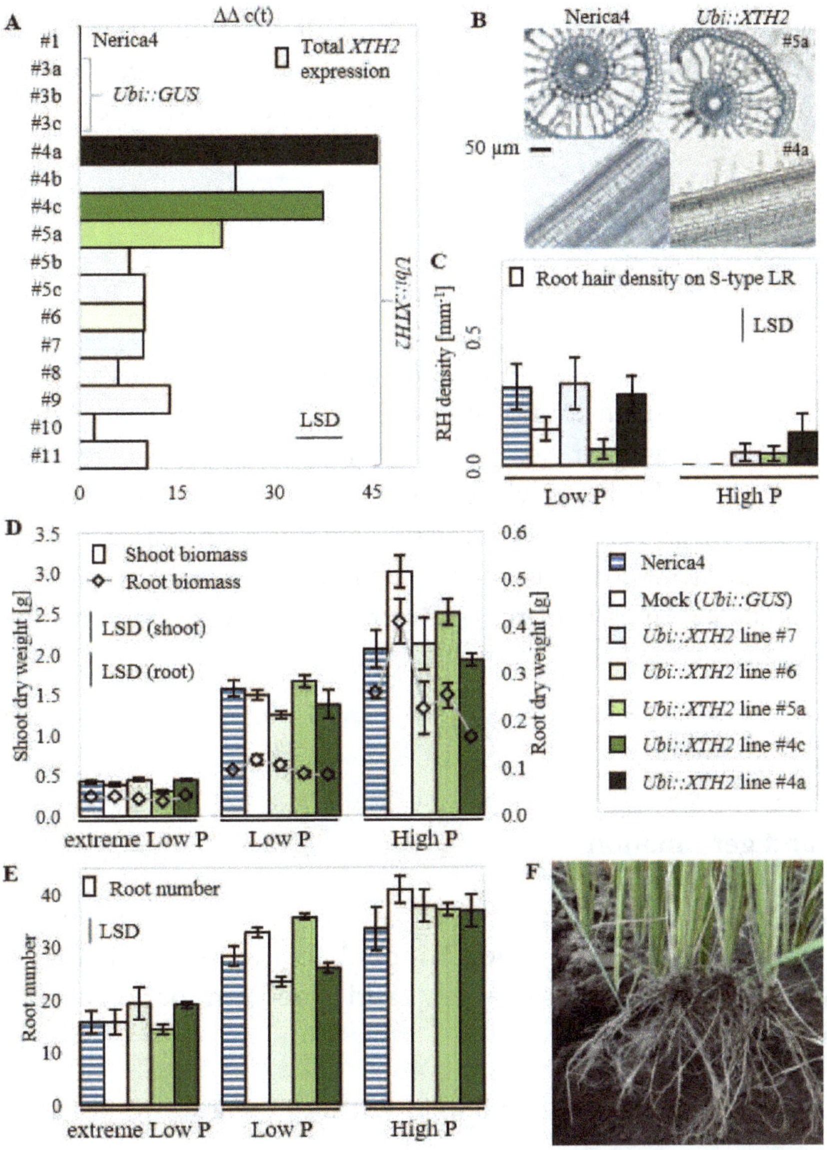

Figure 3. Phenotypic analysis of *Ubi::XTH2* lines. Total *XTH2* expression in a dozen T_2 lines was determined by qRT-PCR (A). Cross and longitudinal vibratome sections (B), and root hair density on S-type lateral roots (LRs) of selected T_2 lines grown in low or high phosphorus containing nutrient solution (C). Shoot and root dry weight (D), and root number (E) of selected T_2 lines grown in extreme low, low (example image in F) or high phosphorus soil. Least significant difference (LSD) values are indicated per graph.

were analyzed (root length and number, lateral root densities, root hair density and length on all root types); root hair density on S-type lateral roots is shown (**Figure 3**). None of the analyzed traits displayed a clear, significant difference between control and all of the *XTH2* lines, but a slightly higher number of root hairs was formed on S-type lateral roots grown in the high P nutrient solution. Also evaluated were shoot and root development after 6 weeks of growth in low or high P soil. Some of the overexpressor lines did produce more roots and a slightly increased shoot biomass in low P soil, but no consistent effect of the overexpression could be detected. In contrast, in high P soil, the overexpressor lines produced lower root biomass than the control lines (**Figure 3**).

Although so *XTH2* had been shown to have higher expression in roots of a P deficiency tolerant genotype in a previous study [53], overexpression in our experiments did not lead to better root and shoot growth or increased root hair production. In another study, tolerance to P deficiency was conferred by overexpression of *PSTOL1*, encoding a protein kinase [76], indicating that transformation can lead to tolerance to P deficiency depending on the specific gene.

3. Conclusion

Overall, it can be concluded that a number of responses to P deficiency exist in rice, yet none of these is a general mechanism found in every rice genotype. Also, often direct evidence for a beneficial effect is lacking for rice. To construct and test a rice genotype optimized for P uptake in P deficient conditions it will be necessary to harness superior traits from many sources and integrate them via marker-assisted breeding or transgenic approaches. For a future sustainable food production, it will also be necessary to overcome the dependence on mining rock phosphate as a major source of P fertilizer. This will include an increase in recycling of biomass and wastewater.

4. Experimental details

4.1. Germplasm and germination

Rice varieties Nerica4, DJ123, Taichung native, and Sadri Tor Misri were used for the phenotyping experiments, and Nerica4 and Nipponbare were used for transgenic plant generation. Dormancy break, sterilization, and pre-germination were performed as described earlier [75].

4.2. Growth in the soil, artificial clay, and agar without the addition of nutrients

Pre-germinated seeds were subjected to different conditions in 50 ml incubation tubes. Plants were grown in low P soil (for details see [53]), vermiculite without nutrient supplementation, and water with the addition of 1% agar (Sigma Aldrich). To exclude light, all tubes were aluminum foil-wrapped and one pre-germinated seed added per tube. Five plants per genotype and condition were grown in a growth cabinet with 16 h light (30°C) and 8 h dark (25°C) for 7 days. Root hair formation was analyzed using a light microscope as described earlier [75].

4.3. Growth in agar-nutrient solution

Half-strength Yoshida nutrient solution [77] was prepared without P. Agar (Sigma Aldrich) was added at concentrations of 1, 1.5, 2, 2.5, 3 and 3.5% to the nutrient solution. Preliminary experiments suggested that beyond 4% agar, lateral root as well as root hair growth becomes impaired (Rose, unpublished). The nutrient solution-agar mixture was poured into clear plastic boxes (200 mm high × 100 mm wide × 25 mm deep, wrapped with aluminum foil) with a 15-mm-diameter hole in the top. A duplicate set of boxes were cut open to determine the resistance of each agar concentration using a penetrometer.

Two germinated seeds per box were sown 5 mm deep in the agar and boxes transferred to a growth cabinet set to 14 h light (27°C) and 10 h dark (22°C). From week two, the boxes were watered with deionized water to weight every 3 d until harvest after 23 d.

4.4. Plasmid construction for *pBIHubi::XTH2*

To amplify the *Xyloglucantransferase2* (Os11g33270) sequence, RNA from genotype DJ123 was isolated, transcribed into cDNA and used as PCR matrix with the oligonucleotides (5′–3′) CAACCCCGGGATGGCGACGACGACGG and GATCGAGCTCTCAGGCGTCGCGGTCG, which introduced the restriction endonuclease recognition sites for *Sma*I and *Sac*I, respectively. According to manufacturer's protocols the PCR product and the target vector pBIH [78] were treated with *Sma*I and *Sac*I (Fermentas, Fast Digest enzymes), the resulting fragments purified (Promega, Wizard PCR clean-up kit), ligated (Roche, Rapid DNA ligation kit), and transformed into DH5α (Promega, library efficient DH5α). The resulting plasmid contains *XTH2* under the control of the *Ubiquitin* promoter. After *pBIHubi::XTH2* sequence confirmation (using the oligonucleotides, 5′–3′: GATGGTGGTGGCAATGTCG and CGGTCGTCGCAGTAGTTGTA) one clone (termed *Ubi::XTH2*) was selected for rice transformation.

4.5. Rice transformation and T_2 selection

Genetic transformation of rice varieties NERICA4 and Nipponbare were conducted by *Agrobacterium*-methods using immature embryos [79]. T_2 plants possessing a single copy of transgene as homozygote were selected [80] and subjected to further experiments.

4.6. Phenotyping of *Ubi::XTH2* T_2 lines in nutrient solution

Pre-germinated T_2 seedlings were grown in water supplemented with iron (12 μM) and calcium (0.1 mM) for 7 days followed by an additional 7 days in 1/3 strength Yoshida solution [77]. At 16 DAS roots were harvested and used for RNA extraction, cDNA production, and qRT-PCR analysis as described earlier [56] and for vibratome sectioning. For detection of the *XTH2* transcript, the oligonucleotides (sequences in 5′–3′) TACCACTCCTACTCCGTCCT and TGGAGTAGAGCTTCATCGGC were used. Cross and longitudinal sections of 75 μm were prepared with a vibratome (Microslicer DTK-1000, DSK) by embedding 5 mm root segments in 4% agarose followed by slicing with a frequency of 8 and cutting speed of 5–7. The sections were then stained for 1 min with 0.05% toluidine blue, briefly washed with water and mounted with 50% glycerol for light microscope (Olympus BX50, Olympus) imaging.

Selected *Ubi::XTH2* lines with very high, high, and moderate ectopic overexpression as well as control (untransformed Nerica4 and Mock transformed) lines were grown in nutrient solution with low (1 μM) or sufficient (100 μM) P nutrition (NaH_2PO_4 $2H_2O$). The nutrient solution was changed and increased from 1/3 to 1/2 and finally to full-strength Yoshida solution [77] weekly, while pH was adjusted regularly to 5.7. At 35 DAS root hair parameters were determined as described previously [75].

4.7. Phenotyping of *Ubi::XTH2* T_2 lines in the soil

Soil containing three levels of P: extreme low P (80% low P soil mixed with 20% subsoil), low P soil, and P-replete soil (fertilized) were used. The soil was sieved and filled into boxes while softly compacted to simulate field conditions. Fertilizer was supplied in the equivalent amount to 30-0-30 or 30-30-30 kg ha^{-1} (N, P_2O_5, K_2O, respectively).

Four germinated seeds (with similar size) were sown directly in soil and thinned to two plants after 1 week of germination. Water was supplied regularly to field capacity to simulate the wet/dry cycle in upland condition. Plants were grown in a greenhouse with temperature and relative humidity varying between 25 and 32°C and 30–50%. At 40 DAS plants were harvested to evaluate plant height, number of leaves, number of tillers, main root length, root number, shoot and root dry matter.

Acknowledgements

Parts of the presented research were supported by post-doctoral scholarships from the Japanese Society for the Promotion of Science (JSPS) awarded to JK (nèe Nestler) and TR.

Author details

Josefine Kant[1]*, Takuma Ishizaki[2], Juan Pariasca-Tanaka[3], Terry Rose[4], Matthias Wissuwa[3] and Michelle Watt[1]

*Address all correspondence to: j.kant@juelich.de

1 Institute for Bio- and Geosciences, Plant Sciences (IBG-2), Forschungszentrum Juelich, Juelich, Germany

2 Tropical Agricultural Research Front, Japan International Research Center for Agricultural Sciences (JIRCAS), Tsukuba, Japan

3 Crop, Livestock and Environment Division, Japan International Research Center for Agricultural Sciences (JIRCAS), Tsukuba, Japan

4 Southern Cross Plant Science, Southern Cross University, Lismore, Australia

References

[1] World Food Situation. Food and Agriculture Organization of the United Nations [Internet]. 2016. Available from: www.fao.org/worldfoodsituation/csdb/en/ [Accessed: 2017-08-17]

[2] Schachtman DP, Reid RJ, Ayling SM. Phosphorus uptake by plants: From soil to cell. Plant Physiology. 1998;**116**:447-453

[3] Lynch JP. Root phenes for enhanced soil exploration and phosphorus acquisition: Tools for future crops. Plant Physiology. 2011;**156**:1041-1049. DOI: 10.1104/pp.111.175414

[4] Raven JA, Taylor R. Macroalgal growth in nutrient enriched estuaries: A biogeochemical perspective. Water, Air, Soil Pollution Focus. 2003;**3**:7-26

[5] Cordell D, Drangert JO, White S. The story of phosphorus: Global food security and food for thought. Global Environmental Change. 2009;**19**:292-305

[6] Rose TJ, Impa SM, Rose MT, Pariasca-Tanaka J, Mori A, Heuer S, Johnson-Beebout SE, Wissuwa M. Enhancing phosphorus and zinc acquisition efficiency in rice: A critical review of root traits and their potential utility in rice breeding. Annals of Botany. 2013;**112**:331-345

[7] Hinsinger P. Bioavailability of soil inorganic P in the rhizosphere as affected by root-induced chemical changes: A review. Plant and Soil. 2001;**237**:173-195

[8] Lambers H, Bishop JG, Hopper SD, Laliberte E, Zuniga-Feest A. Phosphorus-mobilization ecosystem engineering: The roles of cluster roots and carboxylate exudation in young P-limited ecosystems. Annals of Botany. 2012;**110**:329-348

[9] Lambers H, Shane MW, Cramer MD, Pearse SJ, Veneklaas EJ. Root structure and functioning for efficient acquisition of phosphorus: Matching morphological and physiological traits. Annals of Botany. 2006;**98**:693-713. DOI: 10.1093/aob/mcl114

[10] Richardson AE, Simpson RJ. Soil microorganisms mediating phosphorus availability. Plant Physiology. 2011;**156**:989-996

[11] Menezes-Blackburn D, Paredes C, Zhang H, et al. Organic acids regulation of chemical-mircobial phosphorus transformations in soils. Environmental Science and Technology. 2016;**50**:11521-11531

[12] Ryan PR, James RA, Weligama C, et al. Can citrate efflux from roots improve phosphorus uptake by plants? Testing the hypothesis with near-isogenic lines of wheat. Physiologia Plantarum. 2014;**151**:230-242

[13] Smith SE, Smith FA. Roles of arbuscular mycorrhizas in plant nutrition and growth: New paradigms from cellular to ecosystem scales. Annual Reviews of Plant Biology. 2011;**62**:227-250

[14] Koide RT. Nutrient supply, nutrient demand and plant response to mycorrhizal infection. New Phytologist. 1991;**117**:365-386

[15] Sawers RJH, Gutjahr C, Paszkowski U. Cereal mycorrhiza: An ancient symbiosis in modern agriculture. Trends in Plant Science. 2008;**13**:93-97

[16] Yang SY, Gronlund M, Jakobsen I, et al. Nonredundant regulation of rice arbuscular mycorrhizal symbiosis by two members of the phosphate transporter1 gene family. The Plant Cell. 2012;**24**:4236-4251. DOI: 10.1105/tpc.112.104901

[17] Breuillin F, Schramm J, Haijirezaei M, et al. Phosphate systemically inhibits development of arbuscular mycorrhiza in petunia hybrid and represses genes involved in mycorrhizal functioning. The Plant Journal. 2010;**64**:1002-1017. DOI: 10.1111/j.1365-313X.2010.04385.x

[18] Parnell JJ, Berka R, Young HA, Sturino JM, Kang YW, Barnhart DM, DiLeo MV. From the lab to the farm: An industrial perspective of plant beneficial microorganisms. Frontiers in Plant Science. 2016;**7**:1110. DOI: 10.3389/fpls.2016.01110

[19] Jacoby R, Peukert M, Succurro A, Koprivova A, Kopriva S. The role of soil microorganisms in plant mineral nutrition-current knowledge and future directions. Frontiers in Plant Science. 2017;**8**:1617. DOI: 10.3389/fpls.2017.01617

[20] Knief C. Analysis of plant microbe interactions in the era of next generation sequencing technologies. Frontiers in Plant Science. 2014;**5**:216. DOI: 10.3389/fpls.2014.00216

[21] Pii Y, Mimmo T, Tomasi N, Terzano R, Cesco S, Crecchio C. Microbial interactions in the rhizosphere: Beneficial influences of plant growth-promoting rhizobacteria on nutrient acquisition process. A review. Biology and Fertility of Soils. 2015;**51**:403-415

[22] Foo JL, Ling H, Lee YS, Chang MW. Microbiome engineering: Current applications and its future. Biotechnology Journal. 2017;**12**:1600099. DOI: 10.1002/biot.201600099

[23] Shropshire JD, Bordenstein SR. Speciation by symbiosis: The microbiome and behavior. MBio. 2016;**7**:e01785-e01715

[24] Arber A. Water Plants, a Study of Aquatic Angiosperms. Cambridge, UK: Cambridge University Press; 1920

[25] Evans DE. Aerenchyma formation. New Phytologist. 2003;**161**:35-49

[26] Jackson MB, Armstrom W. Formation of aerenchyma and the processes of plant ventilation in relation to soil flooding and submergence. Plant Biology. 1999;**3**:274-287

[27] Konings H, Verschurren G. Formation of aerenchyma in roots of *Zea mays* in aerated solutions, and its relation to nutrient supply. Physiologia Plantarum. 1980;**49**:265-270

[28] Drew MC, He C-J, Morgan PW. Decreased ethylene biosinthesis, and induction of aerenchyma, by nitrogen-starvation or phosphate-starvation in adventitious roots of *Zea mays* L. Plant Physiology. 1989;**91**:266-271

[29] Bouranis DL, Chorianopoulou SN, Siyiannis VF, Protonotarios VE, Hawksford MJ. Aerenchyma formation in roots of maize during sulphate starvation. Planta. 2003; **217**:382-391

[30] Bouranis DL, Chorianopoulou SN, Kollias C, Maniou P, Protonotarios VE, Siyiannis VF, Hawkesford MJ. Dynamics of aerenchyma distribution in the cortex of sulfate-deprived adventitious roots of maize. Annals of Botany. 2006;**97**:695-704

[31] Fan M, Zhu J, Richards C, Brown KM, Lynch JP. Physiological roles for aerenchyma in phosphorus-stressed roots. Functional Plant Biology. 2003;**30**:493-506

[32] Zhu J, Brown KM, Lynch JP. Root cortical aerenchyma improves the drought tolerance of maize (*Zea mays* L.). Plant Cell & Environment. 2010;**33**:740-749

[33] Lenochová Z, Soukup A, Votrubová O. Aerenchyma formation in maize roots. Biologica Plantarum. 2009;**53**:263-270

[34] Burton AL. Genetic basis and high throughput phenotypic evaluation of anatomical and architectural root traits for resource acquisition in *Zea mays*. PhD dissertation: The Pennsylvania State University; 2010

[35] Lynch JP, Brown KM. Regulation of root architecture by phosphorus availability. In: Lynch JP, Deikman J, editors. Current Topics in Plant Physiology. Phosphorus in Plant Biology: Regulatory Roles in Molecular, Cellular, Organismic, and Ecosystem Processes. Rockville, MD: American Society of Plant Physiology; 1998. pp. 148-156

[36] Lynch JP, Brown KM. Root strategies for phosphorus acquisition. In: White PJ, Hammond JP, editors. The Ecophysiology of Plant–Phosphorus Interactions. Dordrecht: Springer; 2008. pp. 83-116

[37] Schneider H, Wojciechowski T, Postma JA, Brown KM, Lücke A, Zeisler V, Schreiber L, Lynch JP. Root corical senescence decreases root respiration, nutrient content, and radial water and nutrient transport in barley. Plant, Cell & Environment. 2017;**40**:1392-1408

[38] Parker JS, Cavell AC, Dolan L, Roberts K, Grierson CS. Genetic interactions during root hair morphogenesis in Arabidopsis. The Plant Cell. 2000;**12**:1961-1974

[39] Leavitt RG. Trichomes of the root in vascular cryptogams and angiosperms. Proceedings of the Boston Society of Natural History. 1904;**31**:273-313

[40] Gahoonia TS, Care D, Nielsen NE. Root hairs and phosphorus acquisition of wheat and barley cultivars. Plant and Soil. 1997;**191**:181-188

[41] Gahoonia TS, Nielsen NE. Direct evidence on participation of root hairs in phosphorus (32P) uptake from soil. Plant and Soil. 1998;**198**:147-152

[42] Stetter MG, Schmid K, Ludewig U. Uncovering genes and ploidy involved in the high diversity in root hair density, length and response to local scarce phosphate in *Arabidopsis thaliana*. PLoS One. 2015;**10**:e0120604. DOI: 10.1371/journal.pone.0120604

[43] Mackay AD, Barber SA. Effect of soil moisture and phosphate level on root hair growth of corn roots. Plant and Soil. 1985;**86**:321-331

[44] Carminati A, Passioura JB, Zarebanadkouki M, Ahmed MA, Ryan PR, Watt M, Delhaize E. Root hairs enable high transpiration rates in drying soils. New Phytologist. 2017;**216**:771-781. DOI: 10.1111/nph.14715

[45] Timsina J, Connor DJ. Productivity and management of rice-wheat cropping systems: Issues and challenges. Field Crops Research. 2001;**69**:93-132

[46] Yamauchi A, Kono Y, Tatsumi J. Comparison of root system structures of 13 species of cereals. Japanese Journal of Crop Sciences. 1987;**56**:618-631

[47] Mori A, Fukuda T, Vejchasarn P, Nestler J, Pariasca-Tanaka J, Wissuwa M. The role of root size versus root efficiency in phosphorus (P) acquisition of rice. Journal of Experimental Botany. 2016;**67**:1179-1189. DOI: 10.1093/jxb/erv557

[48] Hoffland E, Wei C, Wissuwa M. Organic anion exudation by lowland rice (*Oryza sativa* L.) at zinc and phosphorus deficiency. Plant and Soil. 2006;**283**:155-162

[49] Bhattacharyya P, Das S, Adhya TK. Root exudates of rice cultivars affect rhizospheric phosphorus dynamics in soils with different phosphorus statuses. Communications in Soil Science and Plant Analysis. 2013;**44**:1643-1658

[50] Tawaraya K, Horie R, Saito A, Shinano T, Wagatsuma T, Saito K, Oikawa A. Metabolite profiling of shoot extracts, root extracts and root exudates of rice plant under phosphorus deficiency. Journal of Plant Nutrition. 2013;**36**:1138-1159

[51] Kirk GJD, Santos EE, Findenegg GR. Phosphate solubilization by organic anion excretion from rice (*Oryza sativa* L.) growing in aerobic soil. Plant and Soil. 1999;**211**:11-18

[52] Begg CBM, Kirk GJD, MacKenzie AF, Neue H-U. Root-induced iron oxidation and pH changes in the lowland rice rhizosphere. New Phytologist. 1994;**128**:469-477

[53] Pariasca-Tanaka J, Satoh K, Rose T, Mauleon R, Wissuwa M. Stress response versus stress tolerance: A transcriptome analysis of two rice lines contrasting in tolerance to phosphorus deficiency. Rice. 2009;**2**:167-185

[54] Vallino M, Fiorilli V, Bonfante P. Rice flooding negatively impacts root branching and arbuscular mycorrhizal colonization, but not fungal viability. Plant Cell and Environment. 2014;**37**:557-572

[55] Wissuwa M, Ae N. Further characterization of two QTLs that increase phosphorus uptake of rice (*Oryza sativa* L.) under phosphosrus deficiency. Plant and Soil. 2001;**237**:275-286

[56] Nestler J, Wissuwa M. Superior root hair formation confers root efficiency in some, but not all, rice genotypes upon P deficiency. Frontiers in Plant Science. 2016;**7**:1935

[57] Liesack W, Schnell S, Revsbech NP. Microbiology of flooded rice paddies. FEMS Microbiology Reviews. 2000;**24**:625-645

[58] Islam MR, Madhaiyan M, Boruah HPD, Yim W, Lee G, Saravanan VS, Fu QL, Hu HQ, Sa T. Characterization of plant growth-promoting traits of free-living diazotrophic bacteria and their inoculation effects on growth and nitrogen uptake of crop plants. Journal of Microbiology and Biotechnology. 2009;**19**:1213-1222

[59] Sahoo RK, Ansari MW, Dangar TK, Mohanty S, Tuteja N. Phenotypic and molecular characterisation of efficient nitrogen-fixing Azotobacter strains from rice fields for crop improvement. Protoplasma. 2014;**251**:511-523

[60] Maksimov IV, Abizgil'dina RR, Pusenkova LI. Plant growth promoting rhizobacteria as alternative to chemical crop protectors from pathogens (review). Applied Biochemistry and Microbiology. 2011;**47**:333-345

[61] Lavakush, Yadav J, Verma JP, Jaiswal DK, Kumar A. Evaluation of PGPR and different concentration of phosphorus level on plant growth, yield and nutrient content of rice (*Oryza sativa*). Ecological Engineering. 2014;**62**:123-128

[62] Rose MT, Phuong TL, Nhan DK, Cong PT, Hien NT, Kennedy IR. Up to 52% N fertilizer replaced by biofertilizer in lowland rice via farmer participatory research. Agronomy for Sustainable Development. 2014;**34**:857-868

[63] Yanni YG, Rizk RY, Abd El-Fattah FK, et al. The beneficial plant growth-promoting association of *Rhizobium leguminosarum* bv. trifolii with rice roots. Australian Journal of Plant Physiology. 2001;**28**:845-870

[64] Hardoim PR, Nazir R, Sessitsch A, Elhottova D, Korenblum E, van Overbeek LS, van Elsas JD. The new species *Enterobacter oryziphilus* sp. Nov. and *Enterobacter oryzendophyticus* sp. Nov. are key inhabitants of the endosphere of rice. BMC Microbiology. 2013; **13**:164

[65] Jha M, Chourasia S, Sinha S. Microbial consortium for sustainable rice production. Agroecology and Sustainable Food Systems. 2013;**37**:340-362

[66] Sessitsch A, Hardoim P, Döring J, et al. Functional characteristics of an endophyte community colonizing rice roots as revealed by metagenomic analysis. Molecular Plant-Microbe Interactions Journal. 2011;**25**:28-36

[67] Panhwar QA, Radziah O, Rahman AZ, Sariah M, Razi IM, Naher UA. Contribution of phosphate-solubilizing bacteria in phosphorus bioavailability and growth enhancement of aerobic rice. Spanish Journal of Agricultural Research. 2011;**9**:810-820

[68] Gusain YS, Kamal R, Mehta CM, Singh US, Sharma AK. Phosphate solubilizing and indole-3-acetic acid producing bacteria from the soil of Garhwal Himalaya aimed to improve the growth of rice. Journal of Environmental Biology. 2015;**36**:301-307

[69] da Costa EM, de Lima W, Oliveira-Longatti SM, de Souza FM. Phosphate-solubilising bacteria enhance *Oryza sativa* growth and nutrient accumulation in an oxisol fertilized with rock phosphate. Journal of Ecological Engineering. 2015;**83**:380-385

[70] Chen X, Miché L, Sachs S, Wang Q, Buschart A, Yang H, Vera Cruz CM, Hurek T, Reinhold-Hurek B. Rice responds to endophytic colonization which is independent of the common symbiotic signaling pathway. New Phytologist. 2015;**208**:531-543

[71] Vejchasarn P, Lynch JP, Brown KM. Genetic variability in phosphorus responses of rice root phenotypes. Rice. 2016;**9**:29. DOI: 10.1186/s12284-016-0102-9

[72] Abiko T, Obara M. Enhancement of porosity and aerenchyma formation in nitrogen-deficient rice roots. Plant Science. 2014;**215-216**:76-83

[73] Chen CC, Dixon JB, Turner FT. Iron coatings on rice roots: Morphology and models of development. Soil Science Society of America Journal. 1980;**44**:1113-1119

[74] McCormack ML, Dickie IA, Eissenstat DM, et al. Redefining fine roots improves understanding of below-ground contributions to terrestrial biosphere processes. New Phytologist. 2015;**205**:505-518. DOI: 10.1111/nph.13363

[75] Nestler J, Keyes SD, Wissuwa M. Root hair formation in rice (*Oryza sativa* L.) differs between root types and is altered in artificial growth conditions. Journal of Experimental Botany. 2016;**67**:3699-3708. DOI: 10.1093/jxb/erw115

[76] Gamuyao R, Chin JH, Pariasca-Tanaka J, Pesaresi P, Catausan S, Dalid C, Slamet-Loedin I, Tecson-Mendoza EM, Wissuwa M, Heuer S. The protein kinase Pstol1 from traditional rice confers tolerance of phosphorus deficiency. Nature. 2012;**488**:535-541. DOI: 10.1038/nature11346

[77] Yoshida S, Forno DA, Cock JH, Gomez KA. Laboratory Manual for Physiological Studies of Rice. 2nd ed. The International Rice Research Institute; 1972. pp. 1-70

[78] Becker D. Binary vectors which allow the exchange of plant selectable markers and reporter genes. Nucleic Acids Research. 1990;**18**:203

[79] Ishizaki T, Kumashiro T. Genetic transformation of NERICA, interspecific hybrid rice between *Oryza glaberrima* and *O. sativa*, mediated by *Agrobacterium tumefaciens*. Plant Cell Reports. 2008;**27**:319-327. DOI: 10.1007/s00299-007-0465-x

[80] Ishizaki T, Kumashiro T. Investigations of copy number of transgene, fertility and expression level of an introduced GUS gene in transgenic NERICA produced by *Agrobacterium*-mediated methods. In Vitro Cellular & Developmental Biology- Plant. 2011;**47**:339-347. DOI: 10.1007/s11627-011-9341-z

8

Heavy Metal and Mineral Element-Induced Abiotic Stress in Rice Plant

Anitha Mani and Kavitha Sankaranarayanan

Abstract

The adverse effect of nonliving factors on living organisms is described as abiotic stress. It includes drought, excessive watering, extreme temperatures, salinity, and mineral toxicity. Rice is an important cereal crop, grown under diverse ecological and agricultural conditions. Heavy metal contamination of agricultural land causes abiotic stress to the crop plant as well as has a drastic effect on humans. Increased metal concentration in plants leads to the production of reactive oxygen species which results in cell death and thus affects the crop production in plants. In addition, increased heavy metal concentration in the plant has deleterious effects on its consumers. Like other organisms, plants have also designed ways to deal with such stress situations. In this chapter, abiotic stress due to metal toxicity in rice plant, which includes uptake and sequestration mechanisms, biochemical changes taking place in the plant and variation in their gene expression is elucidated. Based on several molecular and biochemical studies in various reviews and research papers, the role of different transporters like zinc-regulated transporter (ZIP), natural resistance-associated macrophage protein (NRAMP), copper transporter (COPT), yellow stripe like (YSL), heavy metal ATPase (HMA), metal tolerance protein (MTP) and other vascular transporters involved in the above processes in rice plant will be discussed in this chapter.

Keywords: transporters, reactive oxygen species, vacuolar sequestration capacity, antioxidant system, gene regulation

1. Introduction

Any negative impact of nonliving factors on living organisms in a particular environment can be described as abiotic stress. There are different types of abiotic stress like those due to drought, excessive watering, that is, water-logging/flooding, extreme temperatures (cold, frost and heat), salinity and mineral (metal and metalloid) toxicity. These have a negative impact on seed germination, plant growth, development, yield, and seed quality of crops. Changes in environmental conditions affect the biological and physiological response of plants. This chapter deals with abiotic stress due to metal toxicity in rice plants, which includes uptake and sequestration mechanisms and biochemical changes taking place in the plant.

For their overall growth and development, plants require 14 different mineral elements [1]. These elements are present in the soil and are taken up by the roots, translocated to the shoots, and then distributed to different organs and tissues of the plant depending on their needs [1].

Minerals comprise of both metals and metalloids that are toxic to both plants and animals even at a very low concentration. Some of these heavy metals such as arsenic (As), cadmium (Cd), mercury (Hg), lead (Pb) or selenium (Se), do not perform any known physiological function in plants, and are called nonessential metals. Others, such as cobalt (Co), copper (Cu), iron (Fe), manganese (Mn), molybdenum (Mo), nickel (Ni), and zinc (Zn), are essential elements as they are required for the normal growth and metabolism of plants. Essential elements can lead to poisoning when their concentration rises beyond their optimal levels.

Plants growing on metal-contaminated soils are categorized as resistant varieties, which have adapted to this stressed environment. Heavy metal resistance is attained by the plants either by avoidance/tolerance or by both. Plants, which can prevent the entry of metal ions into their cell cytoplasm, are categorized as avoiders, while the plants which can detoxify the metal ions can cross the plasma membrane or organellar membranes are grouped as tolerant varieties. Plants were thus classified into three groups as: metal excluders, indicators and accumulators/hyperaccumulators by Baker and Walker [2] based on the approach used by the plants to grow on metal-contaminated soils. Excluder group of plants restricts the uptake and translocation of metal ions to their shoots thus maintaining low levels of heavy metals in the shoots even when grown over a varied range of metal concentrations in soil. Most of the plants belong to the excluder group. Plants, which are classified as metal indicators, can accumulate metals in their aerial shoot system. The level of metal ions in their aerial biomass generally indicates the concentration of metal in the soil. Plants, which are classified as metal accumulators/hyperaccumulators, can take up, transport and accumulate metals generously in their aerial parts to levels higher than the metal concentration found in the soil [2]. The antioxidant defense system in plants helps to accumulate and tolerate the side effects of high levels of internal metal concentrations. Antioxidant system is activated in order to combat the deleterious effects caused by reactive oxygen species (ROS) generated due to stress [3]. Metals can interfere with mineral nutrition and change the concentration and composition of plant nutrients. Metals can also alter the conformation of proteins, including transporters, or other regulatory proteins [4].

Rice (*Oryza sativa L.*) is one of the most important cereal crops in most part of the world and is cultivated in tropical and temperate regions of the world. Rice is the staple food for half of the world's population majorly for many South East Asian countries. It ranks second next to wheat among the most cultivated cereals in the world to feed the ever-growing population. The genus *Oryza* contains 21 wild and 2 cultivated species (*Oryza sativa* and *O. glaberrima*) with 10 genome types [5, 6] *O. sativa* is composed of two subspecies: *japonica* and *indica*. It is also a model monocotyledon plant as it has a small genome, which has been sequenced (for both the cultivars *japonica* [7] and *indica* [8]).

Plants are sessile and its roots are exposed to various stresses like deficiency and an excess of mineral elements. Heavy metal contamination and accumulation in water, soil, and air due to various reasons has become a serious environmental problem and has greatly affected rice growth and quality. Heavy metals accumulated in rice are toxic to growth, metabolism, and development of plants. Thus, the transfer of heavy metals from soil to plants of commercial agricultural value, such as rice, is of great concern as it may lead to biomagnification via food chain causing several deleterious effects to the consumer. Heavy metals can enter the human body through the food chain, leading to an increased prevalence of chronic diseases, deformities and cancer. The uptake of heavy metal ions by rice plants poses a threat to the consumer's health. Thus, it is important to understand the uptake and sequestration of heavy metal in the rice plant in order to understand the stress caused by it to the plant.

Availability of metals in soil to plants is controlled by three steps: (1) soil conditions (upland or flooded soil and soil solution pH); (2) mineralization (ionization and complex formation) and (3) uptake and efflux transporters.

In addition to genetic variation, metal uptake is sometimes limited by its bioavailability in the soil. The availability of metal ions like Fe, Zn, and Cd for plant uptake varies mainly depending on soil redox potential. Generally, rice is cultivated under flooded conditions. The practice of flooding paddy field increases Fe availability while decreasing Zn and Cd availability while moderate soil drying improves Zn and Cd uptake and but decreases Fe uptake [9]. In both drained and flooded soils, Zn mainly exists as Zn^{2+}, while some of it binds to organic substances, and is immobilized as Zn-sulfide (ZnS) in the anaerobic layer of soil [10]. In drained acidic soils, cadmium exists in its ionized state as Cd^{2+} ion, whereas Cd in alkaline paddy field soils is present in the forms of $CdCO_3$ and humic acid-bound Cd [11]. Flooding of soil immobilizes Cd as Cd-sulfide (CdS) and colloidal-bound Cd [12]. Drying of soil converts CdS to Cd^{2+} and increases its availability to plants. In acidic soil, Fe is ionized as Fe^{2+}/Fe^{3+}, and Fe in aerobic alkaline soils is immobilized as $Fe(OH)_3$.

Both Cd and Pb are nonessential elements for plants and are toxic even at a very low concentrations, but are readily transported within the plants. The uptake and sequestration of Cd and Pb by crops is of great concern, due to their accumulation in the edible parts of the plant. The uptake of Cd and Pb, their transport, and accumulation by plants are strongly influenced by soil properties and vary with plant species [13, 14]. Cd is freely taken up by the plants and its uptake increases with increased external concentration. The amount of Cd accumulated after its uptake and its translocation to different organs varies with species and with cultivars within the species. The ability for uptake and sequestration of metals in different parts of the plants varies between different plants. There exists a huge difference in metal uptake and translocation between plant species and even between different cultivars of the same species [15]. Roots of most cereal crop plants are present at a depth of 25 cm from the soil surface from where the heavy metals are absorbed by the plant [16].

Roots are the primary target for the accumulation of metals, and metals like Cd and Pb are accumulated mainly in roots [14]. Plants have developed several strategies to decrease metal ion toxicity, one of which is the cellular transport system. There are several groups of metal transporter proteins identified in plants.

Rice has a distinct root system. Anatomically, rice root is characterized by the two casparian strips present on the exodermis and endodermis. The apoplastic flow of water and movement of mineral elements is prevented in between the cell layers by casparian strips, which act as a barrier [17]. Also, the mature rice root has a well developed aerenchyma, which has well developed vascular bundles for the upstream translocation of metals and other mineral elements from the roots to the shoots. Apoptosis of cortical cells creates the aerenchyma, which has an apoplastic space and no symplastic connections. The apoplastic space is connected by remaining spoke-like connections between the exodermis and the endodermis [18]. Thus to reach the stele, the mineral nutrients in rice roots have to be transported via the symplast of both the exodermis and the endodermis and also through the apoplast of the aerenchyma. The root tips, which lack aerenchyma and casparian strips, can also accumulate mineral elements but in small percentage due to their highly undeveloped vascular system [19]. For efficient translocation of mineral nutrition from the soil solution to the stele in rice, roots require cooperation between both the influx and efflux transporters. Transporters involved in this uptake process have been identified for some mineral elements, but most of them remain unknown. Thus plants take up heavy metals through their roots majorly via various transporters from the soil and accumulate them in their aerial parts.

Metal ions are transported from the soil into the root and then distributed throughout the plant, after crossing both cellular and organellar membranes [20]. All the characterized plant transporter proteins, which are responsible for metal homeostasis, are membrane proteins that mediate heavy metal movement through membranes. Transporters in plants either act at the plasma membrane to move metals into the cytoplasm or at the intracellular organellar membrane to re-circulate metals from intracellular compartment into the cytoplasm. They are classified into different families such as natural resistance-associated macrophage protein (NRAMP), zinc-regulated transporter (ZIP), yellow stripe 1-like family (YSL), and Ctr/copper transporter (COPT) family of high-affinity Cu uptake proteins. Plant transporters have been identified in *Arabidopsis thaliana* to be involved in metal efflux from the cytoplasm either across the plasma membrane or into the organelles. They are classified into two families namely P_{1B}-ATPase family/CPX- ATPase and cation diffusion facilitator (CDF) family/metal tolerance protein1 (MTP1) [21]. Similar metal transporter families are also present in rice.

Various such transporters have been identified in different plant species, which are involved in the metal uptake and sequestration process. This chapter focuses on rice plant transporters that are involved in the uptake of mineral elements in roots and its sequestration into the vacuoles and also the biochemical changes taking place in the rice plant during abiotic stress due to heavy metals.

2. ZIP family

ZIP family of transporters is named after the first proteins identified ZRT, iron-regulated transporter (IRT) like protein [20]. They are present in bacteria, fungus, plants, and humans. ZIP proteins can transport divalent cations like Fe^{2+}, Zn^{+2}, Mn^{+2} and Cd^{+2} [22]. Zn as such cannot be transported across the cell membrane, it requires specific zinc transporters for its transport into the cytoplasm. Generally, ZIP family of transporter proteins transport cations into the cell cytoplasm and play a role in cellular metal ion homeostasis. ZIP proteins generally localized at the plasma membrane and are involved in either moving metals or to remobilize them from intracellular compartments into the cytoplasm [21]. In rice, 14 putative ZIP family of transporter proteins have been identified. It is divided into two subfamilies based on their amino acid sequence similarities. Zinc is taken up by these transporters as a divalent cation and plays role in cellular activities in the form of tetrahedral complexes as it is neither oxidized nor reduced [20].

IRT1 was the first member of ZIP family to be identified. It is a Fe(II) transporter and is involved in the uptake of iron from the soil. Studies have proved that in addition to Fe, it can transport Mn and Zn. IRT1 mediates Cd accumulation in iron-deficient plants. Depending on the plant species, there are two mechanisms for the uptake of Fe by IRT transporter from soil. The first strategy is used by all dicot and non-gramineous monocot plants. In this process, iron uptake by IRT1 transporter happens after ferric Fe is reduced to ferrous Fe by ferric chelate reductase (FRO2) on the plasma membrane [23, 24]. The second strategy is used by graminaceous plants. It is characterized by the secretion of mugineic acid (MA) and forms mugineic acid-ferric complex followed by its uptake by IRT1 transporter [25]. Though the rice root does not have Fe reductase enzyme activity [26], ferrous Fe is abundantly present in paddy fields due to reductive status in soil. Both OsIRT1 and OsIRT2 are involved in Fe uptake in rice plant. They are expressed mostly in roots and induced by Fe deficiency [27]. OsIRT1 can only uptake Fe but not Cu.

OsZIP1 and OsZIP3 are involved in the transport of Zn but not Fe/Mn. OsZIP4, OsZIP5, and OsZIP8 are functional rice zinc transporters expressed on the plasma membrane [28]. OsZIP1 and OsZIP3 are upregulated in roots and shoots upon Zn deprivation. Zn deficiency upregulates OsZIP2 only in roots [27], whereas Zn deficiency upregulates OsZIP5 and OsZIP7 in rice shoots. In the mature rice plant, OsZIP7a and OsZIP8 are expressed constitutively and weakly in roots, culms, leaves and flowering spikes [29]. Expression of OsZIP9 is also induced by Zn deficiency and it complements the absence of OsZIP5. OsZIP4 is expressed in Zn deficient shoots and roots especially in phloem cells and meristems. Zn deficiency induces the expression of OsZIP8 in both roots and shoots. Expression of OsZIP5 is relatively higher in roots than in shoots and is specific to zinc [27]. OsZIP4 is localized to the plasma membrane. It is a Zn-regulated Zn transporter involved in the transport of Zn. It controls the supply of Zn to developing young leaves and is involved in remobilization of Zn from old to young leaves [28].

In rice Zn transporters OsZIP1, OsZIP3, OsZIP4, and OsZIP5 are induced by Zn deficiency [26, 27, 30]. OsZIP1, OsZIP3, and OsZIP4 are expressed in the vascular bundles in both rice shoot and root and only in the epidermal cells in rice root [26, 30]. OsIRT1, OsZIP5, and OsZIP4 are upregulated in Zn-deficient roots and OsZIP4, OsZIP5 and OsZIP7 are upregulated in Zn deficient shoots [28].

OsZIP4 is expressed in the vascular bundles in rice root and shoot and also root and shoot meristem during Zn deficiency. Basically, OsZIP1 and OsZIP3 are involved in Zn uptake in roots and Zn homeostasis in shoots [30]. OsZIP4 transports Zn, specifically into vascular bundles and meristem [28]. OsZIP6 is transcriptionally activated in the shoot and root tissues in response to the deficiency in Fe^{2+}, Zn^{2+}, and Mn^{2+}. Ion transport by OsZIP6 is pH dependent and enhanced transport is observed at acidic pH. OsZIP6 is involved in iron uptake in roots and transport of Fe, Zn, and Mn in the shoots of rice [31]. OsZIP1 and OsZIP3 transport only Zn^{2+} and not Fe^{2+} or Mn^{2+} [30], OsZIP4 transports Zn^{2+} and not Fe^{2+} [28] and OsIRT1 similarly transports Fe^{2+} and not Cu^{2+} [32]. OsZIP6 transports at least three transition metal ions, namely, Fe^{2+}, Co^{2+}, and Cd^{2+}. Substrate affinity for OsZIP6 is in the order $Co^{2+} > Cd^{2+} > Fe^{2+}$. ZIP transporters show both high and low affinity transport [31]. In rice, zinc transporter OsZIP1 exhibits enhanced transport at pH 4.7, in contrast to OsZIP3, where maximum activity is observed at pH 6.0 [30]. OsZIP6 is expressed in both roots and shoots at maximum tillering and mid-grain filling stages [33]. OsZIP8 is a plasma membrane zinc transporter in rice that functions in Zn uptake and distribution. During Zn deficiency, it is highly upregulated in shoots and roots [27].

In rice seed, Zn is present in the embryo, endosperm, and the aleurone layer. The Zn content is specifically high in the embryo [34]. During germination, Zn content in the endosperm decreases, while Zn content increases in the radicle and leaf primordium. Zn content increases in the scutellum and its vascular bundle after 24 hrs of sowing [34]. During germination, expression of ZIP family of transporter members decreases [35]. In the embryo meristematic tissues, Zn accumulation is limited. For such a partial localization of Zn, a decrease in OsZIP family transcripts is required. Different rice genotypes vary mainly in their efficiency to utilize Zn and in their grain Zn contents [36]. Thus, the wrong expression of ZIPs could lead to the irregular distribution of the essential micronutrients in the plant.

3. Yellow stripe-like (YSL) proteins

Though iron is abundant in the earth's crust, it is mostly unavailable to plants because, at a neutral pH, it forms insoluble ferric oxide complexes in aerobic environment [37]. Graminaceous plants use the chelation-based strategy II. In response to Fe deficiency, these plant cells release phytosiderophores (PSs), which belong to the mugineic acid (MA) family and are derived from the precursor nicotianamine (NA). These molecules bind to Fe(III) and specific plasma membrane transporter proteins to import the Fe(III)-PS complexes [38]. The molecular mechanism controlling Fe(III)-uptake was elucidated by cloning the membrane transporter from the maize yellow stripe 1 (ys1) mutant, which showed characteristic interveinal chlorosis or yellow patches [39]. Because that mutant is deficient in Fe(III)-PS uptake, it has been suggested that YS1 is the Fe(III)-PS transporter. The YS1 protein is upregulated by Fe deficiencies in roots and shoots, and functions as a proton-coupled symporter to transport Fe(III)-PS [40].

Similar to maize YS1 gene sequence, 18 such genes have been putatively identified in rice and named as yellow stripe-like (YSL) genes [41]. Though YSL is part of the larger oligopeptide transporter (OPT) family, which is also present in fungi [42], but YSL transporter family members can only be found in plants [43]. YSL family of transporters cannot transport free metals as such but can transport only metals along with nicotianamine (NA) or its derivatives [42]. Nicotianamine is a precursor of phytosiderophores, which are high-affinity Fe ligands exclusively synthesized by Poaceae species and excreted by roots for the chelation and acquisition of Fe [41]. NA is a non-proteogenic amino acid, synthesized from *S*-adenosyl-methionine by the enzyme NA synthase (NAS) [44]. It is a structural analog of 2′-deoxymugineic acid (DMA), which is formed by NA aminotransferase (NAAT) and DMA synthase (DMAS) [45]. In monocot plants, YSL transporters are associated with metal uptake from the soil, while in both monocots and dicots plants they are involved in long-distance metal translocation [46].

In rice, OsYSL2 transports Fe(II)-NA complex and Mn(II)-NA complex and is mainly expressed in the phloem cells of the vascular bundles, especially in the companion cells of Fe-deficient leaves [47]. It also mediates long-distance transport of manganese, especially to the grain [47, 48]. The expression of OsYSL2 is induced by Fe deficiency in the leaves but not in the roots. OsYSL2 is important for Fe translocation at the early stage of growth [35] also important for long-distance transport during grain filling, particularly for Fe translocation to the endosperm [48].

OsYSL15 is localized to the plasma membrane and mediates the uptake of MA-Fe complex [49]. Fe deficiency induces *OsYSL15* in roots but is unaffected by Zn, Mn or Cu deficiency [49]. Under Fe deficiency, *OsYSL15* is expressed in the epidermis, endodermis, cortex, and vascular bundles of the roots and leaves [49]. The OsYSL15 transporter contribution in paddy soil is little as the secreted MA diffuses out of the rhizosphere. In neutral and alkaline soils, Fe^{3+} binds to mineral and organic substances strongly such that ions are hardly available to plants and the iron is solubilized by forming a complex with a phytosiderophore, DMA [50], which is excreted by rice roots by OsTOM1 (rice DMA effluxer) [51]. Fe(III)-DMA complexes in the soil solution is then taken up by OsYSL15 [49]. Under flooded conditions, rice plants may absorb both Fe(III)-DMA and Fe^{2+} [26].

OsYSL16 is a Cu-NA transporter which delivers Cu to the developing tissues and seeds through phloem transport. During vegetative growth, OsYSL16 is expressed in the roots, leaves, and unelongated nodes and during the reproductive stage, it is highly expressed in the upper nodes [52].

Among the 18 OsYSL genes in rice, OsYSL15 transports Fe(III)-DMA and Fe(II)-NA [49]. OsYSL15 expression is strongly induced in the roots and shoots by Fe deficiency. OsYSL2, induced by Fe deficiency, is localized to the plasma membrane and transports Fe(II)-NA and Mn(II)-NA, but not Fe(III)-DMA [47]. OsYSL18 is a transporter of Fe(III)-DMA but not of

Fe(II)-NA [53]. Its expression in flowers and the phloem of lamina joints indicates that it is involved in translocating Fe to the reproductive organs and phloem joints.

Expression of OsYSL2, OsYSL9 and OsYSL15 genes increases when Fe is limited. OsYSL9 is induced by Fe deficiency in the shoots but not the roots. On the other hand, OsYSL16 is constitutively expressed in both roots and shoots at levels similar to OsYSL2, OsYSL9 and OsYSL15 genes, but the alteration of Fe concentration has not shown any effect on the expression of OsYSL16 [47].

OsYSL13 is mostly expressed in the shoots, and its expression is reduced under Fe deficient conditions. OsYSL14 is expressed in both roots and shoots irrespective of the external Fe concentration. OsYSL15 is expressed only in roots and Fe deficiency highly induces its expression. OsYSL16 is expressed in both roots and shoots, and Fe deficiency slightly increases its expression in roots.

OsYSL2 is localized in the plasma membrane. It is not a Fe(III)-phytosiderophore transporter which is involved in the uptake of Fe from the soil [47]. In rice plant, OsYSL2 is expressed in the root companion cells and leaves phloem. OsYSL2 can also transport manganese-NA complex [47].

OsYSL6 is required for detoxification of excess Mn in rice thus helps in Mn tolerance. Irrespective of the Fe status, OsYSL6 is constitutively expressed in both roots and leaves. While OsYSL6 expression is slightly reduced under Fe-deficiency condition. The expression level increases with leaf age. This pattern is similar for Mn concentration in the different leaves [48].

YSL16 is a plasma membrane-localized transporter and is directly involved in distribution and remobilization of Cu as Cu-NA complex in the developing tissues and rice seed. It loads Cu-NA complex into the phloem, which is required for remobilization of Cu from older leaves to developing tissues like young leaves and seeds [52]. OsYSL16 is expressed in many cell types and is more preferentially expressed in the vascular tissues of roots and leaves. It has been studied that enhanced tolerance to a low-Fe environment can be achieved through over expression of OsYSL16.

OsYSL18 is a Fe(III)-DMA transporter which is involved in Fe distribution mediated by DMA in the reproductive organs, lamina joints, and phloem cells at the leaf sheath base. It is localized in the plasma membrane [54]. The other remaining putative OsYSL transporters in rice need to be functionally characterized in future.

4. Ctr/COPT family: copper transporter

This family of transporters is found only in eukaryotes. In plants, it is known as COPT transporters [55] and in animals and fungi as Ctr [56]. COPT family of proteins are important for copper uptake from soil and its transport to pollen in plants [57]. COPT proteins are localized on the plasma membrane and are involved in the transport of Cu from extracellular spaces into cytosol or vacuoles or are localized on the lysosomal membrane and are involved in the supply of Cu from vacuoles or lysosomes to the cytosol [58].

The COPT family of transporters in rice consists of seven members: COPT1–COPT7. In rice plant, COPT proteins are specifically involved in Cu transport. It can transport only Cu(I) but not other bivalent ions such as Mn, Zn or Fe. In rice plant, COPT1–COPT7 are plasma membrane-localized proteins. As these transporters can form symmetrical homotrimer or heterotrimer structure with a diameter that is only suitable for Cu(I) transport and not other divalent ions [59] or heterocomplex with themselves or each other [60] or heterocomplex with other proteins which are involved in Cu transport [61].

COPT1 and COPT5 can exist as homodimers or a heterodimer. Only COPT1 and COPT5 bind to rice XA13 protein, a protein which is susceptible to pathogenic bacterium *Xanthomonas oryzae pv. Oryzae* (Xoo) [61] and mediate Cu transport in rice plant. In rice plant, all the COPTs except COPT1 and COPT5, function independently or together and mediate Cu transport in different tissues. COPT6 acts as a cofactor and aids the efficient localization of COPT2, COPT3 or COPT4 to the plasma membrane for mediating Cu transport. COPT1 and COPT5 show similar tissue and also develop-specific expression patterns. When compared to the sheath, stem, and panicle expression levels of these two genes, they have a higher level of expression in root and leaf tissues. COPT4 has higher expression level in root in comparison to other tissues. COPT1, COPT4, and COPT5 have higher expression level in young leaves than in old leaves, particularly COPT1 and COPT4. COPT2, COPT3, and COPT7 show higher expression levels in old leaves when compared with young leaves. COPT6 is not expressed in root and is highly expressed in leaf than in other tissues. COPT6 is constitutively expressed in different-aged leaves but has a low level of expression in shoot tissue at seedling stage. The expression of rice COPT1 and COPT5 are induced by Cu deficiency and suppressed by excess Cu in both shoot and root tissues. COPT1 and COPT5 together mediate Cu transport in rice plant [61]. The expression of rice COPT2, COPT3, COPT4, COPT6, and COPT7, is also affected by the variation in Cu levels. In both root and shoot tissues, COPT7 shows a similar response as COPT1 and COPT5 to Cu deficiency and overdose. In different-aged leaves of a mature plant, COPT6 is constitutively expressed while in the seedling stage, the shoot tissue has low expression levels. In shoot, COPT6 is induced under Cu deficiency state and suppressed in Cu overdose but no COPT6 expression is detected in root either with or without Cu deficiency. Cu overdose suppresses the expression of COPT2, COPT3, and COPT4 in both root and shoot tissues but their expression is not influenced by Cu deficiency. Expression of COPTs is also influenced by other bivalent cations. Mn deficiency induces expression of COPT1 in root and COPT3 and COPT7 in shoot and slightly suppresses the expression of COPT2 and COPT4 in root. The expression of COPT1, COPT5, and COPT7 is induced by Zn deficiency and COPT4 expression is slightly suppressed in the root. Zn deficiency also induces the expression of COPT5, COPT6, and COPT7 in the shoot. In root, Fe deficiency moderately induces COPT1 and suppresses COPT2 and COPT5 while in the shoot, it induces COPT2, COPT5, COPT6, and COPT7. In root, no COPT6 expression has been found either in the presence or absence of Mn, Zn, or Fe deficiency. COPT family of transporter proteins functions uniquely in different tissues, during various developmental stages, and in different environmental conditions. In rice plant, COPT2, COPT3, COPT4, COPT6 and COPT7 mediate Cu transport either solely or cooperatively with each other. In different tissues of rice plant COPT2, COPT3, or COPT4 function along with COPT6 for Cu transport except in root. Expression of COPT2, COPT3, COPT4, and COPT6 has been observed in stem, sheath, leaf, and panicle tissues. Root shows relatively high expression levels of COPT3 and COPT4 but no expression of COPT6. In leaves, the expression of COPT2, COPT3, and COPT4 is developmentally regulated but not that of COPT6. In rice shoot, Cu deficiency strongly induces the expression of COPT6 but not COPT2, COPT3, and COPT4. COPT7 mediates Cu transport in rice all by itself. Based on its expression pattern, it has been suggested that COPT7 functions in different tissues and is unaffected by Cu deficiency [62].

5. NRAMP family

Natural resistance-associated macrophage protein (NRAMP) family of transporters are found in the three domains of life [63]. NRAMP transporters have a wide range of metal substrates, typically transport Fe^{+2}, Mn^{+2}, Co^{+2}, and Zn^{+2} [63].

The first plant NRAMP genes cloned were from rice [64]. In rice, there are seven Nramp transporters, OsNRAMP1-OsNRAMP7. Though, not all have been functionally characterized [63]. Many of the NRAMP family proteins function as Fe transporters. *OsNRAMP1* is highly upregulated by Fe deficiency. OsNRAMP1 is a plasma membrane-localized transporter and is involved in the transport of Cd and Fe. *OsNRAMP1* expression is observed mainly in roots at the vegetative state and is involved in cellular uptake of Cd and is responsible for high Cd accumulation in rice [65]. The differences observed in Cd accumulation among different rice cultivars are because of differences in *OsNRAMP1* expression levels in roots [65]. *OsNRAMP1* expression is higher during the reproductive stage in leaf blade and stem.

OsNRAMP3 is localized to the plasma membrane and is specifically expressed in vascular bundles, particularly in companion cells of phloem. *OsNRAMP3* is constitutively expressed in the rice node [66]. OsNRAMP3 is a Mn-influx transporter involved in Mn distribution and redistribution to young leaves from old leaf via phloem cells. With leaf aging, the expression of OsNRAMP3 in leaves increases slightly in rice plants. OsNRAMP3 transports Mn from the enlarged vascular bundles to the younger tissues and panicles during Mn deficiency in order to meet its minimal growth requirement. On the other hand, when Mn is in excess, OsNRAMP3 is internalized in vesicles and rapidly degraded. Then, Mn is preferentially loaded into the older leaves, which are directly connected to the enlarged vascular bundles, thereby protecting the developing tissues from Mn toxicity. This indicates the role of post-translational regulation of OsNRAMP3 in response to environmental nutrient availability. Rice plant utilizes OsNRAMP3 to respond to environmental changes to Mn availability. OsNRAMP3 is involved in Mn translocation but not Mn uptake [67].

OsNRAMP4 is also known as Nramp aluminum transporter1 (Nrat1) is the first transporter in this family to be identified as the trivalent Al ion transporter [68]. In contrast to other rice NRAMP members, OsNRAMP4 does not show transport activity for other divalent metal ions, like Zn, Mn, and Fe. It also shares relatively low similarity with the other OsNRAMP members [68]. NRAT1 plays an important role in rice Al tolerance by reducing the level of toxic Al in the root cell wall and transporting Al into the root cell vacuole for sequestration. Rice is the most Al tolerant of all the cereal crops and OsNRAMP4 plays an important role in this [69].

OsNRAMP5 is a plasma membrane protein involved in Mn and Fe transport [70]. OsNRAMP5 gene expression increases slightly in the roots when plants are under Fe or Zn deficiency but varying levels of Mn in the surrounding does not affect it [71]. It is expressed in the mature root zone at the PM of the exodermal and endodermal layers [70, 71]. OsNRAMP5 in rice plant is essential for the uptake of Mn from the soil. In rice plant, OsNRAMP5 is constitutively involved in Fe and Mn uptake, it also plays a role in Fe and Mn transport during flowering and seed development [70].

OsNRAMP5 is highly expressed in hulls. It is also expressed in leaves but the expression level decreases with leaf age. In rice plant, OsNRAMP5 transporter is present in the vascular bundles of roots and shoots particularly the parenchyma cells surrounding the xylem. *OsNRAMP5* is also highly expressed in stele cells especially in the xylem region, thus plays an important role in the xylem-mediated root-to-shoot transport. Thus OsNRAMP5 plays an important role in the uptake, translocation, and distribution of Mn in rice plants. *OsNRAMP5* is highly expressed in stele cells especially in the xylem region, thus plays an important role in the xylem-mediated root-to-shoot transport [72]. OsNRAMP5 is a major transporter for Cd uptake in rice [71].

Recently, OsNRAMP6 has been identified to be involved in uptake of Fe and Mn. It is a plasma membrane-localized protein. It negatively regulates the rice plant immunity as loss of its function results in increased resistance against *M. oryzae* [73].

6. Heavy metal ATPases (HMAs)

The P_{1B}-type ATPases, known as heavy metal ATPases (HMAs) in plants, play an important role in metal transport. HMAs vary in their tissue distribution, subcellular localization, and metal specificity. HMA transporters can be divided into two subgroups based on their metal-substrate specificity, they are Cu/Ag group and Zn/Co/Cd/Pb group. Rice plant has nine such *HMA* genes. OsHMA1–OsHMA3 are members of the Zn/Co/Cd/Pb subgroup in rice. Unlike dicots, only a few reports on HMAs from monocots are available. OsHMA2 plays an important role in root-to-shoot translocation of Zn and Cd and participates in their transport to developing seeds in rice. OsHMA9 phylogenetically belongs to the Cu/Ag subgroup but also plays a role in Zn, Cd, and Pb transport [74]. OsHMA3 transports only Cd and in root cells is involved in the sequestration of Cd into vacuoles [75, 76]. *OsHMA3* has been identified as a responsive gene for quantitative trait loci of Cd concentration in the rice cultivars Anjana Dhan and Cho-kokoku, and loss of function of this protein leads to high Cd accumulation in the shoots [75–77]. There is little information available on the role of OsHMA1 and is thought to be involved in Zn transport. *OsHMA1* expression in shoot tissue is highly upregulated by Zn deficiency [78]. OsHMA1 is suggested to play a role in Zn transport in the plant throughout its growth and developmental stages [74]. Enhanced activity of OsHMA3 is related to increased storage of Cd in roots and its decreased transport to the shoot and the final accumulation in rice grains [74]. OsHMA2 is localized at the root pericycle and plays a major role in of transport of Zn and Cd during xylem loading [74, 75].

OsHMA3 gene selectively sequesters Cd into the vacuoles thus limits the root-to-shoot translocation of Cd [75, 76]. In rice plant, *OsHMA2* gene has also been shown to be involved in the translocation of Cd through xylem from root to shoot [79, 80].

In root cells, OsHMA4 is a vacuolar membrane-localized transporter and is involved in sequestering Cu into the vacuoles. OsHMA4 specifically transports Cu. Increased Cu accumulation in rice grain due to increased root-to-shoot translocation of Cu has been observed when OsHMA4 function is lost. In rice OsHMA4–OsHMA9 are members of the Cu/Ag subgroup of HMAs. OsHMA5 is a Cu transporter, localized to the plasma membrane [81]. In rice, OsHMA5 is involved in transferring Cu into the xylem for its root-to-shoot translocation and/or Cu detoxification in roots [81]. OsHMA4 is induced under long-term exposure of excess Cu and its expression is suppressed by Cu deficiency. In mature root zone, OsHMA4 is localized at the pericycle [81]. OsHMA4 regulates the cellular Cu concentration before loading to the xylem depending on its environmental concentration. OsHMA3 is localized in all root cells [75]. In future, the mechanism responsible for the transporter substrate specificity of the HMAs needs to be studied.

7. CDF/MTP

The cation diffusion facilitator (CDF) family of proteins plays an important role in the maintenance of cation homeostasis in all forms of life from bacteria, yeast, plants to mammals [82]. Generally, CDF proteins are involved in binding and efflux of cations such as Zn, Fe, Co, Cd, and Mn from the cytoplasm either by sequestrating into internal organelles like vacuole or effluxing from the cell [82, 83]. CDF transporters also influence the cation accumulation, metal ion tolerance, signal transduction cascades, oxidative stress resistance, and protein turnover in cells [84]. In plants, CDF members are called Metal Tolerance Proteins (MTPs) and as Solute carrier family 30 (SLC30) in vertebrates [84].

The first plant CDF protein identified was ZAT (zinc transporter of *Arabidopsis thaliana*), because of its role in heavy metal tolerance in Arabidopsis. Later it was renamed as AtMTP1 (Metal Tolerance Protein 1) [85]. MTPs are a group of proteins that play an important role in heavy metal homeostasis in plants [82, 83]. MTP members are present in all the three kingdoms (Archaea, Eubacteria, and Eukaryotes). The plant CDF family can be classified into three subgroups phylogenetically: Zn-CDF, Fe/Zn-CDF, and Mn-CDF [83] based on their main substrate transported: Zn, Zn, and Fe, or Mn [82, 84].

Rice genome has 10 MTP genes [84]. Studies have shown that Rice Metal Tolerance Protein1 (OsMTP1) gene expression is induced by Cd and OsMTP1 belongs to the Zn-CDF subgroup. In rice, there are five Mn-CDFs, three Zn-CDFs, one Fe/Zn- CDF, and one unclassified CDF.

OsMTP1 was characterized recently [86, 87]. In mature leaves and stem, it is highly expressed [86]. Generally, OsMTP1 transports Zn but can also transport Co, Fe, and Cd. Earlier OsMTP1 has been shown to transport Ni [86]. The vacuolar localization of OsMTP1 in the tonoplast, compartmentalizes primarily Zn, but also Co, Fe, and Cd, and serves as a detoxification system when these metals are available in excess. OsMTP1 is expressed constitutively and upregulated by Cd [86]. OsMTP1, OsMTP5, and OsMTP12 belong to the Zn-CDF subgroup [82, 84]. Expression of OsMTP1 in leaves, stems, roots, and flowers is relatively low and spatially and temporally regulated during development of rice. Also, it shows differential response to Cd stress. Transgenic assays in rice have shown that OsMTP1 expression levels can change plant cation absorption and in turn has affect on Zn, Ni, and Cd contents [86].

In rice, OsMTP8.1 is the first Mn-CDF member to be identified. It is localized on to the tonoplast, and its over expression in rice enhances Mn accumulation and tolerance [88]. OsMTP9 is also a Mn-CDF member, which is involved in the uptake and translocation of Mn in rice plant [89]. Mn and other heavy metals induce the expression of *OsMTP11* in rice. In rice plant, *OsMTP1* is involved in Zn and Cd homeostasis/stress and mediates their translocation from roots to the aerial parts [86]. In most rice plant tissues *OsMTP11* is constitutively expressed.

OsMTP8.1 is localized to the tonoplast and involved in the detoxification of manganese by sequestering excess manganese to the vacuoles [90]. In rice root, OsMTP9 is polarly localized at the proximal side of both exodermis and endodermis opposite to Nramp5. The cooperative transport by Nramp5 and MTP9 efficiently transport Mn leading to its high accumulation in rice [91].

The Mn-CDF group in plants is further clustered into two subgroups, Groups 8 and 9 [82]. In the rice genome, there are three members (MTP9/11/11.1) of Group 9. MTP9 shows much higher expression in the roots than in the basal region and shoots. The expression is unaffected by the deficiency of iron, zinc, copper, and manganese. The expression of MTP9 in roots is eightfold higher in the basal parts than that in apical parts. At the reproductive growth stage, MTP9 is also expressed in other organs such as nodes and leaf sheath in addition to the roots. MTP9 is polarly located at the proximal side of the exodermis and the endodermis, which is in opposition to Nramp5 [71]. Therefore, MTP9 at the proximal side of the exodermis releases manganese taken up by Nramp5 to the apoplast of a spoke-like structure in the aerenchyma, whereas MTP9 at the proximal side of endodermis further releases manganese toward the apoplast of stele including xylem vessels. Thus polar localization of transporters plays an important role in the directional transport of minerals. Recently, a number of transporters have been found to show polar localization. However, our understanding of the molecular mechanism underlying polar localization is still very poor. Also, MTP9 is different from other members of the Mn-CDF group as it shows a distinct expression pattern in tissue and subcellular localization. MTP9 is mainly expressed in the roots, but MTP8.1 from rice [90] in the same group is mainly expressed in shoots rather than roots. Different from other members, rice MTP9 is localized to the plasma membrane. These differences are associated with the role of MTP9 in manganese uptake in rice roots. In conclusion, MTP9 is a plasma membrane-localized efflux transporter for manganese uptake and translocation in rice roots.

The polar localization of MTP9 and Nramp5 at both the exodermis and the endodermis leads to efficient and unidirectional flux of manganese from the soil solution to the stele.

8. Root-to-shoot translocation and metal chelation in cytoplasm

Plants that prevent or limit the entry of metals from roots to shoots are categorized as excluders. On the contrary, plants that can transfer metals from root–to-shoot via the xylem along the transpiration stream by increasing the uptake of metals in roots, thus increasing the sequestration of metals in the aerial parts are considered as accumulators/hyperaccumulators. The transfer of metals from the roots to the aerial parts in plants helps to reduce the damage caused by the heavy metals on their root. The translocation of metals from root to the aerial parts is an essential process for the overall growth and development of the plant.

Studies of long-distance root-to-shoot metal transport within plants mainly emphasizes on transporters that are localized to either xylem parenchyma cells or phloem companion cells, as they are directly associated with xylem and phloem loading or unloading thus majorly contribute to the metal redistribution process within the plant. The movement of heavy metals from roots to shoots is facilitated when metals are chelated with ligands such as organic acids, amino acids, and thiols. The movement of metal cations across the xylem cell wall is restricted when the metals are not chelated by ligands due to high cation exchange capability of the xylem wall.

The chelation of metals with NA provides improved tolerance against the restriction by the xylem cell wall. NA facilitates the chelation and transport of divalent ions of metal Ni, Cu, and Zn [92]. Synthesis of NA by trimerization of S-adenosylmethionine is facilitated by nicotianamine synthase (NAS) [93]. In rice plant, increased accumulation of Fe, Zn, and Cu is associated with over expression of the gene *NAS3* [94].

During the process of long-distance transport of metals, some chelators, like nicotianamine [41], glutathione (GSH), and phytochelatins [95], also play a vital role.

Metals pass through xylem unloading process before their distribution and detoxification in the shoot and followed by their redistribution via the phloem. After unloading, the metals either enter into the nearby cells or are symplastically transported or they are apoplastically distributed throughout the leaf tissue [96]. For the symplastic transport of metals across the leaf via the YLS transporter proteins, chelation of metals to NA is required [97].

Excess metals in the plant are sequestered in various aerial plant parts, such as trichomes, leaf epidermal cell vacuole, and mesophyll vacuole. Not much study has been done on the transport of metals through the phloem sap. Nicotianamine is the only molecule to be identified as a phloem metal transporter which is associated with the transport of Fe, Cu, Zn, and Mn [98].

In plants, the vacuole sequestration capacity (VSC) also plays an important role in the long-distance transport and sequestration of metals. Vacuolar metal sequestration capacities are automatically adjusted to the varying availability of metal ions in the environment, due to the cooperation between tonoplast-localized transporters and ion chelators. Hence vacuoles work as a buffering zone.

8.1. Metal sequestration in vacuoles by tonoplast transporters

Plants have developed several defense mechanisms like chelation, excretion, and subcellular compartmentalization to combat heavy metal toxicity as it severely affects its overall growth and development [99].

A large lytic vacuole (LV) is present in most of the plant cells which occupies about 80% of the cell volume. In plants, LVs undergo less metabolism and acts as a store house which can accumulate a huge amount of minerals and water thus play a major role in turgor generation. LVs also function as a store house for other xenobiotic and toxic compounds and also reduce their impact in the cytoplasm where several sensitive processes take place. LV also store plant secondary metabolites and proteins involved in plant defense against pathogens and herbivores and release them when subjected to cellular damage. LV has acidic pH around 4–5, and this acidic environment helps in the degradation of both exogenous and endogenous compounds [100].

Plant LVs are equivalent to lysosomes in animal and vacuoles in yeast and acts as degradation and waste storage compartments. The vacuolar tonoplast of higher plants and fungi as well as lysosomes of animal cells share very similar H^+-ATPases that acidify the lumen. Apart from these, inner organellar membranes also contain multidrug resistance-associated protein (MRP) type ATP binding cassette (ABC) transporters, chloride channels (CLC) type ion channels, cation channels, and aquaporins [99].

Storage of nutrient minerals in the LV is important in buffering any variations in the supply of nutrients like when plants are grown in nutrient-rich conditions, they will deposit large quantities of such nutrients in vacuoles of vegetative tissues. This helps the plant to survive during subsequent periods of nutrient deficiency by remobilizing from the vacuolar store. Thus, the solute composition of vacuoles is highly dynamic and reflects changes in the environment and the plant developmental stage [101].

The vacuole is a major organelle in higher plants functioning as a store for metabolites, mineral nutrients, and toxicants. Studies have shown that in addition to its storage role, the vacuole also contributes to long-distance transport of metals, through the modulation of vacuolar sequestration capacity (VSC) which is basically controlled by cytosolic metal chelators and tonoplast-localized transporters, or the interaction between them [102].

VSC regulates the long-distance transport of mineral nutrients in plants. Zhang et al. [103] isolated the two vacuolar membrane-localized metal transporters OsVIT1 and OsVIT2 in rice. Both *OsVIT1* and *OsVIT2* primarily function to sequester Fe/Zn into vacuoles across the vacuolar membrane. In rice plant, flag leaves show a high expression of *OsVIT1* and *OsVIT2*. *OsVIT1* and *OsVIT2* along with VSC play an important role in Fe and Zn long-distance translocation between flag leaves and seeds.

Studies have also suggested that the long-distance transport of nonessential toxic metals and their detoxification is regulated by VSC thus making the plants highly tolerant to metals [102]. Generally, the VSC of certain metal varies between different plant tissues to ensure proper metal distribution.

Nonessential metals like Cd and As have adverse effects on plants, either through oxidative stress or competitive inhibition of essential mineral nutrient involved in any biological pathways. In order to protect the aerial parts which are associated with important biological processes like photosynthesis, plants have developed several mechanisms to regulate metal distribution between roots and shoots, VSC is one such mechanism.

In rice plant, *OsHMA3* is involved in Cd accumulation in the rice shoots. In rice root, OsHMA3 is localized to the vacuolar membrane, and it mediates the transport of Cd into vacuoles. In rice roots, VSC along with OsHMA3 plays a major role in the long-distance transport of Cd from roots to shoots [75, 76].

9. Biochemical processes modulated by heavy metal stress

Both abiotic and biotic stresses have several effects on plant growth as well as productivity. Plant vigor and crop yields are strikingly influenced by these abiotic stresses. In order to combat these stresses, plants have evolved many responses. Plants have developed and used various strategies to cope with and also to adapt to these stress conditions. It depends on variation in protein relative abundance of stress-responsive proteins, resulting in changes in the whole proteome, transcriptome, and metabolome levels [104].

Expression patterns of these protein and transcript levels are influenced by the intensity and duration of stress apart from the usual post-translational regulatory mechanisms such as RNA stability and protein degradation [105]. In addition, the intensity and duration of stress can have a substantial effect on the complexity of the stress response. Recent progress in different areas of rice research such as analysis of interactome analysis, transcriptome, metabolome etc., have given us a better insight of abiotic stress response in rice plant [106–109]. The study of these differential changes in proteome profiles in response to abiotic stress is an approach to better understanding the physiology and molecular mechanisms that underlie rice stress responses. Most common response to all stresses is the induction of oxidative stress [110] and modulation of gene expression.

9.1. ROS production and modulation of antioxidant system during heavy metal stress

In an environment of metal toxicity, the elevated activities of antioxidant enzymes and nonenzymatic constituents play important role in the plant tolerance to stress. Metal tolerance is enhanced by the plant's antioxidant resistant mechanisms. The harmful effects of heavy metals in plants are due to the production of ROS and induction of oxidative stress. Increased levels of reactive oxygen species such as singlet oxygen ($^1O_2$), superoxide radical (O^{-2}), hydrogen peroxide (H_2O_2), and hydroxyl radical (OH^-) result in oxidative stress [111]. ROS are strong oxidizing agents which cause oxidative damage to biomolecules, like lipids and proteins and can eventually lead to cell death [112]. It has been shown that plant tolerance to metals is correlated with an increase in antioxidants and activity of radical scavenging enzymes [113]. Plants respond to oxidative stress by activating antioxidative defense mechanisms, which involves enzymatic and nonenzymatic antioxidants. The enzymatic components include superoxide dismutase (SOD), catalase (CAT), ascorbate peroxidase (APX), and enzymes of ascorbate glutathione cycle, while the nonenzymatic antioxidants include ascorbate and glutathione and atocoperol [3, 113]. These antioxidants play an important role in the elimination and destruction of the reactive oxygen species [3].

An imbalance between the detoxification of the ROS products and the antioxidative system results in oxidative damage [113]. The tolerance of deleterious environmental stresses, such as heavy metals, is associated with the increased capacity to scavenge or detoxify activated oxygen species [113].

Studies using comparative analysis suggest that each heavy metal is accumulated differentially in root tissues. Heavy metal stress induces production of reactive oxygen species (ROS) which promotes cell death by apoptosis, necrosis, or mechanisms with both features. Methods like H_2O_2 staining are used to measure production of ROS.

SOD, APX, and glutathione peroxidase (GPX) are the ROS scavenging antioxidant enzymes. They play a very important role in scavenging ROS like superoxide radical, H_2O_2, hydroxyl radical, peroxy radical, and singlet oxygen species. SOD acts as the first line of defense against ROS as it converts superoxide radical to H_2O_2 [114]. Ascorbate, which is present inside the cell, plays an important role as an antioxidant either by participating in ascorbate-glutathione cycle or by directly quenching the ROS [115]. Ascorbate is also utilized by APXs as a reducing agent to catalyze the conversion of H_2O_2 to water and GPX breaks down H_2O_2 to H_2O and O_2. Similar modulation of the antioxidant system upon exposure to heavy metal stress has been observed in studies of different rice cultivars [110, 116].

Glutathione is an important antioxidant in plant cells which is involved in scavenging of free radicals and H_2O_2, which are formed as a result of oxidative stress along with ascorbate [117]. Glutathione (GSH) is the precursor of phytochelatins (PC), which bind above the optimal concentrations of heavy metals [118]. It also serves as a substrate for GSTs which catalyzes the conjugation of GSH with xenobiotics like herbicides [119]. In silico studies and over expression of GSTs have shown to provide tolerance toward different heavy metals [120, 121]. Glutathione levels in plant tissues are known to accelerate under stress induced by heavy metals.

In rice plants, the root tissue shows variable response of antioxidant enzymes during growth with different heavy metals of varying concentration.

9.2. Differential expression and modulation of genes during heavy metal stress

Heavy metal stress in plants severely modulates the gene expression pattern. In plants, many genes are downregulated due to heavy metal stress. They are related to the energy metabolism, carbohydrate metabolism, lignin biosynthesis, phenylalanine metabolism, cell growth and death, lipid metabolism, biodegradation of xenobiotics, amino acid metabolism, etc. Among the upregulated genes, the majority of affected genes are associated with the biosynthesis of secondary metabolites, specially flavonoid biosynthesis, lipid metabolism, amino acid metabolism, carbohydrate metabolism, biodegradation of xenobiotics, ascorbate, and aldarate metabolism, membrane transport especially multidrug resistance protein, major facilitator superfamily, ABC transporters, glutathione metabolism, MAPK (mitogen-activated protein kinases) signaling pathway, a large number of GST, etc. In a study using rice seedlings grown in metal supplemented media in comparison to control, genes which significantly modulated were filtered. It was found that 17 and 83 genes are commonly upregulated and downregulated under different heavy metal stress. One each of cytochrome P450, Proton-dependent oligopeptide transporter (POT) family protein, heat shock protein, and two NAC domain-containing proteins are commonly upregulated during heavy metals stress, they play important role in detoxification of heavy metals. On the other hand, one heavy metal-associated domain-containing protein, zinc finger protein, cytochrome P450, ring-H2 zinc finger protein, and catalase-1 are commonly downregulated during heavy metal exposure [122]. Plants have developed cellular mechanisms to tolerate and regulate the uptake of heavy metals [123]. However, molecular mechanisms and networks involved in the uptake and detoxification of heavy metals remain poorly understood. Phytochelatins (PCs), a class of cysteine-rich heavy metal-binding peptides, bind to heavy metals, and detoxify by vacuolar sequestration [123]. Sulfur homeostasis in plants results in the increased production of S-rich metal-binding peptides (such as GSH, PCs), which provide metal tolerance [116].

From the expression data, it was demonstrated that the upregulation of a unique cytochrome P450s in different heavy metal stresses is a major detoxification mechanism. In plants, cytochrome P450s plays a major role in the metabolism of several biosynthetic pathways such as flavonoids, coumarins, anthocyanins, isoflavonoids, phytoalexins, salicylic acid, jasmonic acid, and many others [124]. Previously, it has been reported that cytochrome P450 is involved in the metabolism of toxic compounds, indicating their role in heavy metal detoxification. These results indicate that metabolism of plant biosynthetic pathways are very much affected during metal exposure and different cytochrome P450s are involved in the metabolism of different heavy metals. A large number of transporter genes are differentially up- and downregulated under different heavy metal stresses, which include major facilitator genes, sulfate transporters, peptide transporters, nitrate transporters, ABC transporters, multidrug resistance proteins, zinc transporters, Nramp6, and multidrug and toxic compound extrusion (MATE) efflux family proteins. One of the essential nutrient required for plant growth is sulfur that enters the cell via sulfate transporters as inorganic sulfate, it may induce the production of S-rich metal-binding peptides (such as GSH, PCs) and thus provide defense against heavy metal stress [125]. It is clear from their study that each heavy metal has induced specific sulfate transporters. The peptide transporters [126] have been shown to transport nitrate and tripeptides such as glutathione which is a major component in sulfur metabolism and plant defense during stress [127]. It is suggested that nitrate transporter plays a role in root-to-shoot translocation of nitrate thus plays a role in Cd toxicity [128]. ABC transporter proteins play an important role in the transport of various substances like lipids, phytohormones, carboxylates, heavy metals, chlorophyll catabolites, etc. across various biological membranes [129]. It has been shown that Nramp proteins are conserved bivalent metal transporters [130]. NRAMP3 and NRAMP4 are reported to be responsible for Cd^{2+} efflux from the vacuole [131]. MATE proteins bind to a variety of potentially toxic compounds and function as proton-dependent efflux transporters to remove toxic compounds from the cell [132]. Various methyltransferases are differentially modulated under different stresses. In plants, O-methyltransferases constitute a large family of enzymes that are involved in stress tolerance as reported by Lam et al. [133]. Specific methyltransferases catalyze the transfer of methyl groups which are involved in several pathways that lead to the accumulation of methylated inositols, quaternary amines, and tertiary sulfonium species, which play a significant role in stress tolerance [134]. Therefore, modulation of these transcripts must play a secondary role in different heavy metal toxicity.

During different heavy metal stresses, these metals induce damage to the thylakoid membrane leading to increased lipid peroxidation and thus cause downregulation of peroxidases. These specific peroxidase family genes might play a key role in the enzymatic defense of plant cells by scavenging ROS during stress conditions [135]. Heat shock proteins (HSPs) in particular play important role in protecting plants against stress by re-establishing normal protein conformation and thus cellular homeostasis [136].

It has been observed that various transcription factors like WRKY, MYB family, zinc finger protein, RING-H2 finger protein, and basic leucine zipper (bZIP) are differentially expressed under heavy metal stress [122]. In plants, WRKY transcription factors are linked to various processes associated with different biotic and abiotic stresses and regulation of differential transcription a response to stress in plants [137]. Similarly, MYB TFs play very important roles in many physiological processes under normal or unfavorable growth conditions [138] and also in defense and stress responses [139]. During heavy metal stress various stress-related genes are transcriptionally regulated such as GSTs, dehydrin, sulfite oxidase (SO), L-ascorbate peroxidase, L-ascorbate oxidase, and germin-like proteins. GSTs are a superfamily of multifunctional, dimeric enzymes. It induces the conjugation of GSH a tripeptide glutathione to electrophilic xenobiotics and this is followed by sequestration of this complex into the vacuole for detoxification [140]. Recently, it has been reported that a particular class of GST gene family, that is, Lambda GST plays an important role during heavy metal stress [120, 121]. During drought stress, cold stress, and other defense processes dehydrins are produced in plants [141]. SO catalyzes the transformation of sulfites to the nontoxic sulfate. It has been reported earlier that sulfur is an essential nutrient that is taken up as sulfate by plants and chemical compounds which contain S, such as glutathione (GSH), phytochelatins (the polymers of GSH) play a prominent role in arsenic detoxification [142]. Similarly, L-ascorbate peroxidase plays an important role in defense against oxidative stress as it has been studied that APX is an important antioxidant enzyme which detoxifies H_2O_2 by converting into water. Germin-like proteins have been reported to play a significant role in germination and defense response [143] during Cd toxicity.

Thus transcription factors play a significant role during different heavy metal stress response and indirectly modulate several genes responsible for stress. Further study of these TFs would help to understand the difference in the network of pathways during different heavy metal stresses.

10. Conclusion and future prospective

Though different heavy metals are detoxified through similar mechanism their uptake from soil by root system differs for each metal. There are several families of transporters, which play important role in metal ion uptake from the soil as well as redistribution within the plant and its sequestration in various organelles and plant parts.

Apart from the mineral elements described above, rice requires several other mineral elements for its growth. The transporters associated with the uptake of those mineral elements are yet to be identified in rice. Rice plant because of its distinct root anatomical characteristic requires a pair of influx and efflux transporters for the transport of mineral elements from the soil solution to the stele, thus help to surpass the two casparian strips present in the root exodermis and endodermis. As described above, some of the rice plant transporters associated with mineral uptake have been studied and characterized but there exists no clear understanding about their influx-efflux transporter pairs. More transporters associated with metal uptake and sequestration have to be identified and characterized using different ways like genetics (both forward and reverse), expression pattern, functional characterization in yeast or oocytes, phenotypic analysis using mutants, and so on in future. Each heavy metal modulates specific pathways in addition to common networks such as expression of a specific member of gene families including several transporters, different members of cytochrome P450 and transcription factors are modulated in different heavy metal stresses. In future, these complex responses have to be elucidated using various functional genomic approaches along with proteomic and metabolomic analyses. Also, many plant vacuolar membrane transporters and channels have been identified. Still, there is a dearth of knowledge about the regulation of these networks and how all these transporters and channels interact with each other in order to maintain a cytosolic ion homeostasis. Our understanding about the response of the rice plant to abiotic stress needs to be further refined.

Therefore, further studies on mineral transporters, antioxidant system, and differential gene expression regulated by heavy metals in rice is required to get deeper insight of the abiotic stress caused by heavy metals on rice plant which will in turn help to reduce the abiotic stress caused by heavy metals on rice plant ultimately leading to increased and better crop production for the benefit of mankind.

Acknowledgements

We express our appreciation to Anna University for providing Anna Centenary Research Fellowship (ACRF) to Anitha Mani.

Author details

Anitha Mani and Kavitha Sankaranarayanan*

*Address all correspondence to: skavitham@yahoo.com

Ion Channel Biology Laboratory, AU-KBC Research Centre, MIT Campus of Anna University, Chennai, Tamil Nadu, India

References

[1] Marschner P. Mineral Nutrition of Higher Plants. 3rd ed. London: Academic Press; 2012

[2] Baker AJM, Walker PL. Ecophysiology of metal uptake by tolerant plants: Heavy metal uptake by tolerant plants. In: Shaw AJ, editor. Evolutionary Aspects. Boca Raton: CRC; 1990

[3] Solanki R, Dhankhar R. Biochemical changes and adaptive strategies of plants under heavy metal stress. Biologia. 2011;**66**:195-204

[4] Clemens CS. Molecular mechanisms of plant metal tolerance and homeostasis. Planta. 2001;**212**:475-486. DOI: 10.1007/s004250000458

[5] Ammiraju JS, Luo M, Goicoechea JL, Wang W, Kudrna D, et al. The Oryza bacterial artificial chromosome library resource: Construction and analysis of 12 deep-coverage largeinsert BAC libraries that represent the 10 genome types of the genus Oryza. Genome Research. 2006;**16**:140-147

[6] Jacquemin J, Bhatia D, Singh K, Wing RA. The international Oryza map alignment project: Development of a genus-wide comparative genomics platform to help solve the 9 billion-people question. Current Opinion in Plant Biology. 2013;**16**:147-156

[7] International Rice Genome Sequencing Project. The map-based sequence of the rice genome. Nature. 2005;**436**:793-800

[8] Yu J, Hu S, Wang J, Wong GK, Li S, Liu B, Deng Y, Dai L, Zhou Y, Zhang X, Cao M, Liu J, Sun J, Tang J, Chen Y, Huang X, Lin W, Ye C, Tong W, Cong L, Geng J, Han Y, Li L, Li W, Hu G, Huang X, Li W, Li J, Liu Z, Li L, et al. A draft sequence of the rice genome (*Oryza sativa* L. ssp. Indica). Science. 2002;**296**:79-92

[9] Slamet-Loedin IH, Johnson-Beebout SE, Impa S, Tsakirpaloglou N. Enriching rice with Zn and Fe while minimizing Cd risk. Frontiers in Plant Science. 2015;**6**:121. DOI :10.3389/fpls.2015.00121

[10] Shiratori K, Suzuki T, Miyoshi H. Chemical forms of soil zinc in the salt-rich reclaimed fields, where zinc-deficiency symptoms were observed in rice plants. Japanese Society of Soil Science and Plant Nutrition. 1972;**43**:291-296

[11] Khaokaew S, Chaney RL, Landrot G, Ginder-Vogel M, Sparks DL. Speciation and release kinetics of cadmium in an alkaline paddy soil under various flooding periods and draining conditions. Environmental Science & Technology. 2011;**45**:4249-4255

[12] Hashimoto Y, Yamaguchi N. Chemical speciation of cadmium and sulfur K-edge XANES spectroscopy in flooded paddy soils amended with zerovalent iron. Soil Science Society of America Journal. 2013;**77**:1189-1198

[13] Zhang FQ, Zhang HX, Wang GP, Xu LL, Shen ZG. Cadmium induced accumulation of hydrogen peroxide in the leaf apoplast of Phaseolus aureus and Vicia sativa and the roles of different antioxidant enzymes. Journal of Hazardous Materials. 2009;**168**:76-84

[14] Kumar A, Prasad MNV, Sytar O. Lead toxicity, defense strategies and associated indicative biomarkers in Talinum triangulare grown hydroponically. Chemosphere. 2012;**89**:1056-1065

[15] Kurz et al. Selection of cultivars to reduce the concentration of cadmium and thallium in food and fodder plants. Journal of Plant Nutrition and Soil Science. 1999;**162**:323-328

[16] Ross SM. Toxic Metals in Soil-Plant Systems. Chichester: John Wiley and Sons Inc; 1994

[17] Enstone DE, Peterson CA, Ma F. Root endodermis and exodermis: Structure, function, and responses to the environment. Journal of Plant Growth Regulation. 2002;**21**:335-351

[18] Kawai M, Samarajeewa PK, Barrero RA, Nishiguchi M, Uchimiya H. Cellular dissection of the degradation pattern of cortical cell death during aerenchyma formation of rice roots. Planta. 1998;**204**:277-287

[19] Yamaji N, Ma JF. Spatial distribution and temporal variation of the rice silicon transporter Lsi1. Plant Physiology. 2007;**143**:1306-1313

[20] Guerinot ML. The zip family of metal transporters. Biochimica et Biophysica Acta. 2000;**1465**:190-198

[21] Colangelo EP, Guerinot ML. Put the metal to the petal: Metal uptake and transport throughout plants. Current Opinion in Plant Biology. 2006;**9**(3):322-330

[22] Gaither LA, Eide DJ. Eukaryotic zinc transporters and their regulation. Biometals. 2001;**14**:251-270. PMID: 11831460

[23] Eide D, Broderius M, Fett J, Guerinot ML. A novel iron-regulated metal transporter from plants identified by functional expression in yeast. Proceedings of the National Academy of Sciences of the United States of America. 1996;**93**:5624-5628

[24] Robinson NJ, Procter CM, Connolly EL, Guerinot ML. A ferricchelate reductase for iron uptake from soils. Nature. 1999;**397**:694-697

[25] Ma JF, Nomoto K. Effective regulation of iron acquisition in graminaceous plants. The role of mugineic acids as phytosiderophores. Physiologia Plantarum. 1996;**97**:609-617

[26] Ishimaru Y, Suzuki M, Tsukamoto T, Suzuki K, Nakazono M, Kobayashi T, Wada Y, Watanabe S, Matsuhashi S, Takahashi M, et al. Rice plants take up iron as an Fe^{3+}-phytosiderophore and as Fe^{2+}. The Plant Journal. 2006;**4**:335-346

[27] Lee S, Kim SA, Lee J, Guerinot ML, An G. Zinc deficiency-inducible OsZIP8 encodes a plasmamembrane localized zinc transporter in rice. Molecules and Cells. 2010;**29**(6): 551-558. DOI: 10.1007/s10059-010-0069-0

[28] Ishimaru YS, Kobayashi M, Takahashi T, Nakanishi M, Mori H, Nishizawa S, Naoko K. OsZIP4, a novel zinc-regulated zinc transporter in rice. Journal of Experimental Botany. December 2005;**56**(422):3207-3214. DOI: 10.1093/jxb/eri317

[29] Yang X, Huang J, Jiang Y, Zhang HS. Cloning and functional identification of two members of the ZIP (Zrt, Irt-like protein) gene family in rice (*Oryza sativa* L.). Molecular Biology Reports. 2009;**36**:281-287

[30] Ramesh SA, Shin R, Eide DJ, Schachtman DP. Differential metal selectivity and gene expression of two zinc transporters from rice. Plant Physiology. 2003;**133**:126-134

[31] Kavitha PG, Kuruvilla S, Mathew MK. Functional characterization of a transition metal ion transporter, OsZIP6 from rice (*Oryza sativa* L.). Plant Physiology and Biochemistry. 2015;**97**:165-174. DOI: 10.1016/j.plaphy.2015.10.005

[32] Bughio N, Yamaguchi H, Nishizawa NK, Nakanishi H, Mori S. Cloning an iron-regulated metal transporter from rice. Journal of Experimental Botany. 2002;**53**:1677-1682

[33] Banerjee S, Sharma DJ, Verulkar SB, Chandel G. Use of in silico and semiquantitative RT-PCR approaches to develop nutrient rich rice (*Oryza sativa* L.). Indian Journal of Biotechnology. 2010;**9**:203-212

[34] Takahashi M, Nozoye T, Kitajima B, Fukuda N, Hokura N, Terada Y, et al. In vivo analysis of metal distribution and expression of metal transporters in rice seed during germination process by microarray and X-ray fluorescence imaging of Fe, Zn, Mn, and cu. Plant and Soil. 2009;**325**:39-51

[35] Nozoye T, Inoue H, Takahashi M, Ishimaru Y, Nakanishi H, Mori S, Nishizawa NK. The expression of iron homeostasis-related genes during rice germination. Plant Molecular Biology. 2007;**64**:35-47

[36] Yang X, Ye ZQ, Shi CH, Zhu ML, Graham RD. Genotypic differences in concentrations of iron, manganese, copper, and zinc in polished rice grains. Journal of Plant Nutrition. 1998;**21**:1453-1462

[37] Guerinot ML, Yi Y. Iron – Nutritious, noxious, and not readily available. Plant Physiology. 1994;**104**:815-820

[38] Römheld V, Marschner H. Evidence for a specific uptake system for iron phytosiderophores in roots of grasses. Plant Physiology. 1986;**80**:175-180

[39] Curie C, Panaviene Z, Loulergue C, Dellaporta SL, Briat JF, Walker EL. Maize yellow stripe 1 encodes a membrane protein directly involved in Fe(III)uptake. Nature. 2001;**409**:346-349. DOI: 10.1038/35053080

[40] Schaaf G, Ludewig U, Erenoglu BE, Mori S, Kitahara T, von Wirén N. ZmYS1 functions as a proton-coupled symporter for phytosiderophore and nicotianamine-chelated metals. The Journal of Biological Chemistry. 2004;**279**:9091-9096

[41] Curie C, Cassin G, Couch D, Divol F, Higuchi K, Le Jean M, Misson J, Schikora A, Czernic P, Mari S. Metal movement within the plant: Contribution of nicotianamine and yellow stripe 1-like transporters. Annals of Botany (Lond). 2009;**103**:1-11

[42] Lubkowitz,M. The oligopeptide transporters: A small gene family with a diverse group of substrates and functions? Molecular Plant. 2011;**4**:407-415. DOI: 10.1093/mp/ssr004

[43] Curie C, Cassin G, Couch D, Divol F, Higuchi K, LeJean M, et al. Metal movement within the plant: Contribution of nicotianamine and yellow stripe1-liketransporters. Annals of Botany. 2008;**103**:1-11. DOI: 10.1093/aob/ mcn207

[44] Higuchi K, Suzuki K, Nakanishi H, Yamaguchi H, Nishizawa N-K, Mori S. Cloning of nicotianamine synthase genes, novel genes involved in the biosynthesis of phytosiderophores. Plant Physiology. 1999;**119**:471-480. OI: 10.1104/pp.119.2.471

[45] Bashir K, Inoue H, Nagasaka S, Takahashi M, Nakanishi H, Mori S, et al. Cloning and characterization of deoxymugineic acid synthase genes from graminaceous plants. The Journal of Biological Chemistry. 2006;**281**(43):32395-32402

[46] Conte SS, Walker EL. Transporters contributing to iron trafficking in plants. Molecular Plant. 2011;**4**:464-476. DOI: 10.1093/mp/ssr015

[47] Koike S, Inoue H, Mizuno D, Takahashi M, Nakanishi H, Mori S, Nishizawa NK, et al. OsYSL2 is a rice metal-nicotianamine transporter that is regulated by iron and expressed in the phloem. The Plant Journal. 2004;**39**:415-424

[48] Ishimaru Y et al. Rice metal-nicotianamine transporter, OsYSL2, is required for the long-distance transport of iron and manganese. The Plant Journal. 2010;**62**:379-390

[49] Inoue H, Kobayashi T, Nozoye T, Takahashi M, Kakei Y, Suzuki K, Nakazono M, Nakanishi H, Mori S, Nishizawa NK. Rice OsYSL15 is an iron-regulated iron(III)-deoxymugineic acid transporter expressed in the roots and is essential for iron uptake in early growth of the seedlings. Journal of Biological Chemistry. 2009;**284**:3470-3479

[50] Takagi S, Nomoto K, Takemoto S. Physiological aspect of mugineic acid, a possible phytosiderophore of graminaceous plants. Journal of Plant Nutrition. 1984;**7**:469-477

[51] Nozoye T, Nagasaka S, Kobayashi T, Takahashi M, Sato Y, Uozumi N, Nakanishi H, Nishizawa NK. Phytosiderophore efflux transporters are crucial for iron acquisition in graminaceous plants. The Journal of Biological Chemistry. 2011;**286**:5446-5454

[52] Zheng L, Yamaji N, Yokosho K, Ma JF. YSL16 is a phloem- localized transporter of the copper-nicotianamine complex that is responsible for copper distribution in rice. The Plant Cell. 2012;**24**:3767-3782

[53] Aoyama T, Kobayashi T, Takahashi M, Nagasaka S, Usuda K, Kakei Y, Ishimaru Y, Nakanishi H, Mori S, Nishi-Zawa NK. OsYSL18 is a rice Fe(III)-deoxymugineic acid transporter specifically expressed in reproductive organs and phloem of lamina joints. Plant Molecular Biology. 2009;**70**:681-692

[54] Aoyama T, Kobayashi T, Takahashi M, Nagasaka S, Usuda K, Kakei Y, Ishimaru Y, Nakanishi H, Mori S, Nishizawa NK. OsYSL18 is a rice iron(III)-deoxymugineic acid transporter specifically expressed in reproductive organs and phloem of lamina joints. Plant Molecular Biology. 2009;**70**:681-692. DOI: 10.1007/s11103-009-9500-3

[55] Sancenón V, Puig S, Mira H, Thiele D, Peñarrubia L. Identification of a copper transporter family in *Arabidopsis thaliana*. Plant Molecular Biology. 2003;**51**:577-587. DOI: 10.1023/A:1022345507112

[56] Kim H, Wu X, Lee J. SLC31(CTR) family of copper transporters in health and disease. Molecular Aspects of Medicine. 2013;**34**:561-570. DOI: 10.1016/j.mam.2012.07.011

[57] Sancenón V, Puig S, Mateu-Andres I, Dorcey E, Thiele DJ, Peñarrubia L. The *Arabidopsis* copper transporterCOPT1functions in root elongation and pollen development. Journal of Biological Chemistry. 2004;**279**:15348-15355. DOI: 10.1074/jbc.M313321200

[58] Rees EM, Lee J, Thiele DJ. Mobilization of intracellular copper stores by the Ctr2 vacuolar copper transporter. The Journal of Biological Chemistry. 2004;**279**:54221-54229

[59] Nose Y, Rees EM, Thiele DJ. Structure of the Ctr1 copper trans'PORE'ter reveals novel architecture. Trends in Biochemical Sciences. 2006;**31**:604-607

[60] Beaudoin J, Laliberte J, Labbe S. Functional dissection of Ctr4 and Ctr5 amino-terminal regions reveals motifs with redundant roles in copper transport. Microbiology-Sgm. 2006;**152**:209-222

[61] Yuan M, Chu Z, Li X, Xu C, Wang S. The bacterial pathogen *Xanthomonas oryzae* overcomes rice defenses by regulating host copper redistribution. Plant Cell. 2010;**22**:3164-3176

[62] Yuan M, Li X, Xiao J, Wang S. Molecular and functional analyses of COPT/Ctr type copper transporter-like gene family in rice. BMC Plant Biology. 2011;**11**:69

[63] Nevo Y, Nelson N. The NRAMP family of metal-ion transporters. Biochimica et Biophysica Acta. 2006;**1763**:609-620

[64] Belouchi A, Kwan T, Gros P. Cloning and characterization of the OsNramp family from Oryza sativa, a new family of membrane proteins possibly implicated in the transport of metal ions. Plant Molecular Biology. 1997;**33**:1085-1092. DOI: 10.1023/A: 1005723304911

[65] Takahashi R, Ishimaru Y, Nakanishi H, Nishizawa NK. Role of the iron transporter OsNRAMP1 in cadmium uptake and accumulation in rice. Plant Signaling & Behavior. 2011;**6**(11):1813-1816

[66] Yamaji N, Xia J, Mitani-Ueno N, Yokosho K, Ma JF. Preferential delivery of zinc to developing tissues in rice is mediated by P-type heavy metal ATPase OsHMA2. Plant Physiology. 2013;**162**:927-939

[67] Yang M, Zhang W, Dong H, Zhang Y, Lv K, et al. OsNRAMP3 is a vascular bundles-specific manganese transporter that is responsible for manganese distribution in rice. PLoS One. 2013;**8**(12):e83990. DOI: 10.1371/journal.pone.0083990

[68] Xia J, Yamaji N, Kasai T, Ma JF. Plasma membrane localized transporter for aluminum in rice. Proceedings of the National Academy of Sciences of the United States of America. 2010;**107**:18381-18385

[69] Famoso AN et al. Development of a novel aluminum tolerance phenotyping platform used for comparisons of cereal aluminum tolerance and investigations into rice aluminum tolerance mechanisms. Plant Physiology. 2010;**153**(4):1678-1691

[70] Ishimaru Y, Bashir K, Nakanishi H, Nishizawa NK. OsNRAMP5, a major player for constitutive iron and manganese uptake in rice. Plant Signaling & Behavior. 2012;**7**(7): 763-766. DOI: 10.4161/psb.20510

[71] Sasaki A, Yamaji Y, Yokosho K, Ma JF. Nramp5 is a major transporter responsible for manganese and cadmium uptake in rice. The Plant Cell. 2012;**24**:2155-2167

[72] Yang M, Zhang Y, Zhang L, Jintao H, Zhang X, Kai L, Dong H, Wang D, Zhao F-J, Huang C-F, Lian X. OsNRAMP5 contributes to manganese translocation and distribution in rice shoots. Journal of Experimental Botany. 2014;**65**(17):4849-4861. DOI: 10.1093/jxb/eru259

[73] Peris-Peris C, Serra-Cardona A, Sánchez-Sanuy F, Campo S, Ariño J, Segundo BS. Two NRAMP6 isoforms function as iron and manganese transporters and contribute to disease resistance in rice. Molecular Plant-Microbe Interactions. 2017;**30**:385-398. DOI: 10.1094/MPMI-01-17-0005-R

[74] Lee S, Kim Y-Y, Lee Y, An G. Rice P1B-type heavy-metal ATPase, OsHMA9, is a metal efflux protein. Plant Physiology. 2007;**145**:831-842. PMID: 17827266; DOI: 10.1104/pp.107.102236

[75] Ueno D, Yamaji N, Kono I, Huang CF, Ando T, Yano M, et al. Gene limiting cadmium accumulation in rice. Proceedings of the National Academy of Sciences of the United States of America. 2010;**107**:16500-16505. PMID:20823253; DOI: 10.1073/ pnas.1005396107

[76] Miyadate H, Adachi S, Hiraizumi A, Tezuka K, Nakazawa N, Kawamoto T, et al. OsHMA3, a P1B-type of ATPase affects root-to-shoot cadmium translocation in rice by mediating efflux into vacuoles. New Phytologist. 2011;**189**:190-199. PMID: 20840506; DOI: 10.1111/j.1469-8137.2010.03459.x

[77] Ueno D, Koyama E, Yamaji N, Ma JF. Physiological, genetic, and molecular characterization of a high- Cd-accumulating rice cultivar Jarjan. Journal of Experimental Botany. 2011;**62**:2265-2272. PMID: 21127026. DOI: 10.1093/jxb/erq383

[78] Suzuki M, Bashir K, Inoue H, Takahashi M, Nakanishi H, Nishizawa NK. Accumulation of starch in Zn-deficient rice. Rice. 2012;**5**:9. http://dx.doi. org/10.1186/1939-8433-5-9

[79] Satoh-Nagasawa N, Mori M, Nakazawa N, Kawamoto T, Nagato Y, Sakurai K, et al. Mutations in rice (*Oryza sativa*) heavy metal ATPase 2 (*OsHMA2*) restrict the translocation of zinc and cadmium. Plant and Cell Physiology. 2012;**53**:213-224. PMID: 22123790. DOI: 10.1093/pcp/pcr166

[80] Takahashi R, Ishimaru Y, Shimo H, Ogo Y, Senoura T, Nishizawa NK, et al. The OsHMA2 transporter is involved in root-to-shoot translocation of Zn and Cd in rice. Plant, Cell & Environment. 2012;**35**:1948-1957. PMID: 22548273. DOI: 10.1111/j.1365-3040.2012.02527.x

[81] Deng F, Yamaji N, Xia J, Ma JF. A member of the heavy metal P-type ATPase OsHMA5 is involved in xylem loading of copper in rice. Plant Physiology. 2013;**163**:1353-1362

[82] Gustin JL, Zanis MJ, Salt DE. Structure and evolution of the plant cation diffusion facilitator family of ion transporters. BMC Evolutionary Biology. 2011;**11**:76

[83] Sharma SS, Dietz KJ, Mimura T. Vacuolar compartmentalization as indispensable component of heavy metal detoxification in plants. Plant, Cell & Environment. 2016;**39**(5): 1112-1126

[84] Montanini B, Blaudez D, Jeandroz S, Sanders D, Chalot M. Phylogenetic and functional analysis of the cation diffusion facilitator (CDF) family: Improved signature and prediction of substrate specificity. BMC Genomics. 2007;**107**:1471-2164

[85] Van-Der-Zaal BJ, Neuteboom LW, Pinas JE, Chardonnens AN, Schat H, Verkleij JA, Hooykaas PJ. Overexpression of a novel Arabidopsis gene related to putative zinc-transporter genes from animals can lead to enhanced zinc resistance and accumulation. Plant Physiology. 1999;**119**:1047-1055

[86] Yuan L, Yang S, Liu B, Zhang M, Wu K. Molecular characterization of a rice metal tolerance protein, *OsMTP1*. Plant Cell Reports. 2012;**31**:67-79

[87] Lan HX, Wang ZF, Wang QH, Wang MM, Bao YM, Huang J, Zhang HS. Characterization of a vacuolar zinc transporter OZT1 in rice (*Oryza sativa* L.). Molecular Biology Reports. 2012;**40**:1201-1210

[88] Chen Z, Fujii Y, Yamaji N, Masuda S, Takemoto Y, Kamiya T, Yusuyin Y, Iwasaki K, Kato S, Maeshima M, Ma JF, Ueno D. Mn tolerance in rice is mediated by MTP8.1, a member of the cation diffusion facilitator family. Journal of Experimental Botany. 2013;**64**(14):4375±4387

[89] Ueno D, Sasaki A, Yamaji N, Miyaji T, Fujii Y, Takemoto Y, Moriyama S, Che J, Moriyama Y, Iwasaki K, Ma JF. A polarly localized transporter for efficient manganese uptake in rice. Nature Plants. 2015;**1**:15170

[90] Chen Z et al. Manganese tolerance in rice is mediated by MTP8.1, a member of the cation diffusion facilitator family. Journal of Experimental Botany. 2013;**64**:4375-4387

[91] Ueno D, Sasaki A, Yamaji N, Takaaki Miyaji, Yumi Fuii, Takemoto Y, Moriyama S, Che J, Moriyama Y, Iwasaki K, Jian Feng Ma. A polarly localized transporter for efficient manganese uptake in rice. Nature Plants. 2015;**1**. Article number: 15170. DOI: 10.1038/nplants.2015.170

[92] Takahashi M, Terada Y, Nakai I, Nakanishi H, Yoshimura E, Mori S, Nishizawa N. Role of nicotianamine in the intracellular delivery of metals and plant reproductive development. Plant Cell. 2003;**15**:1263-1280

[93] Shojima S, Nishizawa NK, Fushiya S, Nozoe S, Irifune T, Moris S. Biosynthesis of phytosiderophores- invitro biosynthesis of 2′-deoxymugineic acid from L-methionine and nicotianamine. Plant Physiology. 1990;**93**:1497-1503

[94] Kawachi M, Kobae Y, Mori H, Tomioka R, Lee Y, Maeshima M. A mutant strain Arabidopsis thaliana that lacks vacuolar membrane zinc transporter MTP1 revealed the latent tolerance to excessive zinc. Plant and Cell Physiology. 2009;**50**:1156-1170

[95] Mendoza-Cozatl DG, Jobe TO, Hauser F, Schroeder JI. Long-distance transport, vacuolar sequestration, tolerance, and transcriptional responses induced by cadmiumand arsenic. Current Opinion in Plant Biology. 2011;**14**:554-562. DOI: 10.1016/j.pbi.2011.07.004

[96] Marschner H. In: Marschner H, editor. Mineral Nutrition of Higher Plants. 2nd ed. London, San Diego: Academic Press; 1995. p. 899

[97] DiDonato R, Roberts L, Sanderson T, Eisley R, Walker E. Arabidopsis yellow stripe-Like2 (YSL2): A metal-regulated gene encoding a plasma membrane transporter of nicotianamine-metal complexes. Plant Journal. 2004;**39**:403-414

[98] Stephan UW, Schmidke I, Pich A. Phloem translocation of Fe, Cu, Mn, and Zn in ricinus seedlings in relation to the concentrations of nicotianamine, an endogenous chelator of divalent metal-ions, in different seedling parts. Plant and Soil. 1994;**165**:181-188

[99] Barkla BJ, Pantoja O. Physiology of ion transport across the tonoplast of higher plants. Annual Review of Plant Physiology and Plant Molecular Biology. 1996;**47**:159-184

[100] Echeverria E, Jacqueline JK. Vacuolar acid hydrolysis as a physiological mechanism for sucrose breakdown. Plant Physiology. 1989;**90**:530-533

[101] Marschner H. Mineral Nutrition in Higher Plants. London: Academic Press; 1995

[102] Peng J-S, Gong J-M. Vacuolar sequestration capacity and long-distance metal transport in plants. Frontiers in Plant Science. 2014;**5**:19. DOI: 10.3389/fpls.2014.00019

[103] Zhang Y, Xu YH, Yi HY, Gong JM. Vacuolar membrane transporters OsVIT1 and OsVIT2 modulate iron translocation between flag leaves and seeds in rice. Plant Journal. 2012;**72**:400-410. DOI: 10.1111/j.1365-313X.2012.05088.x

[104] Cho K, Shibato J, Agrawal GK, Jung YH, Kubo A, Jwa NS, Tamogami S, Satoh K, Kikuchi S, Higashi T, Kimura S, Saji H, Tanaka Y, Iwahashi H, Masuo Y, Rakwal R. Integrated transcriptomics, proteomics, and metabolomics analyses to survey ozone responses in the leaves of rice seedling. Journal of Proteome Research. 2008;**7**:2980-2998

[105] Raksha Singh and Nam-Soo Jwa. Understanding the responses of Rice to environmental stress using proteomics. Proteome Research. 2013;**12**:4652-4669. DOI: 10.1021/pr400689j

[106] Seo YS, Chern M, Bartley LE, Han M, Jung KH, Lee I, Walia H, Richter T, Xu X, Cao P, Bai W, Ramanan R, Amonpant F, Arul L, Canlas PE, Ruan R, Park CJ, Chen X, Hwang S, Jeon JS, Ronald PC. Towards establishment of a rice stress response interactome. PLoS Genetics. 2011;**7**:e1002020

[107] Wang L, Xie W, Chen Y, Tang W, Yang J, Ye R, Liu L, Lin Y, Xu C, Xiao J, Zhang Q. A dynamic gene expression atlas covering the entire life cycle of rice. The Plant Journal. 2010;**61**:752-766

[108] Redestig H, Kusano M, Ebana K, Kobayashi M, Oikawa A, Okazaki Y, Matsuda F, Arita M, Fujita N, Saito K. Exploring molecular backgrounds of quality traits in rice by predictive models based on high-coverage metabolomics. BMC Systems Biology. 2011;**28**:176

[109] De Sharma R, Vleesschauwer D, Sharma MK, Ronald PC. Recent advances in dissecting stress-regulatory crosstalk in rice. Molecular Plant. 2013;**6**:250-260

[110] Shri M, Kumar S, Chakrabarty D, Trivedi PK, Mallick S, Misra P, Shukla D, Mishra S, Srivastava S, Tripathi RD, Tuli R. Effect of arsenic on growth, oxidative stress, and antioxidant system in rice seedlings. Ecotoxicology and Environmental Safety. 2009;**72**:1102-1110

[111] Salin ML. Toxic oxygen species and protective systems of the chloroplast. Physiologia Plantarum. 1988;**72**:681-689

[112] Gunes A, Soylemezoglu G, Inal A, Bagci EG, Coban S, Sahin O. Antiox- idant and stomatal responses of grapevine (*Vitis vinifera* L.) to boron toxicity. Scientia Horticulturae. 2006;**110**:279-284

[113] Kachout SS, Ben Mansoura A, Leclerc JC, Mechergui R, Rejeb MN, Ouer-ghi Z. Effects of heavy metals on antioxidant activities of Atriplex hortensis and A. rosea. Journal of Food, Agriculture & Environment. 2009;**7**:938-945

[114] Scandalios JG. Oxygen stress and superoxide dismutase. Plant Physiology. 1993;**101**: 712-726

[115] Zechmann B. Subcellular distribution of ascorbate in plants. Plant Signaling & Behavior. 2011;**6**:360-363

[116] Rai A, Tripathi P, Dwivedi S, Dubey S, Shri M, Kumar S, Tripathi PK, Dave R, Kumar A, Singh R, Adhikari B, Bag M, Tripathi RD, Trivedi PK, Chakrabarty D, Tuli R. Arsenic tolerances in rice (*Oryza sativa*) have a predominant role in transcriptional regulation of a set of genes including sulphur assimilation pathway and antioxidant system. Chemosphere. 2010;**82**:1-10

[117] Alscher RG. Biosynthesis and antioxidant function of glutathione in plants. Physiologia Plantarum. 1989;**77**:457-464

[118] Grill E, Winnacker EL, Zenk MH. Phytochelatins, a class of heavy-metal-binding peptides from plants, are functionally analogous to metallothioneins. Proceedings of the National Academy of Sciences of the United States of America. 1987;**84**:439-443

[119] Marrs KA. The functions and regulation of glutathione S-transferases in plants. Annual Review of Plant Physiology and Plant Molecular Biology. 1996;**47**:127-158

[120] Kumar S, Asif MH, Chakrabarty D, Tripathi RD, Dubey RS, Trivedi PK. Differential expression of rice lambda class GST gene family members during plant growth, development, and in response to stress conditions. Plant Molecular Biology Reporter. 2013;**31**:569-580

[121] Kumar S, Asif MH, Chakrabarty D, Tripathi RD, Dubey RS, Trivedi PK. Expression of a rice lambda class of glutathione S-transferase, OsGSTL2, in Arabidopsis provides tolerance to heavy metal and other abiotic stresses. Journal of Hazardous Materials. 2013;**15**:228-237

[122] Dubey S, Shri M, Misra P, Lakhwani D, Bag SK, Asif MH, Trivedi PK, Tripathi RD, Chakrabarty D. Heavy metals induce oxidative stress and genome-wide modulation in transcriptome of rice root. Functional & Integrative Genomics. 2013;**14**:401. DOI: 10.1007/s10142-014-0361-8

[123] Hall JL. Cellular mechanisms for heavy metal detoxification and tolerance. Journal of Experimental Botany. 2002;**53**:1-11

[124] Li L, Cheng H, Gai J, Yu D. Genome-wide identification and characterization of putative cytochrome P450 genes in the model legume Medicago truncatula. Planta. 2007;**226**:109-123

[125] Nocito FF, Lancilli C, Crema B, Fourcroy P, Davidian JC, Sacchi GA. Heavy metal stress and sulfate uptake in maize roots. Plant Physiology. 2006;**141**:1138-1148

[126] Chiang CS, Stacey G, Tsay YF. Mechanisms and functional properties of two peptide transporters, AtPTR2 and fPTR2. The Journal of Biological Chemistry. 2004;**279**: 30150-30157

[127] Noctor G, Arisi ACM, Jouanin L, Kunert KJ, Rennenberg H, Foye CH. Glutathione: Biosynthesis, metabolism and relationship to stress tolerance explored in transformed plants. Journal of Experimental Botany. 1998;**49**:623-647

[128] Li JY, Fu YL, Pike SM, Bao J, Tian W, Zhang Y, Chen CZ, Zhang Y, Li HM, Huang J, Li LG, Schroeder JI, Gassmann W, Gong JM. The Arabidopsis nitrate transporter NRT1.8 functions in nitrate removal from the xylem sap and mediates cadmium tolerance. Plant Cell. 2010;**22**:1633-1646

[129] Kretzschmar T, Burla B, Lee Y, Martinoia E, Nagy R. Functions of ABC transporters in plants. Essays in Biochemistry. 2011;**50**:145-160

[130] Cailliatte R, Lapeyre B, Briat JF, Mari S, Curie C. The NRAMP6 metal transporter contributes to cadmium toxicity. The Biochemical Journal. 2009;**422**:217-228

[131] Thomine S, LeLievre F, Debarbieux E, Schroeder JI, Barbier-Brygoo H. AtNRAMP3, a multispecific vacuolar metal transporter involved in plant responses to iron deficiency. The Plant Journal. 2003;**34**:685-695

[132] Brown MH, Paulsen IT, Skurray RA. The multidrug efflux protein NorM is a prototype of a new family of transporters. Molecular Microbiology. 1999;**31**:394-395

[133] Lam KC, Ibrahim RK, Behdad B, Dayanandan S. Structure, function, and evolution of plant O-methyltransferases. Genome. 2007;**50**:1001-1013

[134] Bohnert JH, Nelson DE, Jensen RG. Adaptations to environmental stresses. Plant Cell. 1995;**7**:1099-1111

[135] Gülen H, Çetinkaya C, Kadıoğlu M, Kesici M, Cansev A, Eriş A. Peroxidase activity and lipid peroxidation in strawberry (Fragaria X ananassa) plants under low temperature. Journal of Biological and Environmental Sciences. 2008;**2**:95-100

[136] Wang KK, Jiang L, Yi YX, Liu KE, Tang DL, Wang JS, Shi YZ, Wang QP, Xiao XZ. Effect of heat shock response on the cleavage of nucleolin induced by oxidative stress. Zhong Nan Da Xue Xue Bao. Yi Xue Ban. 2004;**29**:504-508

[137] Chen L, Song Y, Li S, Zhang L, Zou C, Yu D. The role of WRKY transcription factors in plant abiotic stresses. Biochimica et Biophysica Acta. 2012;**1819**:120-128

[138] Jin H, Martin C. Mutifunctionality and diversity within the plant MYB gene family. Plant Molecular Biology. 1999;**41**:577-585

[139] Misra P, Pandey A, Tiwari M, Chandrashekar K, Sidhu OP, Asif MH, Chakrabarty D, Singh PK, Trivedi PK, Nath P, Tuli R. Modulation of transcriptome and metabolome of tobacco by Arabidopsis transcription factor, AtMYB12, leads to insect resistance. Plant Physiology. 2010;**152**:2258-2268

[140] Edwards R, Dixon DP, Walbot V. Plant glutathione S-transferases: Enzymes with multiple functions in sickness and in health. Trends in Plant Science. 2000;**5**:193-198

[141] Close TJ. Dehydrins: Emergence of a biochemical role of a family of plant dehydration proteins. Physiologia Plantarum. 1996;**97**:795-803

[142] Zhong L, Hu C, Tan Q, Liu J, Sun X. Effects of sulfur application on sulfur and arsenic absorption by rapeseed in arseniccontaminated soil. Plant, Soil and Environment. 2011;**57**:429-434

[143] Lane BG, Grzelczak ZF, Kennedy TD, Kajioka R, Orr J, Agostino SD, Jaikaran A. Germin: Compartmentation of two forms of the protein by washing growing wheat embryos. Biochemistry and Cell Biology. 1986;**64**:1025-1037

9

Improving Rice Grain Quality by Enhancing Accumulation of Iron and Zinc While Minimizing Cadmium and Lead

Lei Gao and Jie Xiong

Abstract

Iron (Fe) and zinc (Zn) are important trace elements for people's health around the globe. A lot of people, especially children and woman, are suffering from malnutrition caused by Fe and/or Zn deficiency. The deficiency is more pronounced in some parts of Africa and Asia due to low income, which makes it difficult to afford meat or sea foods that are rich in Fe and Zn. Biofortification of Fe and Zn in rice is the most economical and convenient way to supplement these important minerals in the diet of poor people. However, besides Fe and Zn, rice also can accumulate heavy metals, such as cadmium (Cd) and lead (Pb), which are harmful to people, especially for kids' health. Previous researches have shown that there are connections and discrepancies for metal absorption, translocation, and accumulation in rice. So it is imperative to review these issues. This chapter compares the physiological and molecular mechanisms of Fe, Zn, Cd, and Pb uptake, mobilization, and accumulation in rice and discusses the progress and strategies for not only increasing Fe/Zn but also decreasing Cd/Zn accumulation in rice.

Keywords: biofortification, heavy metal, iron, zinc, cadmium, lead, stress

1. Introduction

Metal elements, such as Fe, Zn, Mn and Cu, are essential for living organisms and humans' growth. Various metal nutrients supplied by food contributed to maintain metabolism normally. Unfortunately, iron (Fe) and zinc (Zn), as the most important metal elements, are present in low quantities of staple food, such as rice and wheat [1]. What is worse is that, in some parts of Africa and Asia, people even cannot afford enough food for their kids and families. Fe deficiency is one of the leading risk factors of disability and death worldwide. It is estimated to affect two billion people in the world [2]. Also, it has been estimated that about 30% population of the world suffers from Zn deficiency [3]. Fe-deficiency anemia can impair cognitive and physical development in children and the reduction of daily productivity in adults. Adequate Zn nutrition is also important for the growth of children, immune function, and neurobehavioral development [4]. Thus, Fe and Zn deficiency has emerged as a major and common problem for the health of humans [5, 6].

Rice is the staple food and provides energy to almost half of the world's population, especially in Asia and Africa. Thus, increasing Fe/Zn content of rice has a great potential to mitigate widespread Fe/Zn deficiency problem in humans [7]. Therefore, it is essential to understand the mechanisms through which rice uptakes, mobilizes, and accumulates Fe/Zn. In response to Fe deficiency, higher plants have developed two strategies for acquiring Fe from the rhizosphere [8, 9]. Strategy I is employed in nongraminaceous plants, and Fe(III) is reduced to soluble Fe(II) through activating membrane-bound Fe(III)-chelate reductases, then the reduced Fe(II) was transported into cytoplasm via Fe(II) transporters [10]. In contrast, Strategy II is only applied by graminaceous plants, such as rice and wheat. The root of strategy II plants secretes phytosiderophores (PSs) to rhizosphere and chelates Fe(III) to form Fe(III)-PS complexes. Subsequently, the Fe(III)-PS complexes are transported via specific plasma membrane transporters [7]. Rice not only employs strategy II to acquire Fe from rhizosphere but also utilizes strategy I-like system to uptake Fe(II) directly [10]. Besides Fe uptake, mugineic acid (MA) family also plays crucial roles in chelating Zn from rhizosphere, followed by uptake of Zn-PS complexes via specific plasma membrane transporters [7]. Moreover, Zn can be ionized as Zn(II) and directly enters into root [11]. In spite of rice can apply specific strategies to acquire Fe and Zn, these mechanisms have limited accessibility to resource poor people faced with Fe and Zn deficiency from certain areas of the world. To deal with limited Fe/Zn and improve human Fe/Zn nutritional status, rice with enhanced Fe/Zn absorption will be an effective method for populations consuming rice as their staple foods.

Besides necessary metal elements, such as Fe and Zn, rice also absorbs and accumulates toxic metals such as cadmium (Cd) and lead (Pb), which are harmful for both rice and humans. Cd enters into environment, such as soil and river, mainly through industrial activities or fertilizers [12]. As a highly mobile and soluble metal [13, 14], Cd exposure causes crops yield reduction and does harm to humans' health even at low concentrations [15]. Due to daily consumption, Cd in rice grains poses a latent health problem to humans through food chains and leads to chronic toxicity. The outbreak of "Itai-Itai disease" in the mid-twentieth century in Japan is due to consumption of Cd-contaminated rice [16]. A person with "Itai-Itai" has symptoms of weakness and softening of the bones. Even in recent years, Cd exposure in general Japanese population can be as high as 3–4 mg kg^{-1} body weight every week [17]. The directly observable toxic symptoms of Cd on plants are as follows: reduced rate of transpiration and photosynthesis, growth retardation and declining metabolic activities [15]. In response to Cd toxicity, plants also have evolved several protective mechanisms against Cd toxicity, including avoidance and tolerance strategies [18]. Plants can prevent Cd from entering into plant cells, which is refered to as avoidance strategy. Cell walls serve as the first barrier against Cd entrance [15, 19]. Root exudates, which are majorly consisted of sugars, proteins and organic acids are secreted from root to soil to combine with Cd, keeping Cd apart from root [20, 21]. After entered into the cells, the abilities of resistance to Cd stress are referred to as tolerance strategy [15]. In rice, phytochelatins (PC) acting as Cd chelator plays a key role in Cd detoxification [22]. PC chelates Cd in the cytosol and forms complexes with Cd. Then the Cd-PCs complexes are sequestered in the vacuoles via specific transporters located at tonoplast [23, 24].

Pb after arsenic (As) is ranked as the second most harmful element due to its occurrence, toxicity, and exposure potential [25]. Although Pb occurs naturally only in small amounts within the Earth's crust [26], a large amount of industrial activities are the primary sources of Pb in soil [27]. Pb is released into soil in general forms of Pb(II), lead oxide and lead-metal oxyanion, among which Pb(II) is the most common form [27]. Pb stress can produce reactive oxygen species (ROS) and trigger some antioxidative enzymes, accompanied by the increased level of lipid peroxidation [25, 27]. In rice, Pb toxicity reduces leaf chlorophyll and nitrogen content and increases antioxidative enzymes. Its subsequent translocation to grain causes a great threat to humans' health [28, 29]. High levels of Pb can cause brain and kidney damage, accompanied with central nervous prostration [27]. Therefore, higher plants possess preventive mechanisms against Pb. Similar to Cd detoxification, cell walls, and compartmentalization in vacuoles have been suggested as an important detoxification mechanism [27].

Many researches show that the mechanisms of Fe/Zn/Cd/Pb uptake and accumulation in rice share commons in some aspects as a result of similar entry routes (transporters) within rice cells. However, an increasing number of studies discovered distinct pathways and mechanisms of Fe/Zn/Cd/Pb uptake and accumulation in rice recently. In this chapter, we mainly systematically elaborate and compare physiological and cellular mechanisms of Fe/Zn/Cd/Pb uptake and accumulation in rice. In addition, we also review the mechanisms of maintaining Fe/Zn homeostasis and Cd/Pb detoxification in rice. Effects of fertilizers on Fe/Zn/Cd /Pb accumulation in rice are also discussed. Finally, we enumerate various approaches for reducing grain Cd/Pb accumulation and enhancing Fe/Zn content in rice.

2. Uptake of Fe/Zn and Cd/Pb from rhizosphere to root in rice

Mobility and availability of metals from soil are controlled by three following factors: (1) soil conditions (upland or flooded soil, soil solution pH); (2) mineralization (ionization and complex formation); and (3) uptake transporters [11].

Although Fe in acidic soil is ionized as Fe^{2+}/Fe^{3+} and easily utilized by plants, Fe in aerobic alkaline soil is immobilized as $Fe(OH)_3$. Rice absorbs Fe^{3+} via strategy II. In strategy II, *S*-adenosyl-l-*L*-methionine (SAM) is catalyzed by nicotianamine synthase (NAS) and produces nicotianamine (NA), which is an intermediate for the biosynthesis of MA family and a vital substance of nicotianamine aminotransferase (NAAT) [6]. Three rice NAS genes, *OsNAS1*, *OsNAS2*, and *OsNAS3*, have been identified to play different roles in Fe uptake and translocation in rice several years ago [30]. NAAT is a critical enzyme converting NA to 2′-deoxymugineic acid (DMA). Six rice *NAAT* genes (*OsNAAT1–6*) have been identified, but *OsNAAT1* was the only one highly upregulated under Fe deficiency, suggesting that *OsNAAT1* rather than *OsNAAT2–6* encodes the unique functional enzyme possessing NAAT activity [31]. Cheng et al. [10] demonstrated that *NAAT1* mutant was not able to produce DMA and take up Fe(III) efficiently.

In rice, gene that encodes DMA efflux transporters (OsTOM1) is highly expressed under Fe deficiency stress [32]. TOM1 transporter localizes at plasma membrane and mediates DMA secretion to rhizosphere, followed by Fe(III)-DMA complexes formation [32]. *Yellow stripe 1* (*YS1*) gene that encodes Fe(III)-MAs transporters was first acquired in maize [33], and *YS1*-like (*OsYSL*) genes in rice have been subsequently identified over decades. *OsYSL15* has been demonstrated to be upregulated in rice root and shoot under Fe deficiency to transport Fe(III)-DMA complexes [34]. In addition, *OsYSL* genes that encode transporters are also involved in Fe translocation within rice [35, 36]. Once transported into cytosol, Fe(III)-DMA is reduced by ascorbate to form Fe(II)-NA [37]. NA is not only an important intermediate for the biosynthesis of MAs but also a significant metal chelator that can take part in translocation of Fe within plants [38]. Fe may be excreted to the xylem in the form of Fe(II)–NA and shift to make complexes predominantly with citrate (Fe_2Cit_2, Fe_3Cit_3) and some with DMA [39]. The excretion of citrate from the root cells to the xylem is partly operated by OsFRDL1 (rice ferric reductase defective1-like) to enhance Fe-transport in the xylem as Fe(III)–citrate complexes [40].

In addition to Fe(III)-DMA uptake, rice also absorbs Fe(II) via iron-regulated transporter 1 (OsIRT1) and natural resistance-associated macrophage protein 1 (OsNRAMP1) under flooded conditions [16]. Seven rice *NRAMP* genes have been identified so far [41]. Recent research indicated that plasma membrane-localized protocatechuic acid (PCA) transporter and phenolic efflux zero1/2 (PEZ1/2) also participated in Fe uptake [42]. Such transporters played a role in absorbing apoplasmic precipitated Fe by secreting phenolics like PCA or caffeic acid. Suppression of *PEZ1/2* expression resulted in reduced Fe concentrations [42, 43].

Zn in both drained and flooded soil is largely ionized as (Zn^{2+}) though some Zn may be bound to organic substances and immobilized as Zn-sulfide (ZnS) in the anaerobic layer [7]. Zn-regulated transporters and iron-regulated transporters like protein (ZIP) family participate in Zn uptake [44]. OsZIP1-8 transporters have been characterized for Zn uptake and translocation in rice [7]. OsZIP1, OsZIP3, and OsZIP4 are required for Zn acquisition from rhizosphere [45]. OsZIP1 and OsZIP3 are mainly located in vascular bundles and epidermal cells, while OsZIP4 is located in apical meristem and phloem cells [45]. Under Zn deficiency stress, expression of *OsZIP1, OsZIP3*, and *OsZIP4* are upregulated [44, 45]. In addition, iron-regulated transporters (OsIRT) are also involved in Zn uptake [11]. OsIRT1 is characterized for Zn uptake in rice besides Fe acquisition [46, 47]. After transported into cytosol, Zn can be sequestered into the vacuoles via transporter OsZIP1 [11]. OsZIP1 is located to tonoplast and mediates influx of Zn into vacuoles [11]. MAs are also characterized for their role in Zn uptake in plants [7].

In comparison with Fe and Zn, Cd in paddy alkaline soil is present in the immobilized forms of $CdCO_3$ and humic acid-bound Cd [48]. Cd is immobilized as Cd-sulfide (CdS) and colloids-bound Cd in flooded soil [49]. Cd in drained acidic soil is ionized as Cd^{2+}, and drainage converts CdS to Cd^{2+} dramatically, increasing its availability to plants [11]. Cd uptake from rhizosphere is a dose-dependent process and exhibits saturable kinetic characteristics in rice [16, 50]. After analyzing the kinetics of Cd uptake by root in rice, Fujimaki et al. [50] suggested that uptake rate of Cd was proportional to Cd concentration in the culture solution within range from 0.05 to 100 nM, demonstrating a linear relationship between uptake rate and Cd concentration in a certain range. Ishikawa et al. [51] suggested that this kinetic characteristic of Cd uptake could be mediated by transporters. In fact, entrance of Cd into root cells via transporter OsNRAMP5 or OsIRT1 has been proved, and OsNRAMP5 is predominantly applied [52, 53]. *OsNRAMP5* expression is identified in root epidermis, exodermis, and outer layers of the cortex as well as in tissues around the xylem [54]. Knockout of *OsNRAMP5* reduces Cd accumulation both in straw and grains slightly [4]. Slamet-Loedin et al. [4] also proposed that downregulation of *OsNRAMP5* is a preferential strategy to decrease Cd uptake by root. OsNRAMP5 not only mediates Cd uptake but also regulates manganese (Mn) uptake and has relatively a minor effect on Fe uptake under Fe starvation [54]. In addition, higher expression of *OsNRAMP1* in root could enhance Cd accumulation in shoot of rice, indicating that OsNRAMP1 was also related with Cd uptake and transport [55].

When exposed to Cd contamination, rice supplied with Fe^{3+} generally represents weaker toxic symptoms than rice supplied with Fe^{2+}. The phenomenon is largely attributed to divalent metal transporters that are nonselective for Fe^{2+} uptake. Rice can uptake either Fe^{2+} or Cd^{2+} consequently (**Figure 1**). In contrast, Fe^{3+} transporters are selective for Fe^{3+} with no affinity for other divalent cations, which decreases Cd entrance into rice root to a great extent and reduces Cd toxicity accordingly (**Figure 1**).

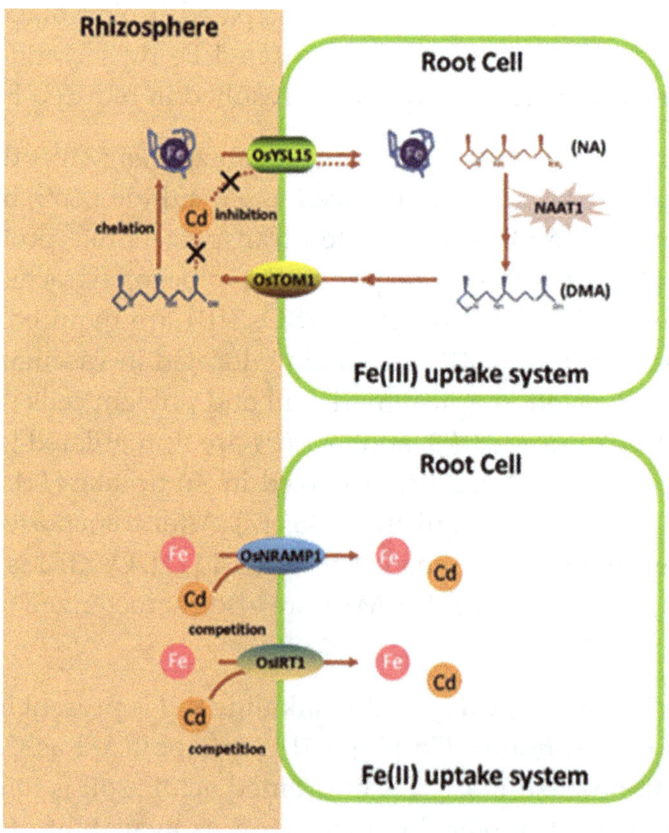

Figure 1. Schematic diagram of Fe and Cd uptake mechanisms in rice root. OsNRAMP1 and OsIRT1 can uptake either divalent ion Fe^{2+} or Cd^{2+}. NA is catalyzed by NAAT1 and finally forms DMA, which is secreted into rhizosphere via transporter OsTOM1 and chelates with Fe^{3+} to form complex. The complex is transported into cell via transporter OsYSL15.

After influx into cytosol, Cd is sequestered into the vacuoles via transporter OsHMA3 [56] and transiently stored in the form of complexes [15]. This pathway decreases Cd mobility in the cytosol and translocation from root to shoot [15, 57]. Enhancement of OsHMA3 activity has been found to increase storage of Cd in root and decrease the transport of Cd to the shoot and the final accumulation of Cd in rice grains [23]. *OsHMA3* is mainly expressed in root [24], and OsHMA3 localized at tonoplast belongs to P_{1B}-ATPases [58]. A high rate of root-to-shoot transport and subsequent accumulation in the grains of ^{107}Cd, which was administered from a culture solution, was observed in OsHMA3-depleted rice lines [51].

Lead forms various complexes with soil components, and only small parts of the lead present as these complexes in the soil solution are phytoavailable [59]. In soil, Pb may occur as a free metal ion or make complexes with inorganic constituents, such as HCO_3^-, CO_3^{2-}, SO_4^{2-}, and Cl^-. Pb also may exist as organic ligands with amino acids, fulvic acids, and humic acids [59]. Lead behavior in soil is mainly controlled by factors, such as pH [60], redox conditions [61], cation-exchange capacity, and organic and inorganic ligand levels [62]. Once adsorbed onto the rhizoderm roots surface, Pb may enter into the root passively, followed by translocating water streams while the mechanism by which Pb enters into root at the molecular level is still unknown. It is suggested that Pb enters into the root through several pathways, and a particular pathway is through ionic channels [59]. Several authors have demonstrated that Ca^{2+}-permeable channels are the main pathway by which Pb enters into root [63, 64]. Ca^{2+} from rhizosphere will compete with Pb^{2+} for common uptake position [25].

3. Translocation of Fe/Zn and Cd/Pb in rice

Following uptake by root, Fe, Zn and Cd are transported to shoot via xylem and phloem, where a large amount of vascular bundles exist [11]. This radial transport system includes symplasmic and apoplasmic pathways, but the former pathway is predominantly utilized as a result of impediment by Casparian strips occurring in apoplasmic pathway [65]. After Fe(II)-NA formation in the cytosol, Fe(II)-NA is transported to xylem and exchanges NA with citrate, transforming into Fe(III)-citrate preferentially [39, 40]. Fe in the xylem is largely in the form of Fe-citrate and then allocated to all leaves, whereas Fe in the phloem is mainly bound to DMA, citrate, and proteins [11]. The translocation of citrate from root pericycle cells to xylem is mediated by ferric reductase defective 1-like transporter OsFRDL1 [40].

Transportation of metals from plant root to shoot requires movement through the xylem [66] and is probably driven by transpiration [67]. Fe, Zn, Pb, and their chemical forms are in rice xylem and phloem saps, and phloem loading is the first step. OsYSL2 plays a role in Fe distribution in the phloem, localizing at the plasma membrane and is responsible for Fe(II)-NA or Mn(II)-NA transport, but not for Fe(III)-DMA transport [68]. Nozoye et al. [32] proposed that the NA efflux transporters (ENA1/2) are responsible for the efflux of NA into xylem or intracellular compartments in order to redistribute Fe. Under Fe deficiency, both *OsYSL2* and *ENA1* are strongly induced [68, 69]. In addition to transporter OsYSL2, OsYSL15 is considered to transport Fe(III)-DMA for phloem trafficking and expressed in the phloem companion cells

[6, 35, 43]. Thereafter, Fe is delivered to grains via phloem in forms of Fe(III)-DMA or binds to some citrates and proteins [11].

The Zn chemical forms in xylem sap are free ions and Zn partially bound to unidentified chelators [40], while the Zn in phloem sap was dominantly bound to NA [70]. Obata and Kitagishi [71] indicated that some Zn in the xylem (transpiration stream) is transferred to the phloem at the vegetative nodes in addition to the mobilization of Zn from mature leaves in rice. Xylem transfer cells have been found in rice vegetative nodes, and localized metal transporters may support the active xylem-to-phloem transfer of xylem sap Zn [72] and Cd [73].

After entered into root cells, part of Cd present as Cd-phytochelatin (Cd-PC) complexes are sequestered in the vacuoles, and the others are transported to xylem mediated by OsHMA2 transporter in root pericycle cells [11, 74, 75]. In the phloem, Cd primarily bounds to specific proteins and slightly to thiol-compounds [76]. In contrast to Fe translocation that is mainly derived from leaves by remobilization, xylem-to-phloem transfer system of Cd mainly occurs at the nodes [50]. In rice nodes, the diffuse vascular bundles (DVBs) that encircle the enlarged elliptical vascular bundles (EVBs) are connected to the panicle [77]. A study demonstrated that Cd was predominantly transported toward the panicle instead of other tissues at the panicle-initiation stage through the nodes and ultimately reached grains by positron-emitting ^{107}Cd tracer imaging system (PETIS) [50]. Node I, the uppermost node, is connected to both flag leaf and panicles. Yamaguchi et al. [77] found that Cd concentration was higher in node I than in blade, culm and panicle due to the accumulation of Cd. In addition, a low-affinity cation transporter (OsLCT1), which is highly expressed in the node I, also takes part in Cd transport to grains [16].

4. Effects of culture managements on Fe/Zn and Cd/Pb accumulation in rice grains

In rice seeds, Fe localizes to dorsal vascular bundle, aleurone layer and endosperm, and it localizes to the scutellum and vascular bundle of the scutellum of embryo [78]. Zn is distributed to all parts of the seed with a significantly high value for the aleurone layer and embryo [79]. Low Fe and Zn contents in rice are often restricted due to low available pools of Fe or Zn in soil. Enriching Fe or Zn concentration in grains through either fertilization or water management is referred to as agronomic biofortification, which is a short-term strategy for complementing the breeding programs.

Fe is abundant in mineral soil, but Fe deficiency still can occur in aerobic condition [80]. The major problem with Fe uptake is solubility. Fe in the soil (usually in the form of Fe^{2+}, either chelated or as a sulfate salt) is easily converted to unavailable Fe^{3+} under aerobic condition. Thus, application of Fe as fertilizer is not an effective strategy for increasing rice seed Fe [81]. Otherwise, foliar application is a better option to overcome Fe deficiency, increasing grains Fe and its bioavailability in rice [82]. In contrast, as soil changes from aerobic to anaerobic conditions after flooding, Fe-oxides are dissolved when the Fe^{3+} is reduced to Fe^{2+}, which weakens the oxide stability and increases its water solubility [83]. In fact, irrigation management in rice strongly influences soil redox potential, which affects the availability of Fe, so flooded soil nearly always has sufficient Fe for rice uptake [4].

Zn status and content in soil are the dominant factors restricting Zn content of rice seeds, followed by rice genotypes and fertilizers [84]. In aerobic condition, Zn mainly presents in soil in the form of ion Zn^{2+}. The application of Zn as fertilizer is effective in promoting rice growth and also in the fortification of rice with Zn [85, 86]. However, the availability of Zn decreases with flooding due to precipitation as insoluble zinc sulfide [87] or as insoluble carbonates mixtures [83]. Positive effects of soil Zn fertilization on grains Zn have been noticed primarily with aerobic water management [84]. In addition, foliar Zn application compared to soil Zn fertilization has been more effective in improving grains Zn concentration in flooded condition [88].

Although foliar application of Fe or Zn is more effective than soil application for increasing Fe or Zn content in rice grains, the efficiency of foliar applied Fe or Zn also depends on the application stages [89]. Late season foliar application of Zn or Fe at flowering or at early grain filling stage is more effective in improving grain Zn or Fe, respectively, than early season application [90, 91]. Although the level of Zn in grains is positively related with Fe, research showed that foliar fertilization of combined Fe and Zn fertilizers enhanced both grain Fe and Zn content without any antagonistic effects [82], indicating that fertilization of one element does not affect the grains concentration of the others [82, 92]. Totally, in order to increase both Fe and Zn content in rice grains under anaerobic or flooded conditions, the most effective fertilization strategy is a combination of foliar Zn and Fe spray soon after flowering or at early grain filling stage [4]. HarvestZinc Fertilizer Project started in 2008 and aimed at assessing the potential of Zn fertilizer in order to increase Zn content in cereal grains, especially in wheat and rice.

N fertilizer application has been reported to be related with Fe and Zn content in rice grains. Optimized N fertilizer application could increase grains Fe and Zn content in several crop species, including rice under sufficient Zn supply [92–94]. The reason is suggested as follows: (1) N nutrition promotes protein synthesis, which is a major sink for Fe and Zn [92]; (2) N nutrition enhances the expression of Zn and Fe transporter proteins, such as ZIP family transporters [92]; (3) N nutrition enhances the production of N compounds, such as NA and DMA [95]; (4) N nutrition increases Fe and Zn accumulation time by increasing vegetative growth and grain filling periods [4]. In contrast, N fertilizer can decrease rice grains Zn content under low Zn condition by increased biomass production and enhanced biological dilution [94]. In summary, optimized N fertilization application in rice production is very important to regulate Fe and Zn accumulation.

As a result of similar physical and chemical characteristics of Zn and Cd [96], Cd is mainly present as free Cd^{2+} in soil under aerobic condition regardless of soil redox potential [97], and the effect of flooding on Zn mobilization is indirect rather than direct compared with Fe. Cadmium in acidic soil is ionized as Cd^{+} [48] and moves toward root system and translocates within plants, resulting in Cd accumulation eventually [98]. In previous reports, phosphate (P) fertilizer was thought to increase rice Cd accumulation [99, 100]. Because Cd emerges in the rock phosphate used for P fertilizer production, P fertilizers generally contain significant amounts of Cd [98]. Nowadays, these relatively high-Cd phosphate rock sources have been avoided in the fertilizer. Sarwar et al. [98] reported that mono-ammonium-phosphate (MAP) could enhance Cd solubility and uptake by lowering soil pH. However, Bolan et al. [12] reported that P fertilizer can reduce Cd solubility by insoluble Cd formation such as $Cd(OH)_2$ or $Cd_3(PO_4)_2$. Yang et al. [101] proposed that P deprivation decreased rice Cd uptake by competitively increasing

Fe uptake and accumulation. P deprivation also enhances the sensitivity to Cd in rice plants by inhibiting biomass accumulation and reducing PCs synthesis deprivation.

An increasing number of evidences show that different N fertilizer forms and content affect Cd accumulation in rice. Sarwar et al. [98] reported that enhanced N application increased biomass production and reduced Cd toxicity to some extent due to dilution effect. N application increased soluble protein that could bind mobile Cd to immobile form. Different N fertilizer forms also have relationships with Cd uptake and accumulation in rice [102]. NH_4^+-containing fertilizer is considered to contribute to enhance Cd uptake [98]. NH_4^+-containing fertilizers acidify rhizosphere by proton excretion from root cells, exchanging with NH_4^+ and leading to low pH in soil [103]. In low pH soil, Cd moves toward root system and translocates within plants, resulting in Cd accumulation. In addition, NH_4^+ can trigger cell membrane depolarization and lead to influx of NH_4^+ into root cells, which accelerates translocation of Cd from root to shoot though this mechanism reduces Cd uptake in a certain way [98, 103]. In contrast, NO_3^--containing fertilizer causes simultaneous NO_3^- and proton absorption by root cells, leading to high pH [99], and cell membrane polarization caused by nitrate can produce Cd detoxification mechanism [98]. Nevertheless, Xie et al. [104] found that plants supplied with NO_3^- accumulated more Cd than NH_4^+ treatment by *Thlaspi caerulescens* in hydroponic experiment, suggesting that effects of NH_4^+ and NO_3^- on Cd uptake are not simply attributed to rhizosphere pH transformation or charge distribution of cell membrane. Yang et al. [105] found that rice fed with excess NO_3^- not only enhanced Fe uptake but also increased Cd uptake by upregulating the expression of *OsIRT1*. In addition, Wangstrand et al. [106] once proposed that application of N fertilizer was dependent on different growth stages and recommended that more N fertilizer should be applied at the vegetative stage while less N doses should be applied during the grain filling stage.

Besides P and N, other fertilizers are also related with Cd accumulation in rice. Fe is reported to remarkably increase Cd concentration in root and shoot of rice [107]. In contrast, a peculiar mechanism against Cd stress by application of Fe fertilizer is iron plague (IP) formation [96]. IP can serve as a barrier and prevent Cd from entering into root cells, resulting in reduced Cd accumulation while enhanced Fe concentration in rice [108]. Si application can reduce mobilization of Cd due to increased pH in soil [98], and complexes formation of Si with Cd is another mechanism for alleviating Cd toxicity in rice [96]. Application of S fertilizer may decrease Cd toxicity in the form of insoluble CdS, by which reduces mobility of Cd in soil [109]. S also participates in GSH and PCs biosynthesis. S increases Cd tolerance by forming Cd-PCs complexes and being transported into vacuoles in rice cells [96, 98, 110].

Pb accumulation in rice was wildly reported in Southeast Asian countries, such as China. Many researchers reported the toxic effects of Pb on rice growth and mineral absorption [111], but researches on reducing Pb accumulation in rice grains by water and fertilizer managements are still limited. Hu et al. [112] reported that Selenium (Se) application reduced Pb concentration in rice tissues but had no significant effect on Pb accumulation in brown rice grains. Soil remediation methods are applied to reduce Cd/Pb toxicity to some extent, including soil removal, replacement, inversion and flooded condition before and after heading [41, 113], but it is not easy to apply. In summary, more work is still needed to explore the effect of water and fertilizer management on Pb accumulation in rice grains.

5. Breeding and transgenic approaches to increase Fe/Zn and reduce Cd/Pb accumulation in rice grains

Since 1992, researchers at International Rice Research Institute (IRRI) have evaluated the genetic variability of Fe [114] and Zn [115] concentration in rice grains. The range in Fe and Zn contents of 939 varieties tested in one study were 7.5–24.4 μg g^{-1} for Fe, and 13.5–58.4 μg g^{-1} for Zn [116]. Among these 939 varieties, high grains Fe and Zn concentrations were identified, including Jalmagna, Zuchem, and Xua Bue Nuo [116]. HarvestPlus Challenge Program launched in 2004 supports biofortification of staple food crops, including rice for increased Fe, Zn, and vitamin A. However, iron biofortification of rice based on conventional breeding has met with only marginal success. The iron level achieved to date is still too low to address the required target level set by HarvestPlus (around 14 μg g^{-1}), indicating that iron biofortification in rice remains a challenge [117]. In contrast, a number of varieties of biofortified zinc in rice and wheat are now available or being tested in countries all over the world, including India, Bangladesh, and Pakistan. Complementing the traditional breeding efforts, modern transgenic technology provides perspectives for efficiently improving Fe and Zn content of rice grains to dietary significant level for humans' nutrition [118].

Recent attempts on the biofortification of Fe and Zn in rice grains using transgenic techniques have shown some positive results. Overexpression of the barley NA synthase gene *HvNAS1* in rice plants caused an increase in DMA and NA concentrations in root, shoot, and seed, accompanied with enhanced Fe, Zn, and Cu concentrations in grains [119]. Zheng et al. [120] indicated that biofortified rice with NA could efficiently enhance bioavailability by overexpression *OsNAS1* in rice endosperm. Alexander et al. [121] constructed three rice populations overexpressing *OsNAS1*, *OsNAS2* and *OsNAS3*. These constitutive overexpression of the *OsNAS* genes led to increased NA level, positively correlated with enhanced Zn concentration both in unpolished and polished grains, which reduces Zn nutrient loss to some extent due to polishing process. Goto et al. [122] demonstrated that high level of Fe in rice endosperm could be acquired by overexpression of ferritin. Swamy et al. [123] suggested that overexpression of the ferritin gene *OsFer2* in basmati rice (Pusa Sugandh II) was observed to accumulate higher levels of Fe and Zn. Combination of upregulated expression of ferritin with overproduction of NA can significantly enhance Fe and Zn content [88]. In addition, manipulation of specific transporters involved in Fe/Zn uptake and translocation is also considered to be promising approach for enhancing Fe/Zn content. Ishimaru et al. [68] introduced *OsYSL2* mediated by sucrose transporter (OsSUT1) promoter into rice plants due to location of OsSUT1 around endosperm, resulting in high concentration of Fe in polished rice. Overexpression of the Fe transporter gene *OsIRT1* or *OsYSL15*, the Fe deficiency-inducible bHLH transcription factor OsIRO2, and knockdown of the vacuolar Fe transporter gene *OsVIT1* or *OsVIT2*, were regarded as an effective approaches to increase the Fe concentration of seeds [124]. Overexpression of OsHMA3 enhanced the uptake of Zn by upregulating the ZIP family genes in the root [125]. OsHMA2 was involved in loading of Zn to the developing tissues in rice [75]. Quantitative trait locus (QTL) analysis is a useful approach to identify responsible genes for the respective transport processes [126]. Anuradna et al. [127] identified QTLs and candidate genes for Fe/Zn transport in rice seeds. *OsYSL1* and *OsMTP1* are responsible for Fe transport, while *OsARD2*, *OsIRT1*, *OsNAS1*, *OsNAS2* are responsible for

Zn transport. Ishikawa et al. [128] detected four QTLs (*qGZn9*, *qGZn10*, *qGZn2–1*, *qGZn2–2*) responsible for high Zn accumulation, and *qGZn9* showed the best effect, which provides valuable allele for breeding rice with high Zn level in grains.

Genotype dependence has been well observed for the accumulation of Cd in rice. More Cd accumulated in shoots and grains of *indica* cultivars than *japonica* cultivars [55, 129, 130]. Recently, Cd and Pb contents of 100 top Chinese rice cultivars were determined. The results also showed that *indica* accumulated more Cd than *japonica* [131]. Studies on rice screened for Cd-free, but Fe/Zn-rich cultivars have been an important issue in agricultural field. Significant efforts have been made on breeding of low-Cd accumulating rice cultivars in Japan, where Cd accumulation in rice grains has long been recognized as serious agricultural issues [41, 132]. Ishikawa et al. [132] identified and screened three low-Cd mutants (*lcd-kmt1*, *lcd-kmt2*, and *lcd-kmt3*) with *japonica* rice cultivar, Koshihikari, which acted as parent by the way of carbon ion-beam irradiation, showing that there were lower Cd concentration in grains of the three mutants than Koshihikari wide type (WT). Such three low-Cd mutants were attributed to mutations of *OsNRAMP5* responsible for Cd transport in rice by sequence analysis [132]. The three low-Cd mutants have different mutation sites in *OsNRAMP5*. An insertion of transposon *mPingA1*, which was activated by ion beam and preferred to insert into exon of *OsNRAMP5*, was identified in *lcd-kmt1*, resulting in nonfunction of OsNRAMP5 and decreased Cd accumulation in grains [132]. Similar results were observed in *lcd-kmt2* and *lcd-kmt3* due to a single-base pair deletion and a large deletion in *OsNRAMP5*, respectively [132]. Meanwhile, Ishikawa et al. [132] proposed that *lcd-kmt1* and *lcd-kmt2* were more promising for breeding program according to agronomic traits as a consequence of earlier heading and smaller plant size than Koshihikari WT in *lcd-kmt3* [132]. In addition, Abe et al. [133] developed a novel population composed of 46 chromosome segment substitution lines (CSSLs), in which LAC23 served as donor segments and were substituted into Koshihikari. LAC23 could result in lower grain-to-straw ratio than Koshihikari [133]. Therefore, cultivars containing LAC23 performed low Cd content in grains [133]. As for breeding Fe/Zn-rich cultivars, Olive et al. [134] bred high level of ferritin cultivars with rice mega variety IR64 as background and introducing ferritin into endosperm increased Fe content in grains [132, 133]. IR64 mutants obtained from sodium azide treatment were reported to have high Zn level [123]. Booyaves et al. [135] expressed Arabibopsis IRT1 (*AtIRT1*) in high-iron NFP rice lines, which expressed *NICOTIANAMINE SYNTHASE* (*AtNAS1*) and *FERRITIN*, enhancing Fe contents in both unpolished and polished grains.

In addition, QTL analysis was also applied to identify responsible genes for Cd transport. QTL for Cd concentration in Anjana Dhan (*indica* rice cultivar) is identified on chromosome 7, responsive gene for which is *OsHMA3* [23, 55, 126]. Abe et al. [136] introduced a nonfunctional allele of *OsHMA3* from Jarian (*indica* rice cultivar) into Koshihikari (*japonica* rice cultivar) by marker-assisted selection, and these plants showed reduced Cd uptake from soil. Suppression of *OsLCT1* expression can decrease grains Cd accumulation by RNAi without influencing nutrient accumulation. On the contrary, Fe content in the brown rice is remarkably higher [16], suggesting that RNAi-mediated *OsLCT1* suppression in rice is a promising approach to establish "high Fe but low-Cd-rice." Furthermore, T-DNA-mediated *OsLCD* knockout mutant showed reduced grain Cd accumulation while having no negative effect on grain yield. Thus, the *lcd* mutant might be a probable mutant line for further research [137].

Breeding low Pb cultivars is also considered to reduce Pb contamination. Developing rice cultivars with low Pb mobilization within root and translocation toward aerial parts to the minimum extent may be a better option to cultivate rice in Pb tainted soils. Li et al. [138] screened three cultivars (Tianyou196, Wufengyou2168, and Guinongzhan) with low Pb level in brown rice. Furthermore, Ashraf et al. [139] compared Meixiangzhan (MXZ-2), Xiangyaxiangzhan(XYXZ), Guixiangzhan (GXZ), Basmati-385 (B-385), and Nongxiang-18 (NX-18) to four different Pb concentrations, indicating that GXZ proved better able to tolerate Pb stress than all other rice cultivars, which are therefore suggested for use in future breeding programs for paddy fields contaminated by Pb.

6. Conclusions and perspectives

Fe and Zn are essential nutrients for humans, but Cd and Pb are toxic at high levels for humans. All these metals accumulated in the grains of rice, a staple cereal worldwide. Compared with Pb, significant progress has been made in investigating the mechanisms for Fe, Zn, and Cd uptake and accumulation in rice grains. These basic discoveries provide us with the increasing possibility to establish high Fe/Zn and low Cd/Pb rice. Here, we summarized a strategy scheme for producing biofortified Fe/Zn but low Cd/Pb rice cultivars as follows (**Figure 2**). On the one hand, scientists have screened or bred nontransgenic rice cultivars with high Fe/Zn or/and low

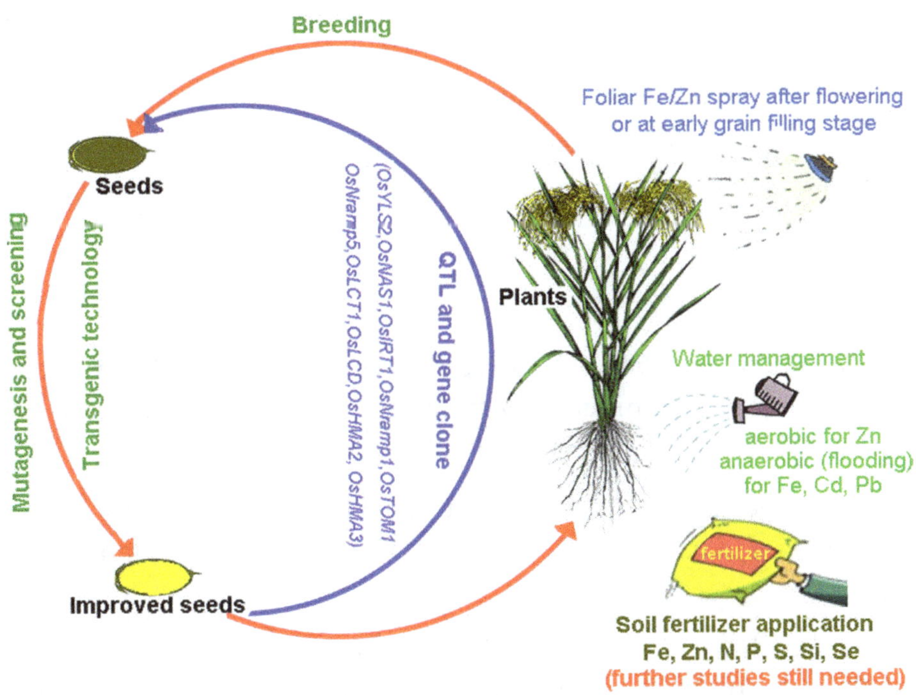

Figure 2. Schematic diagram of strategies for enhancing accumulation of iron and zinc while minimizing cadmium and lead in rice. Based on functional QTLs or genes, scientists screen or breed nontransgenic rice cultivars with high Fe/Zn or/and low Cd/Pb in grains. Modern transgenic technology provides perspectives for efficiently improving Fe/Zn content and decreasing Cd/Pb content in rice grains. Water and fertilizers management are also significantly related with increased Fe/Zn and decreased Cd/Pb in rice grains.

Cd/Pb in grains based on functional QTLs or genes. These cultivars show no agriculturally or economically adverse traits and can be applied sooner. On the other hand, modern transgenic technology provides perspectives for efficiently improving Fe/Zn content and decreasing Cd/Pb content in rice grains to dietary significant levels for humans' nutrition (As to Pb, more researches on QTL and genes still needed). Besides improving rice seeds, water and fertilizer management is also significantly related with increased Fe/Zn and decreased Cd/Pb in rice grains. More studies are still needed to optimize irrigation time, fertilizer categories, dosage, and application stages. In addition, although it is available to establish rice cultivars with high Fe or Zn content, or establish rice cultivars with low Cd or Pb separately, interactions among these metals need to be better understood, and more steps are still needed to cultivate rice with all these merits and without decreasing rice production.

Acknowledgements

This research was supported by Zhejiang Provincial Natural Science Foundation of China under Grant No. LY18C130011 and LY15C130007, Science Foundation of Zhejiang Sci-Tech University, Foundation of Zhejiang Provincial Top Key Discipline of Biology, and Foundation of Zhejiang Provincial Key Discipline of Botany.

Author details

Lei Gao and Jie Xiong*

*Address all correspondence to: jiexiong@zju.edu.cn

College of Life Sciences, Zhejiang Sci-Tech University, Hangzhou, People's Republic of China

References

[1] Gómez-Galera S, Rojas E, Sudhakar D, Zhu C, Pelacho AM, Capell T, Christou P. Critical evaluation of strategies for mineral fortification of staple food crops. Transgenic Research. 2010;**19**:165-180. DOI: 10.1007/s11248-009-9311-y

[2] Sperotto RA, Ricachenevsky FK, Waldow Vde A, Fett JP. Iron biofortification in rice: It's a long way to the top. Plant Science. 2012;**190**:24-39. DOI: 10.1016/j.plantsci.2012.03.004

[3] Welch RM. The impact of mineral nutrients in food crops on global human health. In: Horst WJ et al., editors. Progress in Plant Nutrition: Plenary Lectures of the XIV International Plant Nutrition Colloquium. Developments in Plant and Soil Sciences. Dordrecht: Springer; 2002

[4] Slamet-Loedin IH, Johnson-Beebout SE, Impa S, Tsakirpaloglou N. Enriching rice with Zn and Fe while minimizing cd risk. Frontiers in Plant Science. 2015;**6**:121. DOI: 10.3389/fpls.2015.00121

[5] Alloway B. Soil factors associated with zinc deficiency in crops and humans. Environmental Geochemistry and Health. 2009;**31**:537-548. DOI: 10.1007/s10653-009-9255-4

[6] Kobayashi T, Nakanishi H, Nishizawa NK. Recent insights into iron homeostasis and their application in graminaceous crops. Proceedings of the Japan Academy. Series B, Physical and Biological Sciences. 2010;**86**:900-913. DOI: 10.2183/pjab.86.900

[7] Bashir K, Ishimaru Y, Nishizawa NK. Molecular mechanisms of zinc uptake and translocation in rice. Plant and Soil. 2012;**361**:189-201. DOI: 10.1007/s11104-012-1240-5

[8] Conte SS, Walker EL. Transporters contributing to iron trafficking in plants. Molecular Plant. 2011;**4**:464-476. DOI: 10.1093/mp/ssr015

[9] Kobayashi T, Nishizawa NK. Iron uptake, translocation, and regulation in higher plants. Annual Review of Plant Biology. 2012;**63**:131-152. DOI: 10.1146/annurev-arplant-042811-105522

[10] Cheng L, Wang F, Shou H, Huang F, Zheng L, He F, Li J, Zhao FJ, Ueno D, Ma JF, Wu P. Mutation in nicotianamine aminotransferase stimulated the Fe(II) acquisition system and led to iron accumulation in rice. Plant Physiology. 2007;**145**:1647-1657. DOI: 10.1104/pp.107.107912

[11] Yoneyama T, Ishikawa S, Fujimaki S. Route and regulation of zinc, cadmium, and iron transport in rice plants (*Oryza sativa* L.) during vegetative growth and grain filling: Metal transporters, metal speciation, grain cd reduction and Zn and Fe biofortification. International Journal of Molecular Sciences. 2015;**16**:19111-19129. DOI: 10.3390/ijms160819111

[12] Bolan N, Mahimairaja S, Kunhikrishnan A, Naidu R. Sorption-bioavailability nexus of arsenic and cadmium in variable-charge soils. Journal of Hazardous Materials. 2013;**261**: 725-732. DOI: 10.1016/j.jhazmat.2012.09.074

[13] He S, He Z, Yang X, Stoffella PJ, Baligar VC. Chapter four – Soil biogeochemistry, plant physiology, and phytoremediation of cadmium-contaminated soils. Advances in Agronomy. 2015;**134**:135-225. DOI: 10.1016/bs.agron.2015.06.005

[14] Song W, Chen S, Chen J, Song N, Li N, Liu B. Variation of cd concentration in various rice cultivars and derivation of cadmium toxicity thresholds for paddy soil by species-sensitivity distribution. Journal of Integrative Agriculture. 2015;**14**:1845-1854. DOI: 10.1016/S2095-3119(14)60926-6

[15] Choppala G, Saifullah Bolan N, Bibi S, Iqbal M, Rengel Z, Kunhikrishnan A, Ashwath N, Yong SO. Cellular mechanisms in higher plants governing tolerance to cadmium toxicity. Critical Reviews in Plant Sciences. 2014;**33**:374-391. DOI: 10.1080/07352689.2014.903747

[16] Uraguchi S, Kamiya T, Sakamoto T, Kasai K, Sato Y, Nagamura Y, Yoshida A, Kyozuka J, Ishikawa S, Fujiwara T. Low-affinity cation transporter (OsLCT1) regulates cadmium transport into rice grains. Proceedings of the National Academy of Sciences of the United States of America. 2011;**108**:20959-22064. DOI: 10.1073/pnas.1116531109

[17] Tsukahara T, Ezaki T, Moriguchi J, Furuki K, Shimbo S, Matsuda-Inoguchi N, Ikeda M. Rice as the most influential source of cadmium intake among general Japanese population. Science of the Total Environment. 2003;**305**:41-51. DOI: 10.1016/S0048-9697(02)00475-8

[18] DalCorso G, Farinati S, Furini A. Regulatory networks of cadmium stress in plants. Plant Signaling & Behavior. 2010;**5**:663-667. DOI: 10.4161/psb.5.6.11425

[19] Lang I, Wernitznig S. Sequestration at the cell wall and plasma membrane facilitates zinc tolerance in the moss *Pohlia drummondii*. Environmental and Experimental Botany. 2011;**74**:186-193. DOI: 10.1016/j.envexpbot.2011.05.018

[20] Schwab AP, He Y, Banks MK. The influence of organic ligands on the retention of lead in soil. Chemosphere. 2005;**61**:856-866. DOI: 10.1016/j.chemosphere.2005.04.098

[21] Dong J, Mao WH, Zhang GP, FB W, Cai Y. Root excretion and plant tolerance to cadmium toxicity – A review. Plant, Soil and Environment. 2007;**53**:193-200

[22] Yadav R, Arora P, Kumar S, Chaudhury A. Perspectives for genetic engineering of poplars for enhanced phytoremediation abilities. Ecotoxicology. 2010;**19**:1574-1588. DOI: 10.1007/s10646-010-0543-7

[23] Ueno D, Yamaji N, Kono I, Huang CF, Ando T, Yano M, Ma JF. Gene limiting cadmium accumulation in rice. Proceedings of the National Academy of Sciences of the United States of America. 2010;**107**:16500-16505. DOI: 10.1073/pnas.1005396107

[24] Miyadate H, Adachi S, Hiraizumi A, Tezuka K, Nakazawa N, Kawamoto T, Katou K, Kodama I, Sakurai K, Takahashi H, Satoh-Nagasawa N, Watanabe A, Fujimura T, Akagi H. OsHMA3, a P1B-type of ATPase affects root-to-shoot cadmium translocation in rice by mediating efflux into vacuoles. The New Phytologist. 2011;**189**:190-199. DOI: 10.1111/j.1469-8137.2010.03459.x

[25] Ashraf U, Kanu AS, Mo Z, Hussain S, Anjum SA, Khan I, Abbas RN, Tang X. Lead toxicity in rice: Effects, mechanisms, and mitigation strategies – A mini review. Environmental Science and Pollution Research International. 2015;**22**:18318-11832. DOI: 10.1007/s11356-015-5463-x

[26] Ashraf MA, Wajid A, Mahmood K, Maah MJ, Yusoff I. Removal of heavy metals from aqueous solution by using mango biomass. African Journal of Biotechnology. 2011;**10**: 2163-2177

[27] Gupta DK, Huang HG, Corpas FJ. Lead tolerance in plants: Strategies for phytoremediation. Environmental Science and Pollution Research International. 2013;**20**:2150-2161. DOI: 10.1007/s11356-013-1485-4

[28] Davis JM, Elias RW, Grant LD. Current issues in human lead exposure and regulation of lead. Neurotoxicology. 1993;**14**:15-27

[29] Liu J, Li K, Xu J, Zhang Z, Ma T, Lu X, Yang J, Zhu Q. Lead toxicity, uptake, and translocation in different rice cultivars. Plant Science. 2003;**165**:793-802. DOI: 10.1016/S0168-9452(03)00273-5

[30] Inoue H, Higuchi K, Takahashi M, Nakanishi H, Mori S, Nishizawa NK. Three rice nicotianamine synthase genes, *OsNAS1*, *OsNAS2*, and *OsNAS3* are expressed in cells involved in long-distance transport of iron and differentially regulated by iron. The Plant Journal. 2003;**36**:366-381. DOI: 10.1046/j.1365-313X.2003.01878.x

[31] Inoue H, Takahashi M, Kobayashi T, Suzuki M, Nakanishi H, Mori S, Nishizawa NK. Identification and localisation of the rice nicotianamine aminotransferase gene *OsNAAT1* expression suggests the site of phytosiderophore synthesis in rice. Plant Molecular Biology. 2008;**66**:193-203. DOI: 10.1007/s11103-007-9262-8

[32] Nozoye T, Nagasaka S, Kobayashi T, Takahashi M, Sato Y, Sato Y, Uozumi N, Nakanishi H, Nishizawa NK. Phytosiderophore efflux transporters are crucial for iron acquisition in graminaceous plants. The Journal of Biological Chemistry. 2011;**286**:5446-5454. DOI: 10.1074/jbc.M110.180026

[33] Curie C, Panaviene Z, Loulergue C, Dellaporta SL, Briat JF, Walker EL. Maize *yellow stripe1* encodes a membrane protein directly involved in Fe(III) uptake. Nature. 2001;**409**: 346-349. DOI: 10.1038/35053080

[34] Inoue H, Kobayashi T, Nozoye T, Takahashi M, Kakei Y, Suzuki K, Nakazono M, Nakanishi H, Mori S, Nishizawa NK. Rice OsYSL15 is an iron-regulated iron(III)-deoxymugineic acid transporter expressed in the roots and is essential for iron uptake in early growth of the seedlings. The Journal of Biological Chemistry. 2009;**284**:3470-3479. DOI: 10.1074/jbc.M806042200

[35] Kakei Y, Ishimaru Y, Kobayashi T, Yamakawa T, Nakanishi H, Nishizawa NK. OsYSL16 plays a role in the allocation of iron. Plant Molecular Biology. 2012;**79**:583-594. DOI: 10.1007/s11103-012-9930-1

[36] Koike S, Inoue H, Mizuno D, Takahashi M, Nakanishi H, Mori S, Nishizawa NK. OsYSL2 is a rice metal-nicotianamine transporter that is regulated by iron and expressed in the phloem. The Plant Journal. 2004;**39**:415-424. DOI: 10.1111/j.1365-313X.2004.02146.x

[37] Weber G, von Wirén N, Hayen H. Investigation of ascorbate-mediated iron release from ferric phytosiderophores in the presence of nicotianamine. Biometals. 2008;**21**:503-513. DOI: 10.1007/s10534-008-9137-8

[38] Takahashi H, Kosaka M, Watanabe Y, Nakade K, Fukuyama Y. Synthesis and neuroprotective activity of bergenin derivatives with antioxidant activity. Bioorganic & Medicinal Chemistry. 2003;**11**:1781-1788. DOI: 10.1016/S0968-0896(02)00666-1

[39] Ariga T, Hazama K, Yanagisawa S, Yoneyama T. Chemical forms of iron in xylem sap from graminaceous and non-graminaceous plants. Soil Science and Plant Nutrition. 2014;**60**:460-469. DOI: 10.1080/00380768.2014.922406

[40] Yokosho K, Yamaji N, Ueno D, Mitani N, Ma JF. OsFRDL1 is a citrate transporter required for efficient translocation of iron in rice. Plant Physiology. 2009;**149**:297-305. DOI: 10.1104/pp.108.128132

[41] Uraguchi S, Fujiwara T. Cadmium transport and tolerance in rice: Perspectives for reducing grain cadmium accumulation. Rice. 2012;**5**:5. DOI: 10.1186/1939-8433-5-5

[42] Ishimaru Y, Kakei Y, Shimo H, Bashir K, Sato Y, Sato Y, Uozumi N, Nakanishi H, Nishizawa NK. A rice phenolic efflux transporter is essential for solubilizing precipitated apoplasmic iron in the plant stele. The Journal of Biological Chemistry. 2011;**286**:24649-24655. DOI: 10.1074/jbc.M111.221168

[43] Kobayashi T, Nakanishi Itai R, Nishizawa NK. Iron deficiency responses in rice roots. Rice. 2014;**7**:27. DOI: 10.1186/s12284-014-0027-0

[44] Ishimaru Y, Suzuki M, Kobayashi T, Takahashi M, Nakanishi H, Mori S, Nishizawa NK. OsZIP4, a novel zinc-regulated zinc transporter in rice. Journal of Experimental Botany. 2005;**56**:3207-3214. DOI: 10.1093/jxb/eri317

[45] Ramesh SA, Shin R, Eide DJ, Schachtman DP. Differential metal selectivity and gene expression of two zinc transporters from rice. Plant Physiology. 2003;**133**:126-134. DOI: 10.1104/pp.103.026815

[46] Lee S, An G. Over-expression of OsIRT1 leads to increased iron and zinc accumulations in rice. Plant, Cell & Environment. 2009;**32**:408-416. DOI: 10.1111/j.1365-3040.2009.01935.x

[47] Nouet C, Motte P, Hanikenne M. Chloroplastic and mitochondrial metal homeostasis. Trends in Plant Science. 2011;**16**:395-404. DOI: 10.1016/j.tplants.2011.03.005

[48] Khaokaew S, Chaney RL, Landrot G, Gindervogel M, Sparks DL. Speciation and release kinetics of cadmium in an alkaline paddy soil under various flooding periods and draining conditions. Environmental Science & Technology. 2011;**45**:4249-4255. DOI: 10.1021/es103971y

[49] Hashimoto Y, Yamaguchi N. Chemical speciation of cadmium and sulfur K-edge XANES spectroscopy in flooded paddy soils amended with zerovalent iron. Soil Science Society of America Journal. 2013;**77**:1189. DOI: 10.2136/sssaj2013.01.0038

[50] Fujimaki S, Suzui N, Ishioka NS, Kawachi N, Ito S, Chino M, Nakamura S. Tracing cadmium from culture to spikelet: Noninvasive imaging and quantitative characterization of absorption, transport, and accumulation of cadmium in an intact rice plant. Plant Physiology. 2010;**152**:1796-1806. DOI: 10.1104/pp.109.151035

[51] Ishikawa S, Suzui N, Ito-Tanabata S, Ishii S, Igura M, Abe T, Kuramata M, Kawachi N, Fujimaki S. Real-time imaging and analysis of differences in cadmium dynamics in rice cultivars (*Oryza sativa*) using positron-emitting ^{107}Cd tracer. BMC Plant Biology. 2011;**11**:172. DOI: 10.1186/1471-2229-11-172

[52] Nakanishi H, Ogawa I, Ishimaru Y, Mori S, Nishizawa NK. Iron deficiency enhances cadmium uptake and translocation mediated by the Fe^{2+} transporters OsIRT1 and OsIRT2 in rice. Soil Science and Plant Nutrition. 2006;**52**:464-469. DOI: 10.1111/j.1747-0765.2006.00055.x

[53] Sasaki A, Yamaji N, Yokosho K, Ma JF. Nramp5 is a major transporter responsible for manganese and cadmium uptake in rice. The Plant Cell. 2012;**24**:2155-2167. DOI: 10.1105/tpc

[54] Ishimaru Y, Takahashi R, Bashir K, Shimo H, Senoura T, Sugimoto K, Ono K, Yano M, Ishikawa S, Arao T, Nakanishi H, Nishizawa NK. Characterizing the role of rice NRAMP5 in manganese, iron and cadmium transport. Scientific Reports. 2012;**2**:286. DOI: 10.1038/srep00286

[55] Takahashi R, Ishimaru Y, Senoura T, Shimo H, Ishikawa S, Arao T, Nakanishi H, Nishizawa NK. The OsNRAMP1 iron transporter is involved in cd accumulation in rice. Journal of Experimental Botany. 2011;**62**:4843-4850. DOI: 10.1093/jxb/err136

[56] Takahashi R, Bashir K, Ishimaru Y, Nishizawa NK, Nakanishi H. The role of heavy-metal ATPases, HMAs, in zinc and cadmium transport in rice. Plant Signaling & Behavior. 2012;**7**:1605-1607. DOI: 10.4161/psb.22454

[57] Shahid M, Dumat C, Khalid S, Niazi NK. Cadmium bioavailability, uptake, toxicity and detoxification in soil-plant system. Reviews of Environmental Contamination and Toxicology. 2017;**241**:73-137. DOI: 10.1007/398_2016_8

[58] Williams LE, Mills RF. P_1B-ATPases – An ancient family of transition metal pumps with diverse functions in plants. Trends in Plant Science. 2005;**10**:491-502. DOI: 10.1016/j.tplants.2005.08.008

[59] Pourrut B, Shahid M, Dumat C, Winterton P, Pinelli E. Lead uptake, toxicity, and detoxification in plants. Vol. 213. In: Whitacre D, editor. Reviews of Environmental Contamination and Toxicology (Continuation of Residue Reviews). New York: Springer; 2011. pp. 113-136

[60] Lawal OS, Sanni AR, Ajayi IA, Rabiu OO. Equilibrium, thermodynamic and kinetic studies for the biosorption of aqueous lead(II) ions onto the seed husk of *Calophyllum inophyllum*. Journal of Hazardous Materials. 2010;**177**:829-835. DOI: 10.1016/j.jhazmat.2009.12.108

[61] Tabelin CB, Igarashi T. Mechanisms of arsenic and lead release from hydrothermally altered rock. Journal of Hazardous Materials. 2009;**169**:980-990. DOI: 10.1016/j.jhazmat.2009.04.049

[62] Shahid M, Pinelli E, Pourrut B, Silvestre J, Dumat C. Lead-induced genotoxicity to *Vicia faba* L. roots in relation with metal cell uptake and initial speciation. Ecotoxicology and Environmental Safety. 2011;**74**:78-84. DOI: 10.1016/j.ecoenv.2010.08.037

[63] Wang HH, Shan XQ, Wen B, Owens G, Fang J, Zhang SZ. Effect of indole-3-acetic acid on lead accumulation in maize (*Zea mays* L.) seedlings and the relevant antioxidant response. Environmental and Experimental Botany. 2007;**61**:246-253. DOI: 10.1016/j.envexpbot.2007.06.004

[64] Pourrut B, Perchet G, Silvestre J, Cecchi M, Guiresse M, Pinelli E. Potential role of NADPH-oxidase in early steps of lead-induced oxidative burst in *Vicia faba* roots. Journal of Plant Physiology. 2008;**165**:571-579. DOI: 10.1016/j.jplph.2007.07.016

[65] Enstone DE, Peterson CA, Ma F. Root endodermis and exodermis: Structure, function, and responses to the environment. Journal of Plant Growth Regulation. 2002;**21**:335-351. DOI: 10.1007/s00344-003-0002-2

[66] Verbruggen N, Hermans C, Schat H. Mechanisms to cope with arsenic or cadmium excess in plants. Current Opinion in Plant Biology. 2009;**12**:364-372. DOI: 10.1016/j.pbi.2009.05.001

[67] Liao YC, Chien SW, Wang MC, Shen Y, Hung PL, Das B. Effect of transpiration on Pb uptake by lettuce and on water soluble low molecular weight organic acids in rhizosphere. Chemosphere. 2006;**65**:343-351. DOI: 10.1016/j.chemosphere.2006.02.010

[68] Ishimaru Y, Masuda H, Bashir K, Inoue H, Tsukamoto T, Takahashi M, Nakanishi H, Aoki N, Hirose T, Ohsugi R, Nishizawa NK. Rice metal-nicotianamine transporter,

OsYSL2, is required for the long-distance transport of iron and manganese. The Plant Journal. 2010;**62**:379-390. DOI: 10.1111/j.1365-313X.2010.04158.x

[69] Ogo Y, Kakei Y, Itai RN, Kobayashi T, Nakanishi H, Takahashi H, Nakazono M, Nishizawa NK. Spatial transcriptomes of iron-deficient and cadmium-stressed rice. The New Phytologist. 2014;**201**:781-794. DOI: 10.1111/nph.12577

[70] Nishiyama R, Kato M, Nagata S, Yanagisawa S, Yoneyama T. Identification of Zn-nicotianamine and Fe-2′-Deoxymugineic acid in the phloem sap from rice plants (*Oryza sativa* L.). Plant & Cell Physiology. 2012;**53**:381-390. DOI: 10.1093/pcp/pcr188

[71] Obata H, Kitagishi K. Investigation on pathway of Zn in vegetative node of rice plants by autoradiography: Behavior of zinc in rice plants (III). Japanese Journal of Soil Science and Plant Nutrition. 1980;**51**:297-301

[72] Nishiyama R, Tanoi K, Yanagisawa S, Yoneyama T. Quantification of zinc transport via the phloem to the grain in rice plants (*Oryza sativa* L.) at early grain-filling by a combination of mathematical modeling and ^{65}Zn tracing. Soil Science & Plant Nutrition. 2013;**59**:750-755. DOI: 10.1080/00380768.2013.819774

[73] Rodda MS, Li G, Reid RJ. The timing of grain cd accumulation in rice plants: The relative importance of remobilisation within the plant and root cd uptake post-flowering. Plant and Soil. 2011;**347**:105-114. DOI: 10.1007/s11104-011-0829-4

[74] Takahashi R, Ishimaru Y, Shimo H, Ogo Y, Senoura T, Nishizawa NK, Nakanishi H. The OsHMA2 transporter is involved in root-to-shoot translocation of Zn and cd in rice. Plant, Cell & Environment. 2012;**35**:1948-1957. DOI: 10.1111/j.1365-3040.2012.02527.x

[75] Yamaji N, Xia J, Mitaniueno N, Yokosho K, Ma JF. Preferential delivery of zinc to developing tissues in rice is mediated by P-type heavy metal ATPase OsHMA2. Plant Physiology. 2013;**162**:927-939. DOI: 10.1104/pp.113.216564

[76] White PJ, Broadley MR. Physiological limits to zinc biofortification of edible crops. Frontiers in Plant Science. 2011;**2**:80. DOI: 10.3389/fpls.2011.00080

[77] Yamaguchi N, Ishikawa S, Abe T, Baba K, Arao T, Terada Y. Role of the node in controlling traffic of cadmium, zinc, and manganese in rice. Journal of Experimental Botany. 2012;**63**:2729-2737. DOI: 10.1093/jxb/err455

[78] Bashir K, Takahashi R, Nakanishi H, Nishizawa NK. The road to micronutrient biofortification of rice: Progress and prospects. Frontiers in Plant Science. 2013;**4**:15. DOI: 10.3389/fpls.2013.00015

[79] Takahashi M, Nozoye T, Kitajima N, Fukuda N, Hokura A, Terada Y, Nakai I, Ishimaru Y, Kobayashi T, Nakanishi H. In vivo analysis of metal distribution and expression of metal transporters in rice seed during germination process by microarray and X-ray fluorescence imaging of Fe, Zn, Mn, and cu. Plant and Soil. 2009;**325**:39-51. DOI: 10.1007/s11104-009-0045-7

[80] Zuo Y, Zhang F. Soil and crop management strategies to prevent iron deficiency in crops. Plant and Soil. 2010;**339**:83-95. DOI: 10.1007/s11104-010-0566-0

[81] Briat JF, Ravet K, Arnaud N, Duc C, Boucherez J, Touraine B, Cellier F, Gaymard F. New insights into ferritin synthesis and function highlight a link between iron homeostasis and oxidative stress in plants. Annals of Botany. 2010;**105**:811-822. DOI: 10.1093/aob/mcp128

[82] Wei Y, Shohag MJ, Yang X, Yibin Z. Effects of foliar iron application on iron concentration in polished rice grain and its bioavailability. Journal of Agricultural and Food Chemistry. 2012;**60**:11433-11439. DOI: 10.1021/jf3036462

[83] Kirk G. The Biogeochemistry of Submerged Soils. West Sussex, England: Wiley; 2004

[84] Wang YY, Wei YY, Dong LX, Lu LL, Feng Y, Zhang J, Pan FS, Yang XE. Improved yield and Zn accumulation for rice grain by Zn fertilization and optimized water management. Journal of Zhejiang University. Science B. 2014;**15**:365-374. DOI: 10.1631/jzus.B1300263

[85] Neue HU, Quijano C, Senadhira D, Setter T. Strategies for dealing with micronutrient disorders and salinity in lowland rice systems. Field Crops Research. 1998;**56**:139-155. DOI: 10.1016/S0378-4290(97)00125-1

[86] Yang X, Ye ZQ, Shi CH, Zhu ML, Graham RD. Genotypic differences in concentrations of iron, manganese, copper, and zinc in polished rice grains. Journal of Plant Nutrition. 1998;**21**:1453-1462. DOI: 10.1080/01904169809365495

[87] Bostick BC, Hansel CM, Fendorf S. Seasonal fluctuations in zinc speciation within a contaminated wetland. Environmental Science & Technology. 2001;**35**:3823-3829. DOI: 10.1021/es010549d

[88] Wirth J, Poletti S, Aeschlimann B, Yakandawala N, Drosse B, Osorio S, Tohge T, Fernie AR, Günther D, Gruissem W, Sautter C. Rice endosperm iron biofortification by targeted and synergistic action of nicotianamine synthase and ferritin. Plant Biotechnology Journal. 2009;**7**:631-644. DOI: 10.1111/j.1467-7652.2009.00430.x

[89] Cakmak I. Enrichment of fertilizers with zinc: An excellent investment for humanity and crop production in India. Journal of Trace Elements in Medicine and Biology. 2009;**23**:281-289. DOI: 10.1016/j.jtemb.2009.05.002

[90] Phattarakul N, Rerkasem B, Li LJ, Wu LH, Zou CQ, Ram H, Sohu VS, Kang BS, Surek H, Kalayci M. Biofortification of rice grain with zinc through zinc fertilization in different countries. Plant and Soil. 2012;**361**:131-141. DOI: 10.1007/s11104-012-1211-x

[91] Mabesa RL, Impa SM, Grewal D, Johnson-Beebout SE. Contrasting grain-Zn response of biofortification rice (*Oryza sativa* L.) breeding lines to foliar Zn application. Field Crops Research. 2013;**149**:223-233. DOI: 10.1016/j.fcr.2013.05.012

[92] Cakmak I, Kalayci M, Kaya Y, Torun AA, Aydin N, Wang Y, Arisoy Z, Erdem H, Yazici A, Gokmen O, Ozturk L, Horst WJ. Biofortification and localization of zinc in wheat grain. Journal of Agricultural and Food Chemistry. 2010;**58**:9092-9102. DOI: 10.1021/jf101197h

[93] Kutman UB, YildiZ B, Ozturk L, Cakmak I. Biofortification of durum wheat with zinc through soil and foliar applications of nitrogen. Cereal Chemistry. 2010; **87**:1-9. DOI: 10.1094/CCHEM-87-1-0001

[94] Jin Z, Wu L, Wang M. Can iron and zinc in rice grains (*Oryza sativa* L.) be biofortified with nitrogen fertilisation under pot conditions? Journal of Science and Food Agriculture. 2008;**88**:1172-1177. DOI: 10.1002/jsfa.3194

[95] Haydon MJ, Cobbett CS. Transporters of ligands for essential metal ions in plants. The New Phytologist. 2007;**174**:499-506. DOI: 10.1111/j.1469-8137.2007.02051.x

[96] Imtiaz M, Rizwan MS, Mushtaq MA, Ashraf M, Shahzad SM, Yousaf B, Saeed DA, Rizwan M, Nawaz MA, Mehmood S, Silicon TS. Occurrence, uptake, transport and mechanisms of heavy metals, minerals and salinity enhanced tolerance in plants with future prospects: A review. Journal of Environmental Management. 2016;**183**:521-529. DOI: 10.1016/j.jenvman.2016.09.009

[97] Du Laing G, Rinklebe J, Vandecasteele B, Meers E, Tack FM. Trace metal behaviour in estuarine and riverine floodplain soils and sediments: A review. Science of the Total Environment. 2009;**407**:3972-3985. DOI: 10.1016/j.scitotenv.2008.07.025

[98] Sarwar N, Saifullah MSS, Zia MH, Naeem A, Bibi S, Farid G. Role of mineral nutrition in minimizing cadmium accumulation by plants. Journal of the Science of Food and Agriculture. 2010;**90**:925-937. DOI: 10.1002/jsfa.3916

[99] Eriksson JE. Effects of nitrogen-containing fertilizers on solubility and plant uptake of cadmium. Water, Air, & Soil Pollution. 1990;**49**:355-368. DOI: 10.1007/BF00507075

[100] Williams CH, David DJ. The effect of superphosphate on the cadmium content of soils and plants. Soil Research. 1973;**11**:43-56. DOI: 10.1071/sr9730043

[101] Yang Y, Chen R, Fu G, Xiong J, Tao L. Phosphate deprivation decreases cadmium (cd) uptake but enhances sensitivity to cd by increasing iron (Fe) uptake and inhibiting phytochelatins synthesis in rice (*Oryza sativa*). Acta Physiologiae Plantarum. 2016;**38**:28. DOI: 10.1007/s11738-015-2055-9

[102] Mitchell LG, Grant CA, Racz GJ. Effect of nitrogen application on concentration of cadmium and nutrient ions in soil solution and in durum wheat. Canadian Journal of Soil Science. 2000;**80**:107-115. DOI: 10.4141/S98-085

[103] Zaccheo P, Crippa L, Pasta VDM. Ammonium nutrition as a strategy for cadmium mobilisation in the rhizosphere of sunflower. Plant and Soil. 2006;**283**:43-56. DOI: 10.1007/s11104-005-4791-x

[104] Xie HL, Jiang RF, Zhang FS, Mcgrath SP, Zhao FJ. Effect of nitrogen form on the rhizosphere dynamics and uptake of cadmium and zinc by the hyperaccumulator *Thlaspi caerulescens*. Plant and Soil. 2009;**318**:205-215. DOI: 10.1007/s11104-008-9830-y

[105] Yang Y, Xiong J, Chen R, Fu G, Chen T, Tao L. Excessive nitrate enhances cadmium (cd) uptake by up-regulating the expression of OsIRT1 in rice (*Oryza sativa*). Environmental and Experimental Botany. 2016;**122**:141-149. DOI: 10.1016/j.envexpbot.2015.10.001

[106] Wangstrand H, Eriksson J, Oborn I. Cadmium concentration in winter wheat as affected by nitrogen fertilization. European Journal of Agronomy. 2007;**26**:209-214. DOI: 10.1016/j.eja.2006.09.010

[107] Shao G, Chen M, Wang D, Xu C, Mou R, Cao Z, Zhang X. Using iron fertilizer to control cd accumulation in rice plants: A new promising technology. Science in China. Series C, Life Sciences. 2008;**51**:245-253. DOI: 10.1007/s11427-008-0031-y

[108] Liu H, Zhang J, Christie P, Zhang F. Influence of iron plaque on uptake and accumulation of cd by rice (*Oryza sativa* L.) seedlings grown in soil. Science of the Total Environment. 2008;**394**:361-368. DOI: 10.1016/j.scitotenv.2008.02.004

[109] Hassan MJ, Feng W, Ali S, Zhang G. Toxic effect of cadmium on rice as affected by nitrogen fertilizer form. Plant and Soil. 2005;**277**:359-365. DOI: 10.1007/s11104-005-8160-6

[110] Cobbett CS. Phytochelatins and their roles in heavy metal detoxification. Plant Physiology. 2000;**123**:825-832. DOI: 10.1104/pp.123.3.825

[111] Gopal R, Rizvi AH. Excess lead alters growth, metabolism and translocation of certain nutrients in radish. Chemosphere. 2008;**70**:1539-1544. DOI: 10.1016/j.chemosphere.2007.08.043

[112] Hu Y, Norton GJ, Duan G, Huang Y, Liu Y. Effect of selenium fertilization on the accumulation of cadmium and lead in rice plants. Plant and Soil. 2014;**384**:131-140. DOI: 10.1007/s11104-014-2189-3

[113] Arao T, Kawasaki A, Baba K, Mori S, Matsumoto S. Effects of water management on cadmium and arsenic accumulation and dimethylarsinic acid concentrations in Japanese rice. Environmental Science & Technology. 2009;**43**:9361-9367. DOI: 10.1021/es9022738

[114] Gregorio GB, Senadhira D, Htut H, Graham RD. Breeding for trace mineral density in rice. Food and Nutrition Bulletin. 2000;**21**:382-386. DOI: 10.1177/156482650002100407

[115] Graham JF, Mccague C, Norton PR. Topography and nanomechanical properties of tribochemical films derived from zinc dialkyl and diaryl dithiophosphates. Tribology Letters. 1999;**6**:149-157. DOI: 10.1023/A:1019124026402

[116] Welch RM, Graham RD. Breeding for micronutrients in staple food crops from a human nutrition perspective. Journal of Experimental Botany. 2004;**55**:353-364. DOI: 10.1093/jxb/erh064

[117] Clemens S. Zn and Fe biofortification: The right chemical environment for human bioavailability. Plant Science. 2014;**225**:52-57. DOI: 10.1016/j.plantsci.2014.05.014

[118] Bhullar NK, Boonyaves K, Wang M, Sautter C. Iron and zinc enhancement in rice endosperm by targeted and synergistic action of genes. In: Preedy V, Srirajaskanthan R, Patel V, editors. Handbook of Food Fortification and Health. Nutrition and Health. New York, NY: Humana Press; 2013. pp. 293-230. DOI: 10.1007/978-1-4614-7076-2_23

[119] Masuda H, Usuda K, Kobayashi T, Ishimaru Y, Kakei Y, Takahashi M, Higuchi K, Nakanishi H, Mori S, Nishizawa NK. Overexpression of the barley nicotianamine synthase gene *HvNAS1* increases iron and zinc concentrations in rice grains. Rice. 2009;**2**:155-166. DOI: 10.1007/s12284-009-9031-1

[120] Zheng L, Cheng Z, Ai C, Jiang X, Bei X, Zheng Y, Glahn RP, Welch RM, Miller DD, Lei XG, Shou H. Nicotianamine, a novel enhancer of rice iron bioavailability to humans. PLoS One. 2010;**5**:e10190. DOI: 10.1371/journal.pone.0010190

[121] Alexander ATJ, Bianca Kyriacou, Damien LC, Lorranine C, James S, Enzo L, Mark T. Constitutive overexpression of the *OsNAS* gene family reveals single-gene strategies for effective iron- and zinc-biofortification of rice endosperm. 2011;**6**:e24476. DOI: 10.1371/journal.pone.0024476

[122] Goto F, Yoshihara T, Shigemoto N, Toki S, Takaiwa F. Iron fortification of rice seed by the soybean ferritin gene. Nature Biotechnology. 1999;**17**:282-286. DOI: 10.1038/7029

[123] Swamy BPM, Rahman MA, Inabangan-Asilo MA, Amparado A, Manito C, Chadha-Mohanty P, Reinke R, Slamet-Loedin IH. Advances in breeding for high grain zinc in rice. Rice. 2016;**9**:49. DOI: 10.1186/s12284-016-0122-5

[124] Masuda H, Aung MS, Nishizawa NK. Iron biofortification of rice using different transgenic approaches. Rice. 2013;**6**:40. DOI: 10.1186/1939-8433-6-40

[125] Sasaki A, Yamaji N, Ma JF. Overexpression of OsHMA3 enhances cd tolerance and expression of Zn transporter genes in rice. Journal of Experimental Botany. 2014;**65**:6013-6021. DOI: 10.1093/jxb/eru340

[126] Ishikawa S, Abe T, Kuramata M, Yamaguchi M, Ando T, Yamamoto T, Yano M. A major quantitative trait locus for increasing cadmium-specific concentration in rice grain is located on the short arm of chromosome 7. Journal of Experimental Botany. 2010;**61**:923-934. DOI: 10.1093/jxb/erp360

[127] Anuradha K, Agarwal S, Rao YV, Rao KV, Viraktamath BC, Sarla N. Mapping QTLs and candidate genes for iron and zinc concentrations in unpolished rice of Madhukar×Swarna RILs. Gene. 2012;**508**:233-240. DOI: org/10.1016/j.gene.2012.07.054

[128] Ishikawa R, Iwata M, Taniko K, Monden G, Miyazaki N, Om C, Tsujimura Y, Yoshida S, Ma JF, Ishii T. Detection of quantitative trait loci controlling grain zinc concentration using Australian wild rice, *Oryza meridionalis*, a potential genetic resource for biofortification of rice. Plos One. 2017;**12**:e0187224. DOI: org/10.1371/journal.pone.0187224

[129] Arao T, Ae N. Genotypic variations in cadmium levels of rice grain. Soil Science & Plant Nutrition. 2003;**49**:473-479. DOI: 10.1080/00380768.2003.10410035

[130] Ishikawa S, Ae N, Sugiyama M, Murakami M, Arao T. Genotypic variation in shoot cadmium concentration in rice and soybean in soils with different levels of cadmium contamination. Soil Science & Plant Nutrition. 2010;**51**:101-108. DOI: 10.1111/j.1747-0765.2005.tb00012.x

[131] Xie LH, Tang SQ, Wei XJ, Shao GN, Jiao GA, Sheng ZH, Luo J, PS H. The cadmium and lead content of the grain produced by leading Chinese rice cultivars. Food Chemistry. 2017;**217**:217-224. DOI: 10.1016/j.foodchem.2016.08.099

[132] Ishikawa S, Ishimaru Y, Igura M, Kuramata M, Abe T, Senoura T, Hase Y, Arao T, Nishizawa NK, Nakanishi H. Ion-beam irradiation, gene identification, and marker-assisted breeding in the development of low-cadmium rice. Proceedings of the National Academy of Sciences of the United States of America. 2012;**109**:19166-19171. DOI: 10.1073/pnas.1211132109

[133] Abe T, Nonoue Y, Ono N, Omoteno M, Kuramata M, Fukuoka S, Yamamoto T, Yano M, Ishikawa S. Detection of QTLs to reduce cadmium content in rice grains using LAC23/ Koshihikari chromosome segment substitution lines. Breeding Science. 2013;**63**:284-291. DOI: 10.1270/jsbbs.63.284

[134] Oliva N, Chadha-Mohanty P, Poletti S, Abrigo E, Atienza G, Torrizo L, Garcia R, Dueñas CJ, Poncio MA, Balindong J, Manzanilla M, Montecillo F, Zaidem M, Barry G, Hervé P, Shou H, Slamet-Loedin IH. Large-scale production and evaluation of marker-free indica rice IR64 expressing phytoferritin genes. Molecular Breeding. 2014;**33**:23-37. DOI: 10.1007/s11032-013-9931-z

[135] Boonyaves K, Gruissem W, Bhullar NK. NOD promoter-controlled *AtIRT1* expression functions synergistically with NAS and FERRITIN genes to increase iron in rice grains. Plant Molecular Biology. 2016;**90**:207-215. DOI: 10.1007/s11103-015-0404-0

[136] Abe T, Fumio TS, Kojima Y, Ebitani T, Kuramata M, Yamamoto T, Yano M, Ishikawa S. Detection of a QTL for accumulating cd in rice that enables efficient cd phytoextraction from soil. Breeding Science. 2011;**61**:43-51. DOI: 10.1270/jsbbs.61.221

[137] Shimo H, Ishimaru Y, An G, Yamakawa T, Nakanishi H, Nishizawa NK. Low cadmium (LCD), a novel gene related to cadmium tolerance and accumulation in rice. Journal of Experimental Botany. 2011;**62**:5727-5734. DOI: 10.1093/jxb/err300

[138] Li B, Wang X, Qi XL, Huang L, Ye ZH. Identification of rice cultivars with low brown rice mixed cadmium and lead contents and their interactions with the micronutrients iron, zinc, nickel and manganese. Journal of Environmental Sciences (China). 2012;**24**:1790-1798. DOI: 10.1016/S1001-0742(11)60972-8

[139] Ashraf U, Hussain S, Anjum SA, Abbas F, Tanveer M, Noor MA, Tang XR. Alterations in growth, oxidative damage, and metal uptake of five aromatic rice cultivars under lead toxicity. Plant Physiology and Biochemistry. 2017;**115**:461-471. DOI: 10.1016/j.plaphy.2017.04.019

10

Natural Resistance of Sri Lankan Rice (*Oryza sativa* L.) Varieties to Broad-Spectrum Herbicides (Glyphosate and Glufosinate)

Shyama R. Weerakoon, Seneviratnage Somaratne,
E. M. Sachini I. Ekanayaka and Sachithri Munasighe

Abstract

Since studies on herbicide-resistant rice (HRR) are limited in Sri Lanka, the present study conducted to screen the naturally existing glyphosate and glufosinate resistance in traditional and inbred rice varieties. Six traditional varieties and nineteen inbred lines were selected for the study. Complete randomized design with three pots with 10 replicates for each herbicide concentration was employed. Optimal concentrations of glyphosate (0.5 gl^{-1}) and glufosinate (0.05 gl^{-1}) were applied at 3–4 leaf stages. Varieties ≥50% survival percentage was considered as resistant to respective herbicides. Twelve varieties showed resistance (≥50%) at 0.5 gl^{-1} glyphosate concentration. Survived plants were monitored and agro-morphological and yield characters/parameters were measured. Fifteen varieties were to glufosinate at 0.05 gl^{-1}. Even though no significant differences ($p > 0.05$) were observed in growth parameters across control and treated plants, there was a yield penalty. Nine varieties (At362, Bg352, Bg359, Bg366, Bg369, Bg379-2, Bg403, Bg454, and Pachchaperumal) indicated moderate resistance to both glyphosate and glufosinate. The emerged HRRs indicated varying responses of agro-morphological and yield characters across the type of herbicide and the variety. Glyphosate reduced the growth parameters and yield penalty compared to glufosinate treated varieties. These HRR varieties have a higher potential in rice breeding programs and in developing HR rice varieties in future.

Keywords: glyphosate, glufosinate, herbicide resistance, *Oryza sativa*

1. Introduction

Rice, one of the most important grains, fulfills the carbohydrate requirement of people in the tropical countries and to a lesser extent in subtemperate areas. The cultivated rice belongs to the grass family Gramineae (Poaceae) under the tribe-Oryzeae of the subfamily Pooideae [1]. However, the genus *Oriza* has been divided into several sections and placed *O. sativa* under Series Sativa in Section Sativae [2]. *O. sativa,* an indigenous rice species in Asia, is a diploid species consisting 24 chromosomes. The genomic formula of *O. sativa* is AA [2]. The species *O. sativa* is an annual grass, with round, hollow, jointed culms, rather flat, sessile leaf blades, and a terminal panicle, under favorable conditions. As the other members in the tribe Oryzeae, rice is well-adapted to aquatic and swampy habitats [3].

Rice is cultivated on about 156 million hectares of land to produce about 696 million tons annually in Asian countries which account for 90% of the world's total rice production [4]. There is a growing trend of increasing rice consumptions since 2000s, which surpasses the production. On an annual basis, global rice demand keeps increasing by *ca.* 8 million tons implying that during next 10 years, the rice production need to increase to 80 million tons which is double the present production [5].

The increasing world population especially in tropical countries where rice serves as the staple food, one billion people per year demands an additional rice production (100 million MT) [6]. In future, it is apparent that rice production will continue to grow rapidly as increasing populations attempt to secure food supplies. In order to obtain a good yield in rice, farmers are required to overcome several biotic and abiotic stresses. Among biotic stresses, weeds stand out as the major threat to rice cultivation, which reduce the yield qualitatively and quantitatively. Over the past few decades, climate change has induced transformations in the weed flora of arable ecosystems and the changes in the climate have also influenced weeds indirectly by enforcing adaptations to agronomic practice [7]. Therefore, it is imperative to develop effective weed control strategies while maintaining crop yield [8]. Globally, *ca.*10% loss of rice yield is attributed to weed and specific quantity is more or less closer to 46 million tons (based on 1987 world rough rice production). Depending on the predominant weed flora and on the control methods practiced by farmers, loss of yield caused by weeds varies across countries in the world. In Sri Lanka, a country considered self-sufficient in rice, weeds are the major biotic stress in rice production and account for 30–40% of yield losses [9]. Thus, there is a need to take timely and appropriate measures to preserve the country's rice production.

Rice weeds are the major barriers to rice production because they possess the ability to compete for CO_2, space, moisture, sunlight and nutrients. Under certain conditions, crops fail to successfully compete with weeds [10]. Weed flora varies spatially due to type of rice culture, soil type, hydrology, tillage, cultural practices and irrigation pattern and so on. Approximately, 134 weed species belonging to 32 taxonomic families were identified in rice fields in Sri Lanka, and they were categorized as grasses, sedges and broad leaves [11].

Rice weeds adversely affected on final yield in number of ways. Weed increases the cost of production of rice. The cost of rice weed control, including herbicides, cultural and mechanical practices, and hand weeding, is estimated to be about 5% of world rice production and amount to US$3.5 billion annually. When the10% loss of rough rice grain yield is added to this cost, the world's total estimated cost for rice weeds and their control amounts to 15% of total annual production [12].

Weeds indirectly limit production and act as a host of plant harboring pathogens and pests that adversely affect rice. Furthermore, weeds intervene rice harvesting and increase harvest costs through direct interference with the harvesting operation and by causing lodging. Contamination of rough rice by the seeds of the weeds reduces the grain quality and market value for example weed red rice (*Oryza sativa f. spontanea*) has a pigmented layer that shatters easily and readily contaminates rough rice. Removing all traces of the pigmented layer requires intense milling and results in decreased grain quality and lower milling rates [12]. The drudgery of weeding and labor shortages have made rice farming unattractive. In most tropical countries, farmers spend more time on weeding, by hand or with simple tools, than on any other farming task. Hand weeding of one (01) ha. of rice requires from 100 to 780 labor-hours per crop, depending on the rice culture. Due to these adverse effects, there is a need to improve the present weed control practices.

Herbicides are chemical substances used to kill plants which are often placed under the group of chemicals known as pesticides that prevent, destroy, repel, or mitigate any pest [13]. Herbicides, in general, are classified using different criteria such as activity, timing of application, method of application, mechanism of action and chemical family. Based on the time of application there are three main categories of herbicides recommended for rice. "Pre-plant herbicides" are applied before the crop is planted in order to eliminate weeds that have germinated before planting or were left from following (e.g., glyphosate, glufosinate). "Pre-emergence herbicides" are applied after the crop has been planted but before weeds emerge (butachlor, pretilachloroxadiazon, pendimethalin, oxadiargyl) and "Post-emergence herbicides" are applied after weeds have emerged (bisparybacbispyribac, pentagon, 2,4-D). These herbicides are either broad-spectrum (nonselective) or narrow-spectrum (selective). Some of the most common modes of actions are auxin mimics, mitosis inhibitors, photosynthesis inhibitors, amino acid synthesis inhibitors and lipid biosynthesis inhibitors.

The usage of pre-emergent, broad spectrum herbicide in controlling weeds in rice cultivation has become a popular method among the farmers since it minimizes cost, labor and time. Glyphosate and glufosinate are the most commonly used broad-spectrum herbicides (BSHs) in rice fields and glyphosate usage is comparatively higher. Glyphosate (*N*-(phosphonomethyl) glycine) or $C_3H_8NO_5P$ is a broad spectrum, nonselective systemic herbicide. It is effective in killing all plant types including grasses, perennials and woody plants. Glyphosate is a versatile herbicide used by farmers, land managers and gardeners to simply, safely and effectively control unwanted vegetation. Initially glyphosate was patented and sold by Monsanto

Company in the 1970s under the trade name Roundup and after that glyphosate-based products have become the most commonly used herbicides in the U.S. [14]. This widespread adoption is the result of glyphosate's ability to control a broad spectrum of weeds, its extensive economic and environmental benefits and its strong safety profile. Glyphosate is currently undergoing registration review by the US Environmental Protection Agency (EPA or the Agency) and it is essential that farmers, land managers and gardeners retain access to this important tool for weed control.

As an herbicide, glyphosate is activated by absorbing into the plant mainly though leaves and also through soft stalk tissue. Subsequently, glyphosate is transported throughout the plant where it acts on various enzyme systems inhibiting amino acid metabolism (shikimic acid pathway). Glyphosate inhibits 5-enolpyruvylshikimate-3-phosphate-synthase, the sixth enzyme in the shikimate biosynthetic pathway that produces the essential aromatic amino acids (tryptophan, tyrosine and phenylalanine) and subsequently phenolics, lignins, tannins and other phenylpropanoids [15]. The shikimate pathway is found in all microorganisms and crop plants. This pathway is essential for the biosynthesis of chorismate, the precursor for aromatic amino acids and aromatic secondary metabolites [16] (Priestman *et al.*, 2005). Glyphosate is reported to be causing a significant damage to rice yield with a reduction of yield up to 80% by blocking the shikimate pathway of crop plant [17].

Glufosinate is converted within the plant cell into the phytotoxin named as phosphinothricin (PT). As a structural analogue of glutamic acid, PT inhibits glutamine synthatase—GS (E.C.6.3.1.2.), competitively and irreversibly [18, 19]. GS is an essential ammonia assimilation enzyme found in plants. Inhibition of GS causes a rapid, toxic accumulation of intercellular ammonia resulting in metabolic disruption and inhibition of photosystem I and photosystem II in treated plants [18, 19] (Senseman, 2007; Hensley, 2009). Over 40 monocotyledonous and more than 150 dicotyledonous species are sensitive to PT [20].

In relation to herbicide, usage of the terms "tolerance" and "resistance" are inconsistent among the general public and even weed scientists. Among the members of the weed science community, tolerance and resistance are used interchangeably. Further, herbicide manufacturers/seed companies that develop and/or market HR crop cultivars/varieties generally refer to these as herbicide-tolerant. The present study recognizes the definition of herbicide tolerance and resistance established by the Weed Science Society of America (WSSA) [21].

The official Weed Science Society of America [21] defines herbicide resistance as "the inherited ability of a plant to survive and reproduce following exposure to a dose of herbicide normally lethal to the wild type. In a plant, resistance may be naturally occurring or induced by such techniques as genetic engineering or selection of variants produced by tissue culture or mutagenesis". Herbicide tolerance is defined by WSSA as "tolerance is the inherent ability of a species to survive and reproduce after herbicide treatment. This implies that there was no selection or genetic manipulation to make the plant tolerant; it is naturally tolerant" [21]. Resistance may occur in plants as the result of random and frequent mutation. Through selection, where the herbicide is the selection pressure, susceptible plants are killed while herbicide resistant plants survive.

During the last two decades, considerable effort has been made to breed HR crops and it was expected to relieve the constraints imposed by different combinations of chemicals, overcome problems associated with herbicide residues, expand the range of compounds available for selective use in-crop, simplify the crop management and extend the useful life span of the current nonselective herbicides [22].

Rice cultivars resistant to glufosinate [23], sulfonylureas, imidazolinones and glyphosate have already been developed and are being field-tested, mostly in the USA but also in South America and Japan [24–26]. The main reason for developing HR rice is to attain control of weed species that fail to control rice weeds selectively [27]. In addition, introduction of HR rice improves current cropping systems, with more efficient weed control measures and could reduce the amount of land required to satisfy the global rice needs and fulfill the increase in the future demand of rice. Particularly, HR rice provides the farmer with new efficient chemical options for weed control, for instance, glyphosate and glufosinate target both monocotyledonous and dicotyledonous weeds, which probably allow less herbicide use in terms of amount and number of applications. In relation to HR, both herbicides were post-emergence, which means that doses can be adjusted to actual weed infestation, and spraying can be performed within a wider time frame due to their high efficacy and crop tolerance. Therefore, HR could result in adequate control of hard-to-kill weeds. In addition, weed populations already resistant to currently used herbicides could be controlled with these broad-spectrum herbicides [28].

Many studies focused on optimization of weed management in HR have been conducted with rice resistant to either glufosinate or imidazolinone. Almost complete control of weedy rice and other grasses, including *Echinochloa crus-galli* (L.) Beauv. was achieved in glufosinate-resistant rice in Arkansas (USA) by sequential applications of glufosinate. Initial studies on weed control in Imidazolinone resistant rice (IMI rice) were conducted with imazethapyr, an herbicide proven effective against weedy rice and other rice weeds when applied as a soil or foliar treatment. Imidazolinone resistant rice varieties carrying an insensitive target acetolactatesynthase (ALS) enzyme, which is the target site of these herbicides, were developed through anther culture and backcrossing without exposure to mutagens or genetic transformation [29]. Further, imidazolinone-tolerant rice variety was engineered through mutation of the rice variety AS3510 with EMS. The resulted M2 plants were sprayed with imazethapyr. A single surviving plant was identified, and the progeny of this rice plant showed tolerance to several AHAS-inhibiting herbicides [30]. This mutant line was referred to as 93AS3510, and subsequently two imidazolinone-tolerant rice varieties, CL121 and CL141 were developed with this tolerance trait and were first marketed in the USA in 2001 [31, 32].

Even with such achievements, inadequate weed and pest management practices led to creation of a yield gap in the rice production. Literature on the subject revealed that studies have focused on the importance of controlling weeds including hard-to-control *Echinochloa* spp.

and *Eleusine* spp. [8]. In addition, herbicide resistant (HR) conspecific weeds such as weedy rice with varying dormancy patterns have become more abundant in rice fields in Sri Lanka throughout the cropping season. As a result, Sri Lankan farmers tend to use pre- and postemergent herbicides in land preparation specially to control weedy rice. These pre- and postemergent herbicides include selective and nonselective (broad spectrum) herbicides. As far as selective herbicide usage is considered, the number of application and their amount to control common weeds such as: *Cyperus iria* L. (family: Cyperaceae), *Echinochloa* sp. (family: Gramineae), *Monochoria vaginalis* (family: Pontederiaceae) and weedy rice (*Oryza sativa f. spontanea*; family: Poaceae) has been increased considerably leading to sever threats to the rice growing environment [33]. Thus, it is critically important to evaluate the possibility of applying commonly used broad-spectrum herbicides; glyphosate and glufosinate as postemergent herbicides along with herbicide resistant technology to eliminate hard-to-control weeds. Thus, the objective of the present study was to evaluate the herbicide resistance of Sri Lankan traditional and inbred cultivated rice varieties to pre-emergent herbicide—glyphosate and glufosinate.

2. Methodology

Seeds of twenty-five rice (*Oryza sativa* L.) varieties (**Table 1**) were collected from the Rice Research and Development Institute (RRDI) of Sri Lanka. These lines were maintained in a plant house at the Open University of Sri Lanka, located in low country wet zone of the Western province, with an average temperature of 28–32°C and 65–70% relative humidity.

The selected seeds were pre-soaked overnight and allowed to germinate. One week old seedlings were planted in pots (with 23 cm diameter) filled with puddle soil (5.5 kg per pot) and excess plantlets were thinned out 1 week after planting [34] leaving 10 plants per pot. Fertilizer application and other crop management practices were performed according to the recommendations of the Department of Agriculture, Sri Lanka.

Glyphosate (0.5 gl^{-1}) and glufosinate (0.05 gl^{-1}) [35] were applied at 3–4 leaf stage (Department of Agriculture, Sri Lanka) of plants separately. The research design used was complete randomized design (CRD) with three pots (10 replicates in each pot) for each treatment and nontreated plants served as the control.

The total number of plants and the number of surviving plants were counted for each variety and percentage resistance (PR) was calculated as follows: plants with ≥50% resistance to herbicides were arbitrary considered as resistant varieties [36].

$$PR\ (\%) = \left[\frac{\textit{Number of survivng seedlings in a variety}}{\textit{Total number of seedlings grown in the same variety}}\right] \times 100$$

Agro-morphological characters of resistant plant were measured/evaluated in 2 weeks after sawing and the yield parameters, respectively, by application of respective herbicide.

Selection number	Name	Age (month)	Attributes
1	Bg94-1	3 ½	High yield WP
2	Bg250	2 ½	
3	Bg300	3	Resistant to GM-1, BL, BB, Bph
4	Bg304	3	Resistant to GM 1&2, BL, BB, Bph
5	Bg305	3	Resistant to GM-1 and 11, BPH, BL and BLB
6	Bg352	3 ½	Resistant to BL, BB & GM-1, Bph
7	Bg357	3 ½	Resistant to GM-1& 2, BL, BB, Bph
8	Bg359		Resistant to GM 1 & 2, BL, BB, Bph
9	Bg360	3 ½	Resistant to GM-1, GM-2, BL, Bph
10	At362	3 ½	
11	Bw364	3 ½	
12	Ld3 65	3 ½	Resistant to iron toxicity
13	Bg366	3 ½	
14	Bg369	3 ½	
15	Bg379-2	4 ½	Resistant to Bph and BB
16	Bg403	4	Resistant to BB, BL and Bph
17	Bg450	4 ½	Resistant GM-I
18	Bg454	4 ½	
19	H4	4	Resistant to BL
20	*Kaluheenati*	4	Moderately tolerant Gr. 2
21	*Kuruluthuda*	4	
22	*Suwadal*	5	
23	*Rathhal*	5	
24	*Madel*	5	
25	*Pachchaperumal*	3 ½	

BB: bacterial leaf blight; BL: rice blast disease; GM-1: biotype one of rice gall midge; GM-2: biotype two of rice gall midge; Bph: brown plant hopper; PS: photo period sensitivity.

Source: Jeyawardena *et al.*, 2010, RRDI, Batalagoda, Department of Agriculture, Sri Lanka

Table 1. List of chosen Sri Lankan rice varieties form the results of a previous for the study on natural herbicide resistance.

3. Results and discussion

Evaluation of natural resistance to glyphosate and glufosinate among rice varieties.

The results obtained from the screening for glufosinate resistant and glyphosate resistant varieties revealed that some of the selected traditional rice varieties and inbred lines possess the ability to resist the detrimental effects of those broad-spectrum herbicides (**Figure 1**). Two rice varieties (*Rathal*—2% and Bg305—1%) were found to be lethal to 0.05 gl^{-1} glufosinate concentration whereas no such varieties were observed under the application of 0.5 gl^{-1} glyphosate. Fifteen rice varieties (At362—90%, Bg250—83%, Bg300—96%, Bg352—100%, Bg357—53%, Bg359—100%, Bg360—96%, Bg366—73%, Bg369—83%, Bg379/2—93%, Bg403—100%, Bg450—57%, Bg454—97%, Bg94/1—73%, *Pachchaperumal*—53%) showed natural resistance under glufosinate application and 12 rice varieties (At362—75%, Bg352—50%, Bg359—55%, Bg366—65%, Bg369—60%, Bg379/2—65%, Bg403—60%, Bg454—55%, Ld365—70%, *Kaluheenati*—55%, *Kuruluthuda*—55%, *Pachchaperumal*—70%) were able to survive under glyphosate application. Results indicated that nine varieties (At362, Bg352, Bg359, Bg366, Bg369, Bg379-2, Bg403, Bg454 and *Pachchaperumal*) were resistant for both glyphosate and glufosinate (**Figure 1**).

Very limited studies have been conducted regarding the natural or induced herbicide resistance in Sri Lankan rice varieties [36] and findings of Sri Lankan rice varieties which are able to resist broad-spectrum herbicides (BSHs) such as glufosinate and glyphosate have hardly been recorded. In this study, nine rice varieties with the ability to resistant the application of concentrations, 0.05 gl^{-1}glufosinate and 0.5 gl^{-1} glyphosate have been identified. Among these varieties, only two red grain rice varieties (*Pachchaperumal* and At362) are included indicating that most of the cultivated traditional rice varieties, except *Pachchaperumal* did not possess the ability to resist both glufosinate and glyphosate as observed in inbred rice varieties. However, further studies are required to confirm such findings. Sri Lankans do admire red grain rice such as *Kuruluthuda*, *Kaluheenati* and *Rathal* due to their high nutritive qualities (**Table 1**), and it is important to note that such varieties need to be developed as BSHs resistant varieties in future.

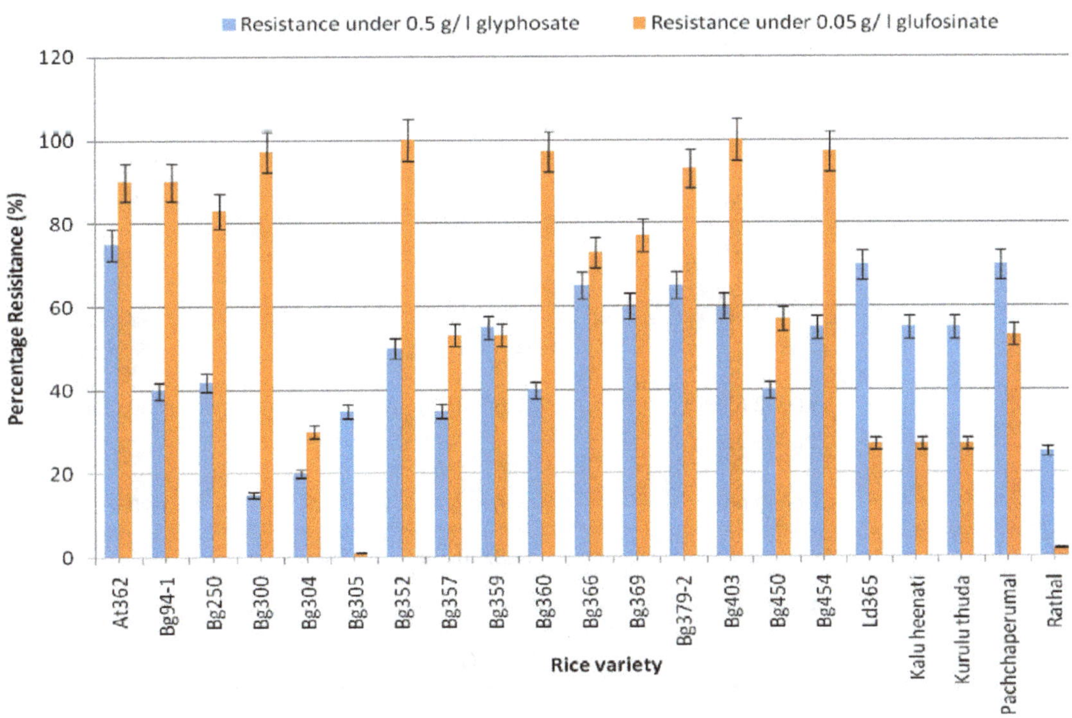

Figure 1. Comparison between natural resistances in selected rice (*Oryza sativa* L.) varieties to glyphosate and glufosinate.

On the other hand, according to the results of the study, relatively high survival percentage toward both BSHs was reported by inbred rice varieties which possess many valuable attributes other than glyphosate- and glufosinate-resistance such as resistant to GM-1, GM-2, BL, BB, Bph and have high yield potential (**Table 1**). These rice varieties could be incorporated in rice breeding programs to strengthen the sustainable cultivation.

Comparison table of plant height and yield parameters of glyphosate treated and untreated rice varieties are shown in **Table 2**. Though rice plant showed considerable HR in general, growth

Rice variety	Plant height (cm)	1000-grain weight (g)	Yield/plant (g)
Control			
At362	66.33 (1.20)	25.00 (0.44)	23.43 (2.69)
Bg359	62.00 (2.08)	22.46 (0.32)	12.38 (0.46)
Bg366	46.67 (2.73)	23.17 (0.28)	5.34 (0.60)
Bg369	34.33 (1.67)		
Bg379-2	64.00 (0.58)	25.67 (0.44)	5.22 (0.70)
Bg403	67.00 (1.15)	21.39 (0.38)	15.00 (1.09)
Bg454	52.00 (2.08)		
Bw364	58.67 (2.33)	19.23 (2.25)	12.51 (2.65)
Ld365	59.00 (2.08)	13.26 (0.29)	4.04 (0.77)
Kaluheenati	73.33 (0.88)	22.89 (1.51)	4.69 (0.75)
Kuruluthuda	70.00 (1.15)		
Pachchaperumal	70.33 (1.20)	31.64 (0.38)	11.31 (1.02)
Treated			
At362	50.33 (0.88)	16.86 (0.32)	12.61 (3.25)
Bg359	52.00 (3.06)	16.37 (0.29)	5.38 (0.94)
Bg366	27.67 (3.67)	18.49 (0.90)	3.19 (0.36)
Bg369	48.67 (2.33)		
Bg379-2	48.67 (3.48)	15.27 (1.41)	1.84 (0.40)
Bg403	49.00 (2.65)	17.56 (0.38)	6.50 (1.45)
Bg454	45.33 (1.45)		
Ld365	48.00 (2.00)	9.85 (0.20)	1.90 (0.38)
Kaluheenati	59.83 (3.68)	14.37 (0.59)	3.11 (0.50)
Kuruluthuda	64.33 (2.33)		
Pachchaperumal	61.00 (3.12)	22.94 (1.34)	3.29 (0.37)

Table 2. Summary of the parametric variables; plant height, 1000-grain weight and yield per plant control and treated with glyphosate (0.5 g/l).

retardation is indicated by the decrease in plant height resulting stunting of glyphosate treated plants. Similarly, the yield parameters such as 1000-grain weight yield per plant also showed apparent decrease in treated plants. However, yield per plant of the treated At362 indicated comparatively less reduction of yield. These findings led to conclude that though most of the rice varieties included in the study were resistant to the glyphosate, there was a considerable yield penalty. Similarly, the response of rice varieties included in the study to glufosinate are summarized in **Table 3** and according to the table, a general trend of decreasing plant height that is stunting growth and yield parameters specially yield per plant was observed. Comparatively, glufosinate treated At362 variety indicated low reduction in yield per plant (**Table 3**).

Rice variety	Plant height (cm)	1000-grain weight (g)	Yield/plant (g)
Control			
At362	66.33 (1.20)	25.00 (0.44)	84.57 (0.59)
Bg359	62.00 (2.08)	22.46 (0.32)	75.37 (0.64)
Bg366	46.67 (2.73)	23.17 (0.28)	72.27 (0.50)
Bg369	34.33 (1.67)		
Bg379-2	64.00 (0.58)	25.67 (0.44)	68.70 (0.29)
Bg403	67.00 (1.15)	21.39 (0.38)	82.43 (0.67)
Bg454	52.00 (2.08)		
Bw364	62.33 (1.45)	24.21 (0.42)	85.17 (0.49)
Ld365	59.00 (2.08)	13.26 (0.29)	81.27 (0.50)
Kaluheenati	73.33 (0.88)	22.89 (1.51)	70.93 (0.54)
Kuruluthuda	70.00 (1.15)		
Pachchapermal	70.33 (1.20)	31.64 (0.38)	80.23 (1.13
Treated			
At362	50.33 (0.88)	16.86 (0.32)	82.87 (0.24)
Bg359	52.00 (3.06)	16.37 (0.29)	69.60 (0.38)
Bg366	27.67 (3.67)	18.49 (0.90)	64.27 (0.62)
Bg369	48.67 (2.33)		
Bg379-2	48.67 (3.48)	15.27 (1.41)	65.20 (0.59)
Bg403	49.00 (2.65)	17.56 (0.38)	80.57 (0.46)
Bg454	45.33 (1.45)		
Bw364	55.00 (3.40)	14.26 (0.55)	70.33 (2.50)
Ld365	48.00 (2.00)	9.85 (0.20)	74.97 (2.98
Kaluheenati	59.83 (3.68)	14.37 (0.59)	68.50 (0.36)
Kuruluthuda	64.33 (2.33)		
Pachchaperumal	61.00 (3.12)	22.94 (1.34)	70.63 (0.30

Table 3. Summary of the parametric variables; plant height, 1000-grain weight and yield per plant of control and treated with glufosinate (0.05 gl^{-1}).

The nonparametric variables of treated and untreated rice varieties with glyphosate and glufosinate are shown in **Tables 4** and **5**, respectively. According to the tables, it is evident that there was no discernible different in growth parameters; however, yield parameters were considerably varied between the herbicide treated plants.

Rice variety	Number of tillers/ plant	Number of leaves/ plant	Number of panicles/ plant	Number of seeds/ panicle
Control				
At362	1	1	2	15
Bg359	0	3	1	14
Bg366	1	2	1	10
Bg369	1	2		
Bg379	1	2	1	10
Bg403	1	5	2	20
Bg454	1	3		
Bw364	1	1	1	15
Ld365	1	5	2	20
Kaluheenati	1	3	2	6
Kuruluthuda	1	3		
Pachchaperumal	1	3	1	15
Treated				
At362	0	2	3	34
Bg359	1	4	2	10
Bg366	2	1	1	5
Bg369	1	2		
Bg379	1	4	1	9
Bg403	1	3	3	20
Bg454	1	2		
Bw364	1	1	2	22
Ld365	0	2	1	15
Kaluheenati	0	1	1	11
Kuruluthuda	1	4		
Pachchaperumal	0	2	2	10

Table 4. Summary of nonparametric variables; number of tillers per plant, number of leaves per plant, number of panicles per plant and number of seeds per panicle of control and treated with glyphosate (0.5 gl^{-1}).

Rice variety	Number of tillers/ plant	Number of leaves/ plant	Number of panicles	Number of seeds/ panicle
Control				
At362	1	1	2	15
Bg359	0	3	1	14
Bg366	1	2	1	10
Bg369	1	2		
Bg379	1	2	1	10
Bg403	1	5	2	20
Bg454	1	3		
Bw364	1	1	1	15
Kaluheenati	1	3	2	6
Ld365	1	5	2	20
Kuruluthuda	1	3		
Pachchaperumal	1	3	1	15
Treated				
At362	0	2	3	34
Bg359	1	4	2	10
Bg366	2	1	1	5
Bg369	1	2		
Bg379	1	4	1	9
Bg403	1	3	3	20
Bg454	1	2		
Bw364	1	1	2	22
Kaluheenati	0	1	1	11
Kuruluthuda	1	4		
Ld365	0	2	1	15
Pachchaperumal	0	2	2	10

Table 5. Summary of nonparametric variables; number of tillers per plant, number of leaves per plant, number of panicles per plant and number of seeds per panicle of control and treated with glufosinate (0.05 gl^{-1}).

3.1. Effect of glufosinate and glyphosate on agro-morphological characters of HR resistant rice varieties

The results of the study suggest that several specific growth parameters of certain glufosinate-resistant varieties at 0.05 gl^{-1}showed no significant difference ($p > 0.05$) compared to control plants. For instance, plant height of Bg379-2 (**Table 3**), leaf blade width of Bg366, and leaf length of Bg352 at 0.05 $gl^{-1}$of glufosinate application showed no significant difference ($p > 0.05$) (data not given). In addition, the control plants and the plants resistant to 0.05 gl^{-1}glufosinate, number of leaves per plant and number of tillers per plant (**Table 5**) were not statistically significant.

Analysis of the variance of yield parameters indicated no significant difference for number of the seeds per panicle at 0.05 gl^{-1} glufosinate application except Bg454, Bg369 and *Kuruluthuda*. Almost all varieties indicated significant differences ($p \leq 0.05$) for flag leaf length, flag leaf width at0.05 gl^{-1}glufosinate application predicting the possibility of glufosinate (at 0.05 gl^{-1}) to cause reduction in flag leaf quality even when applied at 3–4 leaf stage of the plant (data not given). Varieties such as Bg360, Bg357, Bg369, Bg379-2, Bg450, Bg403, Bg250 and Bg 454 reported insignificant differences for thousand seed weight character at 0.05 gl^{-1}. Significant yield reduction was observed for Bg362, Bg359, Bg94-1, Bg358, Bg300 and At362 at 0.05 gl^{-1}.

After application of 0.05 gl^{-1} concentration of glufosinate, injuries were identified (**Figure 2**) as rapid chlorosis (**Figure 2B**) of treated leaves followed by wilting (**Figure 2D**), necrosis (**Figure 2C**) and ultimate death of susceptible plants. Similar symptoms have been reported for different rice varieties [19, 34, 35] and for wheat [37] (Deeds *et al.*, 2006). In addition, brown color lesions (**Figure 2A**) were also observed on leaves, and browning of leaf tips (**Figure 2E**) commonly occurred on all varieties. The injuries were significantly higher after 1 week from herbicide application. Severe chlorosis was observed in rice leaves depending on the susceptibility of the varieties within 3–6 days after herbicide treatment. Within 2 weeks after herbicide application, the observable symptoms were disappeared even in the varieties which were exposed to the highest concentration of glufosinate. Previous studies have been shown that rupture and contortion of inter-venal mesophyll cells with concomitant disorganization of bundle sheath cells herbicide treated plants [38, 39].

Comparatively, the glyphosate treated rice plants indicated that all the yield parameters (number of panicle/plant, number of seeds/panicle and 1000 grain weight) were significant differ from the controls (**Tables 2** and **4**).

After application of 0.5 gl^{-1} glyphosate concentration, a number of visual injuries were observed in individuals of varieties (**Figure 3**). The injuries were promptly observable after 1 week of herbicide application. Among these injuries, general chlorosis in the upper part of the leaves was most abundant. Comparatively, severe chlorosis was observed in rice leaves that depend on the resistance of the varieties within 3–6 days after herbicide treatment. In susceptible varieties, leaf wilting leads to plant death. Newly emerged leaves of survived varieties remained in green color; however, the young emerging leaves which were subjected to treatment were often tightly curled inwardly. Multiple shoots arising from internodes of main stem (**Figure 3A**) were observed, and the secondary shoots and flag leaves were wrinkled or curled in *Kaluheenati* and *Pachchaperumal*. At the booting stage, all the leaves of the variety *Kuruluthuda* were curled and leaf discoloration had occurred. The plants remained in the same stage and indicated no maturity until the harvesting stage (**Figure 3B**). Malformation of inflorescences was also observed in certain varieties at the reproductive stage. The inflorescence of Bg369 and Bg454 was found aborted inside the flag leaf sheath and unable to emerge as a panicle (**Figure 3C**). Meanwhile, panicles of certain varieties were yet to appear in full due

Figure 2. Visual injuries caused by glufosinate: (A) brown color lesions on leaf blade, (B) severe chlorosis on leaf blade after glufosinate treatment, (C) necrotic areas of leaf blade, (D) wilting of susceptible plants, and (E) browning of leaf tips.

Figure 3. Visual injuries caused by glyphosate: (A) multiple shoots and roots that sprouted from the internodes, (B) leaf curling and discoloration, (C) fused panicle to flag leaf, and (D) bleached lemma and Palea.

to the fusion of the flag leaf at the maturity stage. Malformation of inflorescence and developing grains with only bleached lemma and palea were commonly found in Bg366 (**Figure 3D**). In the variety Bg379-2, distorted and crescent-shape spikelet were observed.

4. Conclusions

The rice varieties such as At362, Bg359, Bg366, Bg369, Bg379-2, Bg403, Bg454 and *Pachchaperumal* were resistant to both glyphosate (0.5 gl^{-1}) and glufosinate (0.05 gl^{-1}) applications. Even though the herbicide resistant varieties emerged from the screening, the responses of agro-morphological and yield characters varied across the type of herbicide and the variety. Glyphosate substantially reduced the growth parameters as well as yield compared to glufosinate treated varieties. As far as yield is concerned, there was a significant yield penalty in HR rice varieties. These broad-spectrum HR rice varieties have a higher potential to be utilized in rice breeding programs to breed new HR varieties and can be used to develop HR rice in future.

Acknowledgements

The research grants provided by the National Research Council, Sri Lanka (Grant No. NRC 12-037) and the Faculty of Natural Sciences, The Open University of Sri Lanka are greatly acknowledged. The assistance provided by Ms. Deshani Lakshika is highly appreciated.

Author details

Shyama R. Weerakoon*, Seneviratnage Somaratne, E. M. Sachini I. Ekanayaka and Sachithri Munasighe

*Address all correspondence to: shyamaweerakoon@gmail.com

Department of Botany, The Open University of Sri Lanka, Nawala, Nugegoda, Sri Lanka

References

[1] Ge S, Li A, Lu BR, Zhang SZ, Hong DY. A phylogeny of the rice tribe Oryzeae (Poaceae) based on matK sequence data. American Journal of Botany. 2002;**89**:1967-1972

[2] Khush GS, Brar DS, Hardy B. Rice Genetics IV. Los Baños, The Philippine: Int. Rice Res. Inst; 2001

[3] Bardenas EA, Chang TT. Morphology and Varietal Characteristics of the Rice Plant. Los Baños, The Philippine: The. Int. Rice Res. Inst; 1965

[4] Chang TT, Bardenas EA. The Morphology and Varietal Characteristics of the Rice Plant. Int. Rice Res. No. 36. Los Baños, The Philippine: Int Rice Res. Inst; 1965

[5] GRiSP (Global Rice Science Partnership). Rice Almanac. 4th ed. Los Baños, Philippines: International Rice Research Institute; 2013. p. 283

[6] IRRI (International Rice Research Institute). Annual Report 2010. Los Baños, Philippines: International Rice Research Institute; 2011

[7] Hegde S, Hegde V. Assessment of global rice production and export opportunity for economic development in Ethiopia. International Journal of Science and Research (IJSR), India. 2013;**2**(**6**):257-260

[8] Peters K, Breitsameter L, Gerowitt B. Impact of climate change on weeds in agriculture: A review. Agronomy for Sustainable Development. 2014;**34**:707-721

[9] Labrada R. Weedy and wild rice: Their impact and management. In: Proceedings of 21st Asian Pacific Weed Science Society (APWSS) Conference; Colombo, Sri Lanka. 2007. pp. 8-15

[10] Abeysekara ASK. Management of *Echinochloa* spp. in rice in Sri Lanka. In: FAO Workshop on *Echinochloa* spp. Control; Beijing, China. 2011. p. 13

[11] Singh S, Singh G, Singh VP, Singh AP. Effect of establishment methods and weed management practices on weeds and rice in rice-wheat cropping system. Indian Journal of Weed Science. 2005;**37**:51-57

[12] Department of Agriculture. Crop Recommendation. 2006. Link https://www.doa.gov.lk/rrdi/index.php?option=com_sppagebuilder&view=page&id=42&lang=en

[13] Ampong-Nyarko K. Weed management in rice in Africa. In: Auld BA, Kim KU, editors. Weed Management in Rice. Rome: FAQ; 1996. pp. 183-191

[14] Benbrook CM. Trends in glyphosate herbicide use in the United States and globally. Environmental Sciences Europe. 2016;**28**(1):3. DOI: 10.1186/s12302-016-0070-0

[15] Husnain T. Transformation and transgenic expression studies of glyphosate tolerant and cane borer resistance genes in sugarcane (*Saccharum officinarum* L.). Molecular Plant Breeding. 2015;**6**(12):1-17

[16] Weed Science Society of America (WSSA). Resistance and tolerance definitions. Weed Technology. 1998;**12**:789

[17] Priestman MA, Healy ML, Funke T, Becker A, Schönbrunn E. Molecular basis for the glyphosate-insensitivity of the reaction of 5-enolpyruvylshikimate 3-phosphate synthase with shikimate. FEBS Letters. 2005;**579**(25):5773-5780

[18] Davis BM, Scott RC, Norsworthy JK, Smith KL. Effects of low rates of glyphosate and glufosinate on rice. Arkansas Agricultural Experiment Station Research Series. 2009:581

[19] Senseman SA. Herbicide Handbook. 9th ed. Lawrence, K.S.: Weed Science Society of America; 2007. pp. 81-82

[20] Hensley JB, Webster EP, Harrell DL, Bottoms SL. Herbicide drift effects Louisiana rice production. Louisiana Agriculture. 2009;**52**:6-7

[21] Metz PLJ, Stiekema WJ, Nap JP. A transgene-centered approach to the biosafety of transgenic phosphinothricin-tolerant plants. Molecular Breeding. 1998;**4**(4):335-341

[22] Weed Science Society of America (WSSA). Herbicide Handbook. 9th ed. Lawrence, KS: WSSA; 2007

[23] Senior IJ, Bavage AD. Comparison of genetically modified and conventionally derived herbicide tolerance in oilseed rape: A case study. Euphytica. 2003;**132**(2):217-226

[24] Datta SK, Datta K, Soltanifar N, Donn G, Potrykus I. Herbicide resistant indica rice plants from IRRI breeding line ER72 after PEG-mediated transformation of protoplasts. Plant Molecular Biology. 1992;**20**:619-629

[25] Anon. AgrEvo adds Brazil rice breeding. Chemical Week. 1998;**160**:7

[26] Anon. Transgenic rice. Chemical Week. 1999;**161**:35

[27] Sissell K. Brazil burns AgrEvo's rice. Chemical Week. 1999;**161**:37

[28] Gealy DR, Dilday RH. Biology of Red Rice (*Oyrza sativa* L.) Accessions and their Susceptibility to Glufosinate and Other Herbicides. Lawrence, Kansas: Weed Science Society of America, Allen Press; 1997. p. 34

[29] Wilcut JW, Coble HD, York AC, Monks DW, Duke SO. The niche for herbicide-resistant crops in US agriculture. In: Herbicide-Resistant Crops. Agricultural, Environmental, Economic, Regulatory, and Technical Aspects. 1996. pp. 213-250

[30] Croughan TP. Application of tissue culture techniques to the development of herbicide resistant rice. Louisiana Agriculture. 1994;**37**(3):25-26

[31] Croughan TP. U.S. Patent No. 5,736,629. Washington, DC: U.S. Patent and Trademark Office; 1998

[32] Webster EP, Masson JA. Acetolactate synthase-inhibiting herbicides on imidazolinone-tolerant rice. Weed Science. 2001;**49**:652-657

[33] Gealy DR, Mitten DH, Rutger JN. Gene flow between red rice (*Oryza sativa*) and herbicide-resistant rice (*O. sativa*): Implications for weed management. Weed Technology. 2003;**17**:627-645

[34] Olofsdotter M, Valverde B, Madsen-Hauge K. Herbicide resistant rice (*Oryza sativa* L.) —A threat or a solution. In: FAO Report of the Global Workshop on Red Rice Control; 30 August-3 September; Varadero, Cuba. 1999. pp. 121-145

[35] Ellis JM, Griffin JL, Linscombe SD, Webster EP. Rice (*Oryza sativa*) and corn (*Zea mays*) response to simulated drift of glyphosate and glufosinate. Weed Technology. 2003;**17**:452-460

[36] Davis B, Scott RC, Norsworthy JK, Gbur E. Response of rice (*Oryza sativa*) to low rates of glyphosate and glufosinate. Weed Technology. 2011;**25**(2):198-203

[37] Weerakoon SR, Somaratne S, Wijeratne RGD, Ekanayaka EMSI. Natural herbicide resistance (HR) to broad-spectrum herbicide, glyphosate among traditional and inbred-cultivated rice (*Oryza sativa* L.) varieties in Sri Lanka. Pakistan Journal of Biological Sciences. 2013;**16** (16):796-803

[38] Deeds ZA, Al-Khatib K, Peterson DE, Stahlman PW. Wheat response to simulated drift of glyphosate and imazamox applied at two growth stages. Weed Technology. 2006;**20**:23-31

[39] Bellinder RR, Lyons RE, Scheckler SE, Wilson HP. Cellular alterations resulting from foliar applications of HOE-39866. Weed Science. 1987;**35**:27-35

11

Rice Crop Rotation: A Solution for Weed Management

Ananda Scherner, Fábio Schreiber, André Andres, Germani Concenço, Matheus Bastos Martins and Andressa Pitol

Abstract

The challenges for weed management have increased in rice cultivation due to the high number of cases of herbicide-resistant weeds, especially the widespread distribution of imidazolinone-resistant weedy rice. Therefore, there has been particular interest in preventive, physical, and cultural methods in recent decades. In this context, the adoption of the rice-soybean rotation is reported to be one of the most important factors for weed management in rice fields. Additionally, the use of a diversified crop rotation enables the implementation of a broader herbicide program, which is an important feature influencing weed population dynamics. Rice-soybean rotation has been adopted by farmers to control problematic weed species, reduce seed bank of troublesome weed species, and prevent rice grain yield and quality losses caused by its interference. This crop rotation scheme has brought several benefits when it comes to weed management; however, there are also some drawbacks when adopting this strategy such as the limited productivity of soybean and new weed species becoming problematic, such as *Conyza* species. Thus, this chapter explores the advantages and disadvantages of adopting crop rotation in Brazilian lowlands, and proposes a set of strategies to successfully implement crop rotation in lowland soils as a tool for weed management.

Keywords: rice-soybean rotation, herbicides, residual activity, weed resistance, agriculture

1. Introduction

Weed management strategies are described as biological, cultural, chemical, or mechanical practices employed in an integrated manner to prevent and satisfactorily control weed infestations. Since the introduction of herbicides, after the Second World War, the chemical approach has been the major method of weed control [1] and the reliance on herbicides, with limited diversification of mechanisms of action, has led to the appearance of increased cases of herbicide-resistant weed species. Additionally, the lack of active ingredients with new mechanisms of action [2] and public concern associated with environmental and health hazards, further emphasizes the need to rethink herbicide use [3].

Brazilian rice production has changed considerably in the past decades, partially due to the availability of high-yielding varieties and improved production techniques that have increased productivity by approximately 50% in the Southern region. Considerable progress has been also achieved in terms of weed control with the introduction of the Clearfield® technology, which allowed producers to selectively control weedy rice (*Oryza sativa* L.) by using rice genotypes tolerant to the imidazolinone herbicides. The introduction of these varieties increased the yields by more than 2.5 t/ha, allowing productivity levels to be greater than 10 t/ha in these areas [4]. However, the continued monocropping exerted a selection pressure on the weed community, favoring weed species with phenotypes and phenology that are similar to rice, such as weedy rice and *Echinochloa* spp. Moreover, the intensive use of imidazolinone herbicides concomitantly with minimal alternative cultural practices being adopted, led to the appearance of resistant biotypes of these species.

Facing the widespread distribution of imidazolinone-resistant weedy rice in Brazil, there has been particular interest in preventive, physical, and cultural methods during recent decades. Weed control strategies in general should follow integrated weed management (IWM) principles, relying less on the use of the herbicides and, whenever feasible, including non-chemical methods [5]. IWM practices have not been adopted by all rice producers in Brazil and one of the greatest constraints is the pragmatic solution provided by the use of herbicides as compared to the long-term strategies used in IWM. In practice, IWM strategy is costly in short term and the biggest challenge is to persuade farmers to spend money in preventing problems, such as herbicide resistance, that they still do not have, but probably will face in their own fields in the near future. Herbicide resistance usually evolves due to a poor weed control program, based mainly on the chemical approach, which is largely under the farmer's own control. Thus, the recent cases and obstacles caused by herbicide resistance are changing farmers' perceptions, making them now more positive toward the adoption of non-chemical weed management methods as part of an IWM strategy.

In this context, a very diverse crop rotation is reported to be one of the most important factors in diversifying weed communities and affecting their seed bank dynamics. It is believed that a crop rotation scheme composed of crops with great variability in their biological traits can be the most effective tool for controlling weeds [6] and avoiding weed resistance. The variation of cropping sequences creates an unstable environment, which prevents the annual recurrence of particular weed species [7]. Crop rotation strategies may not eradicate troublesome species, but they can limit their growth and reproduction.

Factors such as the choice of crops and cultivars, plant row spacing, crop seeding rate, sowing date, and use of fertility-building measures have to be taken into account when planning crop sequences. These measures, when properly planned and implemented, can enhance a crop's competitive ability against weeds. Variation in crop sowing dates is one of the best strategies to reduce the seed bank size because it allows for changes in the timing of direct control strategies, such as tillage and herbicide spraying, and disrupts the germination periods of the weed species [8]. Weed germination is affected by crop cultivars and plant spacing due to changes in the canopy and thus the quality of the light that reaches the soil [9]. The incorporation of crops into rotation schemes that release allelopathic substances can be used as a tool to reduce the germination and emergence of some weed species [10]. Therefore, the right choice of crops and the sequence in which they appear is a strong tactic for preventing the establishment of several weed species in the field.

Additionally, the use of a diversified crop rotation enables the implementation of a diversified herbicide rotation scheme. The use of herbicides is considered by some researchers to be the main factor influencing seed bank dynamics [11], as they can drastically reduce weed populations. Based on this scenario, this chapter aims to explore possibilities of crop rotation sequences in Brazilian lowlands, addressing the benefits and drawbacks of each crop sequence when it comes to weed management and crop productivity. Furthermore, the authors aim to propose a set of strategies that can be used to successfully implement crop rotation in lowland soils as a tool for weed management.

2. Weed resistance in Brazilian rice fields

An overview of the current resistant status of herbicide resistance and the efficiency of chemical control against weed species in Southern Brazil adds a new perspective to better understand the need of non-chemical weed management methods, such as crop rotation, as part of an IWM strategy in lowlands. **Table 1** summarizes the herbicides available for rice production in Brazil by mechanism of action, and provides the reader with valuable information about the control efficiency of these compounds against the most troublesome weed species. The problems with weed resistance in rice go far beyond weedy rice, with cases of herbicide-resistant biotypes being reported for *Echinochloa* spp. (*E. crus-galli, E. crus-pavonis,* and *E. colona*), *Eleusine indica, Cyperus* spp. (*C. rotundus* and *C. difformis*), and *Sagittaria* spp. (*S. montevidensis*), which are mainly associated with the intensive use of ALS-inhibiting herbicides, poor crop rotation schemes, and cropping strategies such as irrigation systems. ALS-inhibiting herbicides are known to be highly efficient in low doses against a broad range of weed species and this is probably one of the main reasons associated with the great acceptance of the Clearfield® technology, reducing the use of other herbicides that were widely used before such as, pendimethalin, oxadiazon, oxifluorfen, thiobencarb, bentazon, propanil, and quinclorac.

Echinochloa spp. is also resistant to acetyl-CoA carboxylase (ACCase) inhibitors and quinclorac (AUX, auxin-mimic herbicides), while *Eleusine indica* and *Sagittaria* spp. are also resistant to ACCase and photosystem II (PS II) inhibitors, respectively. Quinclorac was widely used during the 1990s in Brazil to control *Echinochloa* spp. and *Aeschynomene* spp., and some researchers believe [12] that the first case of herbicide resistance in rice cultivation in the country is associated with this herbicide, which selected resistant plants of *Echinochloa* spp.

The current resistance problem evidences the urgent need of alternative management strategies to efficiently control these species and reduce the reliance on chemical control. The occurrence of resistant weed species, such as weedy rice, can reduce rice yields from 5 to 100%

HRAC Group	Mechanism of action	Herbicide	Application Timing	*Echinochloa* spp.	*Oryza sativa*	*Eleusine indica*	*Urochloa plantaginea*	*Digitaria horizontalis*	*Cyperus* spp.	*Aeschynomene* spp.	*Sagittaria* spp.	*Heteranthera reniformis*
A	ACCase	Cyhalofop-p-butyl	POST	R	*	R			*	*	*	*
		Fenoxaprop-p-ethyl	POST		*	R		*	*	*	*	*
B	ALS	Bispyribac-sodium	POST	R	*	*	*	*	R		R	
		Penoxsulam	PRE/POST	R	*	*	*	*	R		R	*
		Imazapyr + Imazapic	PRE/POST	R	R	*		*	R			*
		Imazethapyr	PRE/POST	R	R	*	*	*	R	*	*	*
		Imazethapyr + Imazapic	PRE/POST	R	R	*	*	*	R	*	*	*
		Ethoxysulfuron	POST	*	*	*	*	*			R	*
		Pyrazosulfuron-ethyl	POST	*	*	*	*		R	*	R	
		Metsulfuron-methyl	POST	*	*	*	*	*	*		R	
C	PS II	Bentazon	POST	*	*	*	*	*		*	*R	*
		Propanil	POST		*			*				
D	PS I	Paraquat	NP						*	*	*	*
E	PPO	Carfentrazone-ethyl	POST	*	*	*	*	*		*		
		Saflufenacil	POST	*	*	*	*	*	*			*
		Oxadiazon	PRE						*	*	*	
		Oxyfluorfen	PRE							*	*	*
F_4	DOXP	Clomazone	PRE		*				*		*	*
G	EPSPS	Glyphosate	NP								*	*
K_1	MA	Pendimethalin	PRE		*				*	*	*	*
O	AUX	2,4-D	PRE/POST	*	*	*	*	*	*		*	*
		Quinclorac	POST	R	*	*	*	*	*		*	*
-	PS II + AUX	Propanil + Triclopyr	POST		*	*	*	*	*		*	*

☐Information not available; ■no control; ■control < 50%; ■control 50–70%; ■control 71–95%; ■control > 95%.

*Product not registered to control the weed species. ACCase: lipid synthesis inhibition (inh. of ACCase); ALS: inhibition of ALS (branched chain amino acid synthesis); PS II: inhibition of photosynthesis PS II; PS I: PS I electron diversion; PPO: Inhibition of protoporphyrinogen oxidase; DOXP: Inhibition of DOXP (1-deoxy-d-xylulose-5-phosphate or clomazone) synthase; EPSPS: Inhibition of EPSPS (5-enolpyruvylshikimate-3-phosphate) synthase; MA: Inhibition of microtubule assembly; AUX: Synthetic auxin. Pre: Pre-emergence; Post: Post-emergence; NP: Application on needle point, glyphosate applied over the first-day emerging rice, R: Resistant. HRAC: Herbicide Resistance Action Committee. Font: SOSBAI, 2016 and Agrofit, 2017 < Available at: http://agrofit.agricultura.gov.br/agrofit_cons/principal_agrofit_cons>.

Table 1. Application timing, control levels of the most troublesome weed species in Brazilian Rice.

[13], resulting in large economic losses [14]. Thus, greater use of cultural methods, such as crop rotation, should be taken into account to reduce the weed population, resulting in less dependence on herbicides, selection pressure, and herbicide resistance.

The majority of weed resistance cases in rice are reported for ALS inhibitors, indicating that herbicide use tends to shift to other mechanisms of action to efficiently control ALS-resistant weed species. For example, clomazone and propanil provide a great control (>95%) of *Echinochloa* spp. in Brazil as indicated in **Table 1**; however, biotypes with resistance to those herbicides have been already reported in Arkansas and California due to their frequent use [15, 16]. Thus, it is likely that herbicide resistance might evolve in Brazil for these species as a consequence of the increasing frequency in which they are sprayed.

The future introduction of a new herbicide-tolerant technology for paddy rice in Brazil, the Provisia™ Rice System, includes post-emergence ACCase-inhibiting herbicides as an alternative to improve the control of resistant grass species such as weedy rice [17]. Therefore, this technology tends to increase the use of these herbicides, which is a mechanism of action considered to pose a high resistance risk [18]. In the past 30 years, more than 35 grass species have evolved resistance to ACCase-inhibiting herbicides worldwide, especially due to target-site resistance mechanism [19], which threatens the long-term use of this technology in paddy rice. Therefore, this new technology shows great potential to reduce problems with resistant grass species but to ensure the longevity and optimize its efficiency, it is necessary to carefully follow the recommendations for use.

It is possible to observe that there is a great amount of herbicides registered for weed control in rice (**Table 1**). However, weed resistance has been reported for several molecules, especially for weedy rice and *Echinochloa* spp. There are some herbicides that provide satisfactory control of resistant biotypes when sprayed in pre-emergence, for instance, *Echinochloa* spp. resistant to ALS inhibitors can be controlled with the application of pendimethalin (MA) and clomazone (DOXP), with great control levels being reported (up to 95%) in experimental studies. However, it is likely that some plants will escape pre-emergence control and the herbicide options for post-emergence are quite limited because the species are already resistant to most of them.

Table 1 also shows that herbicides such as oxadiazon and oxyfluorfen (PPO) do not control *Echinochloa* spp. and weedy rice when applied in pre-emergence. However, when applied on a water layer before sowing the crop (label instructions), these herbicides can provide better control of such weeds. The application of these herbicides is quite complex and growers must follow carefully the label instructions to achieve greater control efficiency. Moreover, these herbicides are likely to contaminate the environment and can cause crop injuries, which are some of the reasons for their greatly reduced usage in recent years.

It is also important to mention that the control levels given in **Table 1** for all herbicides are only expressed when they are applied following the instructions of the manufacturer, with specific doses and at the correct development stage of the crop and weed.

Based on the aforementioned facts, the need to include other control strategies, such as crop rotation that would enhance the number of molecules that can be used to control these species is evident. Nevertheless, it is important to mention that weed control levels provided by cultural measures are often meager in comparison to the efficacy of herbicides and, thus, do not reduce their need, at least in the short term. Moreover, the costs and the unpredictability of many cultural strategies are the main reasons why farmers are reluctant to adopt them, and IWM strategy will only be prioritized when the occurrence of resistant weed biotypes causes extreme failures and almost complete lost in herbicide efficacy [20].

3. Crop rotation in Brazilian lowlands

Lowlands in Southern Brazil are mainly cultivated with rice in the summer period and kept uncultivated during the fallow season. Crop residues left on the soil surface can be used for cattle grazing and in some cases, cover crops are sown during the winter. In general, long-term crop rotation is not included in this cropping system due to the introduction of chemical fertilizers and pesticides, mechanization, and improved crop varieties [21]. However, crop rotation is one of the essential practices in sustainable agricultural systems, because of its effects on soil fertility, control of pathogens and pests, including weeds.

Yield reduction due to weed competition in rice cropping is estimated between 10 and 15% of potential production [22, 23]. Nevertheless, it was the widespread distribution of imidazolinone-resistant weedy rice that promoted the introduction of new crops in areas under rice monoculture. This strategy aims to reduce the seed bank of troublesome weed species and prevent rice grain yield and quality losses caused by weed interference. There are several mechanisms responsible for this effect, including allelopathy, changes in fauna, and disturbance patterns, which could diversify selection pressures by influencing seed bank dynamics. Some studies have shown that the seed bank of troublesome species in rice cultivation is greatly reduced when monoculture is abandoned [24, 25]. Rotation also affects species communities by determining the tillage frequency and effects attributed to cropping practices, such as herbicide programs, crop seed rate, and sowing time [26, 27].

Moreover, the introduction of other crops, such as soybean, in lowlands increases soil fertility due to nutrients cycling and reduces some disease pressure. Even though the positive effects of a very diverse crop rotation scheme that includes legumes and cereals are well recognized, the greatest constraints for the introduction of this strategy in lowlands are the drainage problems and the absence of species that endure long periods of water surplus in the soil. The poor natural drainage in these areas is usually the result of the flat relief associated with a shallow soil profile and impermeable sub-surface layer [28]. The physicochemical characteristics of the soils and the low natural fertility are other factors that affect crops performance in these fields [29].

Several studies aimed to evaluate the performance of various summer and winter crops to be used in a rotation scheme with rice in lowlands and will be explored in more detail in the following sub-sections.

3.1. Summer crops

Summer crops such as maize (*Zea mays*), sorghum (*Sorghum bicolor*), and soybean (*Glycine max*) have been explored in a crop rotation scheme with rice. Researchers have been trying to identify cultivars of these crops that can adapt to lowlands [30].

The performance of maize in these soils is quite limited because their physicochemical features do not favor the development and productivity of this crop. Lowlands soils in Southern Brazil are generally acidic, with low pH (ranging from 4.5 to 5.4), and maize plants develop better in soils with pH close to 7. Therefore, liming the soil is an essential practice in these soils to allow maize cultivation [31]. The choice of a maize cultivar with vigorous stalk, adequate height, low spike insertion, reduced lodging, and breaking resistance is another aspect that has to be considered when including this crop in a crop rotation system with rice [30].

On the other hand, sorghum is a species that adapts well in lowland soils because it has a great tolerance to drought periods and water excess when in the advanced stages of development (more than five leaves), producing up to 70 t/ha of biomass that can be used for cattle grazing. Therefore, the introduction of this species in a rotation system with rice can be a tool to reduce the seed bank of troublesome species in these areas (**Figure 1**), though trading of

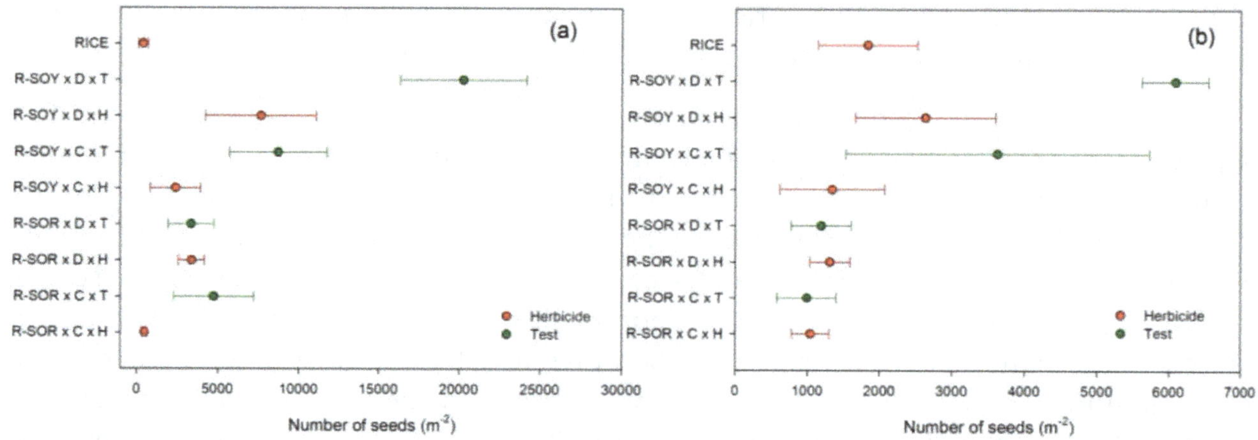

Figure 1. Number of seeds of *Echinochloa* spp. (a) and *Urochloa plantaginea* (b) under two tillage systems (conventional and direct drilling) and herbicide application (H- with or T-without) as a function of two crop rotation schemes: R-SOY—rice-soybean and R-SOR—rice-sorghum, two tillage systems (C- conventional and D- direct drilling) and herbicide application (with or without). Treatment means were compared on the basis of 95% confidence intervals.

the production can be a real obstacle to its wide cultivation. Moreover, the performance of this crop in lowland soils is highly associated with the choice of cultivar, adequate sowing date, and the use nitrogen fertilizers [30].

From the crop rotation point of view, the introduction of maize and sorghum in areas under rice monoculture brings several benefits for weed management; however, the inclusion of a legume species adds much more diversity to this system. Thus, soybean is probably the most promising crop to be used in a crop rotation scheme with rice, allowing farmers to increase their income, control weeds more efficiently by diversifying herbicide mechanisms of action of the herbicides and cultural practices. Moreover, leguminous species increase nitrogen (N) availability in the soil due to symbiotic N_2-fixation, lowering fertilizer needs for the following crop and increasing the yields of cereals grown in succession [31].

Studies evaluating the performance of soybean in Brazilian lowlands showed that this crop may be highly productive in these soils, reaching more than 4.000 kg ha^{-1} [32]. However, soybean is still less profitable than rice in lowland soils because the crop is quite sensitive to water excess, especially during germination and emergence. Water surplus in soil during flowering and grain filling can also affect soybean productivity, though the crop is slightly less sensitive to this stress at those development stages [33]. Thus, it is clear that the feasibility of the rice-soybean rotation depends on the progress of research works for the adaptation of different genotypes to flooding and poor drainage conditions. Moreover, the compaction of lowland soils and reduced nitrogen fixation due to low rhizobium activity are other limiting factors to soybean productivity. Nevertheless, soybean has been shown to be a valuable tool in controlling weedy rice and aquatic weed problems as well as reducing some disease pressure.

In a study conducted at Embrapa Temperate Agriculture (Pelotas-Brazil), evaluating the effects of two crop rotation systems: rice-soybean (R-SOY) and rice-sorghum (R-SOR); two tillage systems: conventional and direct drilling; and herbicide application: with or without; on the seed bank of *Echinochloa* spp. and *Urochloa plantaginea* (**Figure 1**), it was reported that in general, the number of seeds of both species in the seed bank was higher under rice-soybean rotation than in rice-sorghum rotation.

The seed bank of both species under R-SOR rotation was not affected by tillage systems and herbicide treatment. In R-SOY rotation, the number of seeds of *Echinochloa* spp. was higher in direct drilling than in conventional tilling in the control treatments (without herbicides). Moreover, the inclusion of herbicides reduced the seed number of this species in both tillage systems under R-SOY rotation. On the other hand, the soil seed bank of *U. plantaginea* in a R-SOY rotation in the control plots was not affected by tillage, but the inclusion of herbicides reduced the number of seeds per m^2 in direct-drilling plots. These results showed that rice-sorghum rotation is a good option to reduce the seed bank of *Echinochloa* spp. and *U. plantaginea* independently of tillage system and herbicide treatment. The success of a rice-soybean rotation to reduce the seed bank of these species depends on the tillage system and the inclusion of herbicides. When this system is not well manipulated, there is a risk of increasing the number of seeds in the seed bank as seen in the combination of this crop rotation (R-SOY) with direct drilling.

Another study, conducted at the same institution, tested several herbicide treatments that can be considered when soybean is introduced into a crop rotation system with rice to control troublesome species such as weedy rice, *Echinochloa* spp. and *U. plantaginea* (**Figure 2**). The results showed that all treatments efficiently controlled *U. plantaginea*, which was also suppressed in

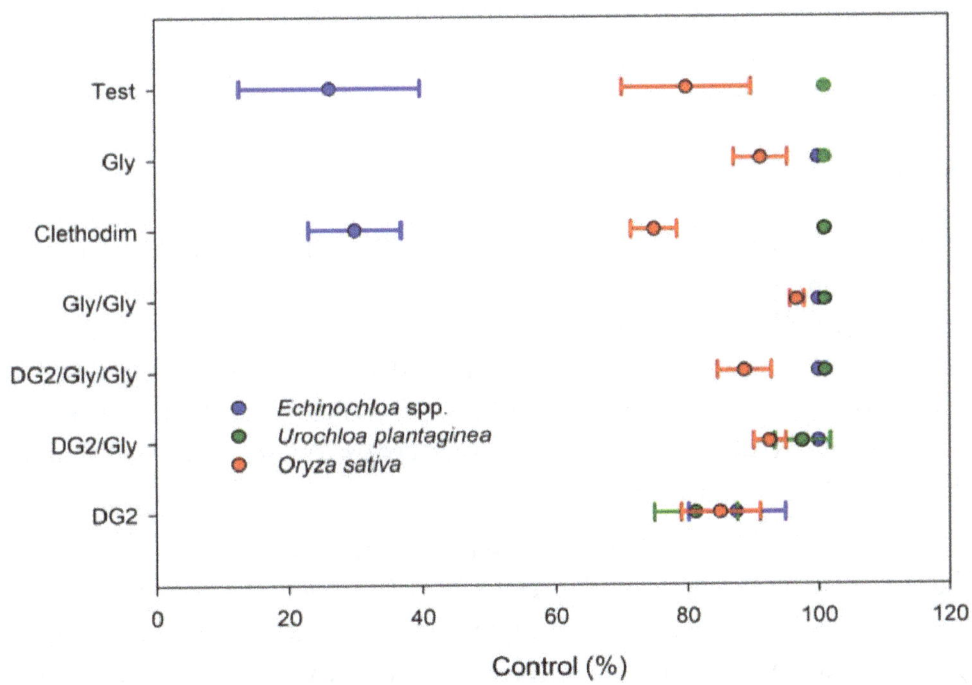

Figure 2. Control (%) of *Echinochloa* spp., *Urochloa plantaginea*, and *Oryza sativa* (weedy rice) with different herbicides treatments, considering a rice-soybean rotation. Gly-one post-emergence application of 3 L ha^{-1} of glyphosate; clethodim—one post-emergence application of 600 mL ha^{-1} of clethodim; Gly/Gly—two post-emergence application of 3 L ha^{-1} of glyphosate; DG2- one pre-emergence application of s-metolachlor (dual gold); DG2/Gly- one pre-emergence application of s-metolachlor and one post-emergence application of 3 L ha^{-1} of glyphosate; DG2/Gly/Gly- one pre-emergence application of s-metolachlor and two post-emergence application of 3 L ha^{-1} of glyphosate. Treatment means were compared on the basis of 95% confidence intervals.

the control plot (test) due to the high infestation of *Echinochloa* spp. The control of *Echinochloa* spp. was satisfactory (> 80%) in all treatments, except when clethodim was sprayed, which showed similar results with the untreated plots (test). Clethodim resulted in lower control percentage for weedy rice as well, which were not statistically different from the control plots. Post-emergence applications of glyphosate demonstrated good control for weedy rice. Moreover, a single application of s-metolachlor (DG2, **Figure 2**) provides more than 80% control for the three species. Thus, to avoid the pressure selection and future cases of weed resistance in soybean, the application of a pre-emergent followed by a post-emergent herbicide is a good control strategy in this scenario. Based on the results of this study, it is possible to observe that the herbicide rotation scheme is made viable by the inclusion of soybean into a crop rotation system in lowland soils, and can greatly reduce weed occurrence and consequently, the seed bank of some troublesome species in rice cultivation.

Nowadays, soybean is considered the best option in a crop rotation scheme with rice in lowland soils, although it presents some obstacles. The variation of cropping sequences with the inclusion of soybean creates an unstable environment for most weeds, which prevents the annual recurrence of particular weed species that are promoted by rice cultivation. Crop rotation, in general, adds more diversity into the systems; however, a monotonous rotation scheme composed only of rice and soybean can exert a selection pressure on the weed community, favoring species most adapted to both crop environments. Therefore, weed species such as *Conyza* spp. that were not problematic in these areas when under rice monoculture, can be favored due to the introduction of soybean in the system. Moreover, *Echinochloa* spp. and weedy rice can become problematic for soybean cultivation if their control is not satisfactory and end up evolving resistance to frequently used herbicides such as glyphosate.

3.2. Winter crops

Winter cover crops, which are grown during an otherwise fallow period, are a possible means of improving weed control in rice cultivation. Cover crops are well known to improve nutrient dynamics, soil organic matter content, microbial activity, water retention, and prevent nitrate leaching [34]. Moreover, returning of crop straws has been suggested to improve overall soil conditions, reduce the requirement for N fertilizers, and support sustainable rice productivity. However, while the benefits of cover crops for nutrient management are well documented, weed effects are less verified.

Rice demands high amount of potassium (K), which is mainly accumulated in the straw residues, and is easily lost by leaching and surface runoff after crop harvesting. Therefore, the inclusion of cover crops composed of grass species that tend to produce a great amount of biomass and absorb nutrients such as nitrogen and potassium, are a great strategy for nutrient cycling, substantially avoiding nutrient losses [35].

Moreover, pertinent choices of cover crop species can suppress the growth of serious weeds and protect the soil during winter, resulting in a better soil structure as opposed to leaving soil bare. Italian ryegrass (*Lolium multiflorum*) is a grass species that has been widely used during winter in paddy soils and is a good option for cattle grazing and as cover crop. The species has great biomass production and high nutritional value for animals, as well as impressive ability to re-grow after grazing, being highly competitive for nutrients, water, and sunlight [36]. Italian ryegrass can naturally establish itself from soil seed bank after the first year of cultivation in a given area, reducing costs with its plantation [37]; besides, it is well adapted to lowland soils. Moreover, when this species is cultivated, weed growth and development are inhibited due to the alellopathic substances released by the crop [38, 39].

Figure 3 shows how important the inclusion of cover crops is to hamper weed infestations. In the left side of the picture, the area was left uncultivated favoring weed establishment, especially *Conyza* spp., whereas in the right side, Italian ryegrass was sown, which greatly reduced the infestation weeds.

Moreover, **Figure 4** gives the density of the weed flora (plants per m^{-2}) in a given area cropped with soybean for 2 years. In winter, the first half of the area was left fallow, while ryegrass was grown in the other half. The weed flora was mainly composed of *Conyza* spp, *Soliva* spp., and *Richardia brasiliensis*. Weed density was assessed in the beginning of spring, showing a clear reduction in the number of plants where the cover crop was established in comparison to leaving the soil bare. It was also possible to observe that weed density in fallow plots was approximately 10 times bigger than the one assessed in the plots with Italian ryegrass. Similar results were obtained by other authors who found that this species is an effective choice of cover crop to suppress weed communities due to its great biomass production and alellopathic properties [38, 40, 41].

It is clear that the introduction of Italian ryegrass into a crop rotation system with rice has many benefits in lowland soils. Nevertheless, the inadequate management of the crop residues can jeopardize the establishment of rice in succession due to the great amount of biomass that is kept on the soil surface. When the amount of crop residues left on the soil surface is greater than 30 t ha^{-1} it is difficult for the seed drill to cut the straw as the residues act as a physical barrier [40], resulting in the poor establishment of the following crop. Moreover, the alellopathic properties of this species can be considered another drawback for its inclusion in a crop rotation scheme, especially in no-tillage systems, affecting rice germination and emergence [41].

Figure 3. Weed infestation in experimental plots with (right) and without (left) Italian ryegrass (*Lollium multiflorum*) in southern Brazil.

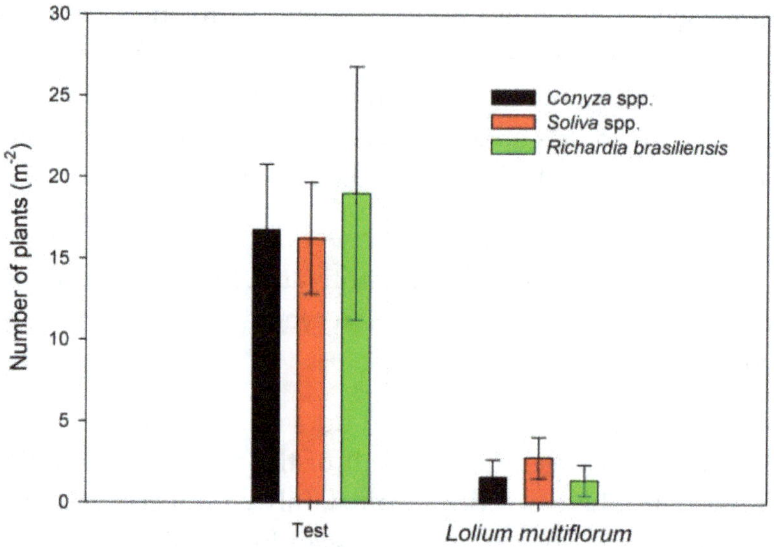

Figure 4. Number of plants (m^{-2}) of *Conyza* spp, *Soliva* spp., and *Richardia brasiliensis* in plots without (fallow) and with *Lollium multiflorum* as a cover crop in Southern Brazil.

Glyphosate is normally used to control this species prior to sowing the summer crop due to its broad spectrum of action, efficiency, and low cost. However, the intensive use of this herbicide has left resistant biotypes of Italian ryegrass, which are becoming more frequent [42]. Other herbicides, such as the popularly known graminicides that inhibit the ACCase enzyme can be used to efficiently control biotypes resistant to glyphosate. These herbicides can be used in a mixture with glyphosate or with another herbicide with broad spectrum of action such as paraquat or ammonium-glufosinate [43, 44]. However, it is important to consider that biotypes of Italian ryegrass resistant to ACCase inhibitors have been already reported and, therefore, require careful use.

The efficiency of these herbicides is highly dependent on the development stage in which they are applied. Nevertheless, the main constraint in the use of these compounds in lowland soils is associated with the negative effects that they can cause in rice plants due to their residual activity. Residual herbicides tend to dissipate slowly in paddy soils due to poor drainage, and when the interval between spraying and sowing rice is short, crop establishment is likely to be affected.

Therefore, it is essential to determine the correct sowing and desiccation time of the cover crops, because the decomposition of crop residues can be quite slow in lowland soils, due to the great soil moisture and physical-chemical features of this type of soil. Thus, the herbicides used can affect the establishment and consequently the yield of the following crop. The ideal timing should be set according to cover crop traits, cultivation density, developmental stage, soil cultivation technique adopted, and level plus type of herbicide used [45].

Another option for winter rotation is the mixture of grass species and legumes, which has high nutritional value for animals if the crop is used for grazing; it can benefit rice cultivation due to the nitrogen (N) input in the soil and a great improvement of the physical-chemical properties of the soil. Among the legumes species that can be introduced in a crop rotation system in lowland soils, common bird's foot trefoil (*Lotus corniculatus* L.) and white clover (*Trifolium repens*) seem to be good alternatives for Brazilian lowland scenarios, as they survive to some degree in soils with poor drainage. However, little is known about the benefits of these species in relation to weed management when introduced into a crop rotation system in lowland areas, due to the lack of more detailed studies.

4. Recommendations to successfully implement crop rotation in lowlands

In the last decade, there has been an increasing interest in the introduction of new crops in lowland soils in Southern Brazil, which are mainly cultivated with rice in summer and used for cattle grazing in winter. This interest is driven by several factors such as to increase the profitability of the production system and reduce the problems that have been caused by herbicide-resistant weeds. As mentioned in the chapter, sorghum, maize, and soybean are the main crop choices to be included in a crop rotation scheme with rice; however, the success of these crops is highly dependent on manipulating the ecosystem according to their needs, especially making sure that poor soil drainage and fertility will not hamper the productivity of these crops. Moreover, ensuring good soil drainage during the winter as well, would allow farmers to introduce other grass species, such as *Avena sativa,* that can be very useful for cattle grazing for instance but are quite sensitive to waterlogging.

There are several strategies that can be adopted to enhance the natural drainage in these areas, even though they are quite flat. The use of furrows works quite well and can be used as an irrigation system as well [46]. Drought periods are quite common over the summer in Southern Brazil, and considering the water requirements of maize and soybean, irrigation might be needed to ensure crops productivity in this region. The digital elevation model (DEM) is another technique that can be used for the design and allocation of drains in the area, enhancing the natural soil drainage. The DEM can be obtained with geodesic data collected by a global navigation satellite system (GNSS), with accuracy improved by a real-time kinematic (RTK) positioning. This system provides a detailed survey of the area and through the analysis of the water flow, the drainage system is designed [46].

On the other hand, instead of only manipulating the environment to meet the needs of sorghum, maize, and soybean plants, the development of crop cultivars that tolerate periods of water surplus in the soil would be a great tool to ensure their adaptation to lowland soils. To date, several genes that control the behavior of plants under water stress have been already identified and characterized. However, the information gathered so far is not enough for the development of crop cultivars that would tolerate water excess in the soil; there is still a long way to go. To enhance this knowledge and amend the strategies that have been used for plant breeding, researchers are developing high-performance sequencers and making use of statistical and transformation techniques [47]. Therefore, farmers should expect in near future the introduction of high-yield crop varieties of soybean and maize that are well adapted to the paddy soils ecosystem. Nevertheless, there are some varieties of soybean and maize available in the market in Southern Brazil that perform better in lowland soils in terms of productivity and could be an option for producers, even though they do not tolerate waterlogging periods. For instance, among others, BMX Apolo, BMX Ícone, BRS Taura RR, BR IRGA 6070 RR, and BR IRGA 1642 IPRO are good choices of soybean cultivars, whereas P30F53H, 2B655Hx, and 2B688Hx are maize cultivars that are most promising to show high yields in lowlands [48]. Even though high-yielding cultivars of these crops are already available in the market, it is important to mention that new cultivars aiming for greater yields and stress tolerance are frequently launched. Thus, producers and professionals in this sector must keep themselves informed.

It is important to mention that there is no magic recipe to ensure the success of crops such as maize, soybean, or sorghum in lowland soils as each field has some peculiar attributes that should be taken into account and climatic conditions change all the time. Moreover, the introduction of a diverse crop rotation system alone is not sufficient to guarantee that the density of troublesome weed species will be reduced. However, the introduction of this strategy allow farmers to diversify herbicides (with different mechanisms of action), soil cultivation type, and timing and sowing dates, that together are capable of disturbing the ecosystem and hampering the establishment of recurrent weed species.

The most important thing to consider in the real world when establishing new cropping systems in a farm is to plan and introduce it in small areas in the beginning, allowing the necessary cultural modifications to be applied. This will make the crop rotation functional and productive, avoiding a possible economical drawback in case of problems in the first tests of the new crop rotation scheme.

Author details

Ananda Scherner[1], Fábio Schreiber[1], André Andres[2*], Germani Concenço[2], Matheus Bastos Martins[1] and Andressa Pitol[1]

*Address all correspondence to: andre.andres@embrapa.br

1 Federal University of Pelotas, Pelotas, Brazil

2 Embrapa Temperate Agriculture, Pelotas, Brazil

References

[1] Owen MDK. Diverse approaches to herbicide-resistant weed management. Weed Science. 2016;(Special Issue):570-584

[2] Green JM. Current state of herbicide in herbicide-resistant crops. Pest Management Science. 2014;**70**:1351-1357

[3] Grundy AC. Predicting weed emergence: A review of approaches and future challenges. Weed Research. 2002;**43**:1-11

[4] Merroto A Jr, Goulart ICGR, Nunes AL, Kalsing A, Markus C, Menezes VG, Wander AE. Evolutionary and social consequences of introgression of nontransgenic herbicide resistance from rice to weedy rice in Brazil. Evolutionary Applications. 2016;**7**:837-836

[5] Swanton CJ, Weise SF. Integrated weed management: The rationale and approach. Weed Technology. 1991;**5**:657-663

[6] Nichols V, Verhulst N, Cox R, Govaerts B. Weed dynamics and conservation agriculture principles: A review. Field Crops Research. 2015;**183**:56-68

[7] Bond W. What is the weed seed bank. In: Weed Management Handbook. 9th ed. London, United Kingdom: British Crop Protection Council; 2002

[8] Bàrberi P, Cascio Lo B. Long-term tillage and crop rotation effects and weed seed bank size andcomposition. Weed Research. 2001;**41**:325-340

[9] Fenner M. Seed Ecology. New York: Chapman and Hall; 1985

[10] Khanh TD, Chung IM, Xuan TD, Tawata S. The exploitation of crop allelopathy in sustainableagricultural production. Journal of Agronomy and Crop Science. 2005;**191**:172-184

[11] Cardina J, Herms CP, Doohan DJ. Crop rotation and tillage system effects on weed seedbanks. Weed Science. 2002;**50**:448-460

[12] Andres A, Concenço G, Melo PTBS, Schmidt M, Resende RG. Detecção da resistência de campim-arroz (Echinochloa sp.) ao herbicida quinclorac em regiões orizicolas do Sul do Brasil. Planta Daninha. **1**:221-226

[13] Burgos NR, Singh V, Tseng TM, et al. The impact of herbicide-resistant rice technology on phenotypic diversity and population structure of United States weedy rice. Plant Physiology. 2014;**166**:1208-1220

[14] Lee S, Jia Y, Jia M, Gealy DR, Oslen KM, Caicedo AL. Molecular evolution of the rice blast resistance gene pi-ta in invasive weedy rice in the USA. PLoS One. 2011;**6**:e26260

[15] Norsworthy JK, Scott R, Smith K, Still J, Estorninos LE Jr, Bangarwa S. Confirmation and management of clomazone-resistant barnyardgrass in rice. In: Proceedings of the 62nd Southern Weed Science Society Meeting; Orlando, Fla, USA, Abstract 210; 2009

[16] Yasuor H, TenBrook PL, Tjeerdema RS, Fischer AJ. Responses to clomazone and 5-ketoclomazone by *Echinochloa phyllopogon* resistant to multiple herbicides in Californian rice fields. Pest Managment Science. 2008;**64**:1031-1039

[17] Webster EP, Linscombe SB, Bergeron EA, McKnight BM, Fish JC. Provisia rice: A future option in rice. 2017. Web page: http://wssaabstracts.com/public/29/abstract-271.html [Accessed: October 6, 2017]

[18] Heap I. International Survey of Herbicide Resistant Weeds. 2006. Web page: http://www.weedscience.org. [Accessed: April 6, 2017]

[19] Délye C, Zhang X-Q, Michel S, Matéjicek S, Powles SB. Molecular bases for sensitivity to acetyl-coenzyme A carboxylase inhibitors in blackgrass. Plant Physiology. 2005;**137**:794-806

[20] Chauvel B, Guillemin JP, Colbach N, Gasquez J. Evaluation of cropping systems for management of herbicide-resistant populations of blackgrass (*Alopecurus myosuroides* Huds.). Crop Protection. 2001;**20**:127-137

[21] Power JF, Follett RF. Monoculture. Scientific American. 1987;**256**:57-64

[22] Moody K. Weed control in wet-seeded rice. Experimental Agronomy. 1993;**29**:393-403

[23] Baltazar AM, Smith RJ Jr. Propanil-resistant barnyardgrass (*Echinochloa crus-galli*) control in rice (*Oryza sativa*). Weed Technology. 1994;**8**:576-581

[24] Avila LA, Marchezan E, Machado SL, Silva RP. Evolução do banco de sementes de arroz vermelho em diferentes sistemas de utilização do solo de várzeas. Planta Daninha. 2002;**18**:217-230

[25] Andres A, Avila LA, Marchezan E, Menezes V. Rotação de culturas e pousio do solo na redução do banco de sementes de arroz vermelho em solo de várzea. Revista Brasileira de Agrociência. 2001;**7**:85-88

[26] Smith RG, Gross KL, Januchowski S. Earthworms and weed seed distribution in annual crops. Agriculture, Ecosystem & Environment. 2005;**108**:363-367

[27] Scherner A, Melander B, Kudsk P. Vertical distribution and composition of weed seeds within the plough layer after eleven years of contrasting crop rotation and tillage schemes. Soil and Tillage Research. 2016;**161**:135-142

[28] Pinto LFS, Miguel P, Pauletto EA. Solos de várzea e terras baixas. In: Cultivo de soja e milho em terras baixas do Rio Grande do Sul. Pelotas -RS, Brazil: Embrapa; 2017. pp. 23-43

[29] Sousa RO, Camargo FA, Vahl LC. 2006. Solos alagados: reação de redox. In: Meurer EJ (Org). Fundamentos de química do solo. 3rd ed. Porto Alegre: Evangraf, p. 185-211

[30] Gomes AS, Magalhães AM Jr. Arroz irrigado no sul do Brasil. Brasília-DF: Embrapa Informação Tecnológica; 2004. pp. 74-95

[31] Parfitt JMB. Produção de milho e sorgo em várzea. Pelotas: Embrapa Clima Temperado; 2000. p. 146

[32] Timm PA, Campos ADS, Bueno MV, Aires T, Silva JT, Schreiber F, Parfitt JMB, Scvittaro WB, Timm LC. Avaliação de cultivares de Soja produzida em sistema de camalhão em terras baixas. In: Proceedings of the X Congress Do Arroz Irrigado, Gramado – RS. Brazil; 2017

[33] Oliveira ACB. Cultivares de Soja. In: Cultivo de soja e milho em terras baixas do Rio Grande do Sul. Pelotas -RS, Brazil: Embrapa; 2017. pp. 127-140

[34] Wittwer RA, Dorn B, Jossi W, van der Heijden MGA. Cover crops support ecological intensification of arable cropping systems. Scientific Reports. 2017;**7**:41911

[35] Scivittaro WB, Sousa RO, Silva LS. Manejo da fertilidade do solo para o cultivo de soja e milho. In: Cultivo de soja e milho em terras baixas do Rio Grande do Sul. Pelotas -RS, Brazil: Embrapa; 2017. pp. 105-125

[36] Lopes MLT, Carvalho PCDF, Anghinoni I, et al. Sistema de integração lavoura-pecuária: desempenho e qualidade da carcaça de novilhos superprecoces terminados em pastagem de aveia e azevém manejada sob diferentes alturas. Ciência Rural. 2008;**38**:178-184

[37] Trezzi MM. Antagonismo das associações de clodinafop-propargyl com metsulfuron-methyl e 2,4-d no controle de azevém (*Lolium multiflorum*). Planta Daninha. 2007;**25**: 839-847

[38] Moraes PVD et al. Manejo de plantas de cobertura no controle de plantas daninhas na cultura do milho. Planta Daninha. 2009;**27**:289-296

[39] Soares GLG, Vieira TR. Inibição da germinação e do crescimento radicular de alface (cv. "Grand rapids") por extratos aquosos de cinco espécies de Gleicheniaceae. Revista Floresta e Ambiente. 2000;**7**:180-197

[40] Rizzardi MA, Silva LF. Influência das coberturas vegetais de aveia-preta e nabo forrageiro na época de controle de plantas daninhas em milho. Planta Daninha. 2006;**24**:669-675

[41] Trezzi MM, Vidal RA. Potencial de utilização de cobertura vegetal de sorgo e milheto na supressão de plantas daninhas em condição de campo: II – efeitos da cobertura morta. Planta Daninha. 2004;**22**:1-10

[42] Powles SB, Yu Q. Evolution in action: Plants resistant to herbicides. Annual Reviews in Plant Biology. 2010;**61**:317-347

[43] Vargas L, Roman ES, Rizzardi MA, Toledo EB. Manejo de azevémresistente ao glyphosate em pomares de maçã com herbicida select (clethodim). Revista Brasileira de Herbicidas. 2006;**1**:30-36

[44] Spader V, Cristiane É, Lopes P, Geraldo E. Residual activity of ACCaseinhibitor herbicides applied at pre-sowing of corn. Revista Brasileira de Herbicidas. 2012;**11**:42-48

[45] Dabney SM, Delgado JA, Reeves DW. Using winter cover crops to improve soil and water quality. Communications in Soil Science and Plant Analysis. 2001;**32**:1221-1250

[46] Parfitt JMB, Winkler AS, Pinto MAB, Silva JT, Timm LC. Irrigação e drenagem para cultivo de soja e milho. In: Cultivo de soja e milho em terras baixas do Rio Grande do Sul. Pelotas -RS, Brazil: Embrapa; 2017. pp. 45-78

[47] Mertz-Henning LM, Nepomuceno AL, Nakayama TJ, Neumaier N, Farias JRB. Avanços biotecnológicos para o desenvolvimento da tolerâncias de soja e milho ao estresse por encharcamento do solo. In: Cultivo de soja e milho em terras baixas do Rio Grande do Sul. Pelotas -RS, Brazil: Embrapa; 2017. pp. 317-336

[48] Emygdio BM. Cultivares de Milho. In: Cultivo de soja e milho em terras baixas do Rio Grande do Sul. Pelotas -RS, Brazil: Embrapa; 2017. pp. 141-162

12

Microwave Weed and Soil Treatment in Rice Production

Muhammad Jamal Khan and Graham Ian Brodie

Abstract

Herbicides resistance has challenged sustainable rice productivity. Consequently, interest in chemical-free weed management has increased to overcome this constraint. This chapter has demonstrated the effect of pre-sowing microwave soil heating as a new alternative to chemicals for confirmed herbicide resistant weeds of the Australian rice production system. Microwave can superheat weed plants, creating micro-steam explosions in the plant structures to kill weeds. This requires the least amount of energy to achieve weed control and can be likened to a 'knock down' herbicide treatment. Considerably, more microwave energy can be applied to the soil to achieve weed seed bank deactivation; however, there is growing evidence that this strategy also changes the soil biota and nutrient profile in favour of substantial increases in crop yield, when crops are planted into this microwave-treated soil. An energy application of approximately 400–500 J cm^{-2} gave approximately 70–80% reduction in weed establishment in three field trials conducted at two agro-ecological zones of the Australia. In addition, there was a 10 times higher nitrogen use efficiency, and a 37% higher water use efficiency was achieved through this aspect of the microwave technology. There is also evidence that the soil treatment strategy provides persistent effects, beyond a single season; therefore, the rice production is better than when using conventional weed control methods.

Keywords: weeds, herbicide resistance, microwave, soil health, crop health

1. Introduction

Rice (*Oryza sativa* L.) is the staple food of 60% of the world's population [1], performs a significant role in the socio-economic constancy of the world, and is grown in a vast range of agro-ecological conditions. In Australia, rice farming is done in the Murray-Darling Basin, on an area of 70,000 ha, with an annual grain production capacity of 0.69 M t [2]. Direct seeding is a common sowing strategy of rice in Australia due to high labour costs associated with transplanted rice systems. Weeds are one of the major biological constraints to increasing rice yield. Oerke [3] estimated that globally 34% crop productivity losses are due to weeds. However, the global decline in the production of rice, due to weeds, is estimated to be 10.2% [4]. Yield loss in a direct seeded rice crops is high, compared to transplanted rice [5]. Productivity losses of rice crops generally depend on climatic conditions, weed types, weed population density, rice variety, sowing methods and weed management practices.

The troublesome weeds of the Australian rice growing belt are barnyard grass (*Echinochloa crus-galli*), dirty dora (*Cyperus difformis*), arrowhead (*Sagittaria montevidensis*), and starfruit (*Damasonium minus*). Among all the weeds, Barnyard grass (*Echinochloa crus-galli* L.) is the major problematic bio-agent of rice [6] and is also considered to be the main weed of several semi-aquatic cropping systems [7]. It follows the C_4 photosynthetic pathway [5] and has indistinguishable morphology to rice at seedling stage, which makes it extremely competitive with the rice crop. A 57% reduction in rice yield was documented, with a barnyard grass population density of 9 plants m^{-2} [8]. Additionally, higher densities of barnyard grass may remove up to 80% of the soil nitrogen, especially during its vegetative growth stages [7]. Seed production is the key element of long-standing weed population dynamics [9]. The average seed production capacity of barnyard grass ranged from 20,000 to 73,000 seeds per plant [10] and 60% of these seed could become part of the weed seed bank. Therefore, effective weed management depends on reducing the soil weed seed bank [11].

1.1. Herbicide resistance

Globally, there are 400 weed species that have developed resistance to herbicides and annually nine new weed biotypes are reported as being herbicide resistant [12]. The overall number of herbicide resistant weed species in various crops is illustrated in **Figure 1**. Cross-resistance in weed flora is described as resistant to two or more weedicides of the same or different chemistry because of one resistant mechanism (RM) [13]. However, multiple resistances in individual weed species are generally characterized by the presence of two or more RMs. These mechanisms might be the mutation at the site of action (SOA) of herbicides (target site) or change in metabolism and translocation (nontarget site), which reduces the phytotoxic effect of herbicides on their SOA [14]. Of particular concern, the numbers of weed species, which have become resistance to glyphosate in Australian agricultural systems are shown in **Figures 2** and **3**. Metabolic resistance is more commonly found in monocot (grasses) than in dicot (broadleaf) weeds [14]. Herbicide resistance in weeds is the greatest threat to sustainable productivity of agricultural commodities in industrialized countries. Therefore, there is a present need of an alternative weed management strategy in exiting cropping system. A series of experiments have been conducted, at Dookie Campus of the University of Melbourne, to assess the effects of microwave energy as an alternative of chemical weed control.

1.2. Water use efficiency

Higher grain production per unit application of water is needed to enhance sustainable rice production for future demands. Australia is the driest inhabited continent on the planet

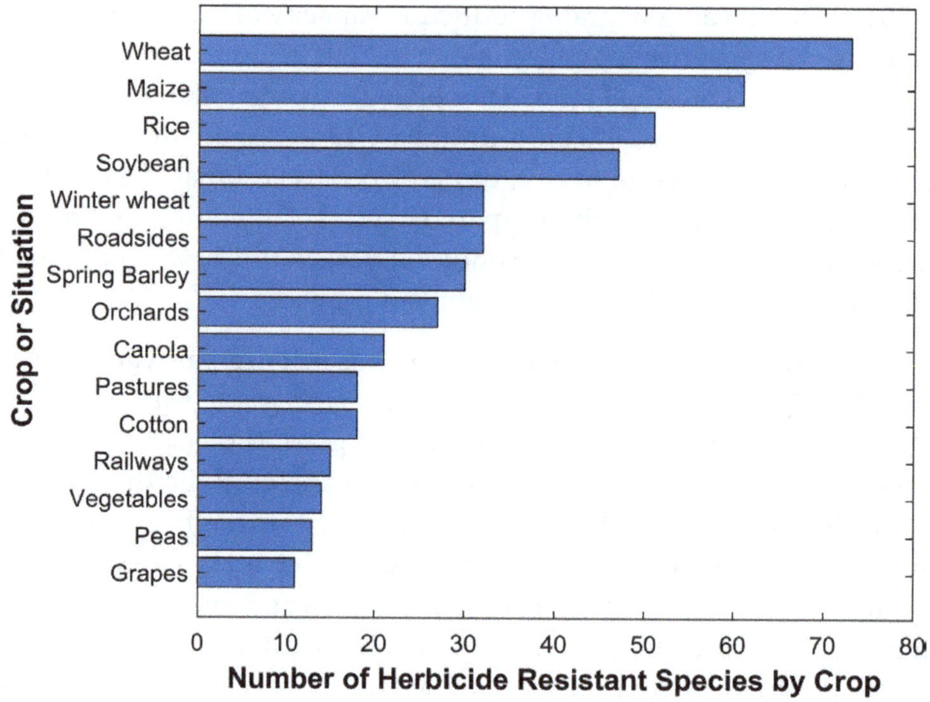

Figure 1. The overall herbicides resistant scenario of weed species in crops. Source: [12].

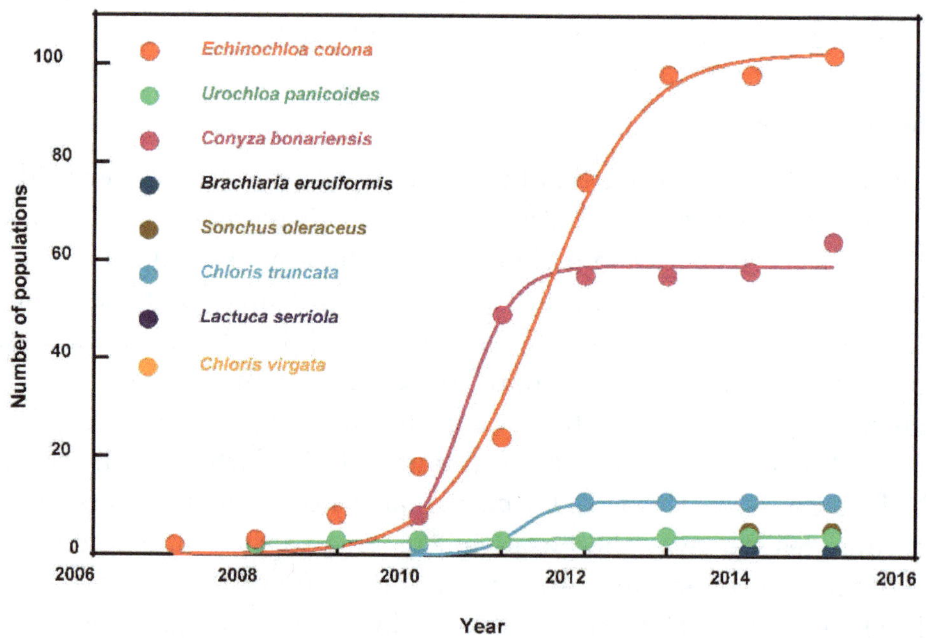

Figure 2. Confirmed glyphosate resistance summer weeds of Australian crop production systems. Source: [15].

and the Australian Academy of Technology and Engineering (ATSE) reported that 62% of Australia's water was consumed by the agriculture sector in 2013–2014. Effective water use, to improve crop yield, can save the sector's water for future generation. The cost of water in Australia is about AU$ 200–300 per Ml, which is consequently increasing the cost of rice production in Australia, independent of direct-drill farming, which postpones permanent flooding of the crop for almost 35 days.

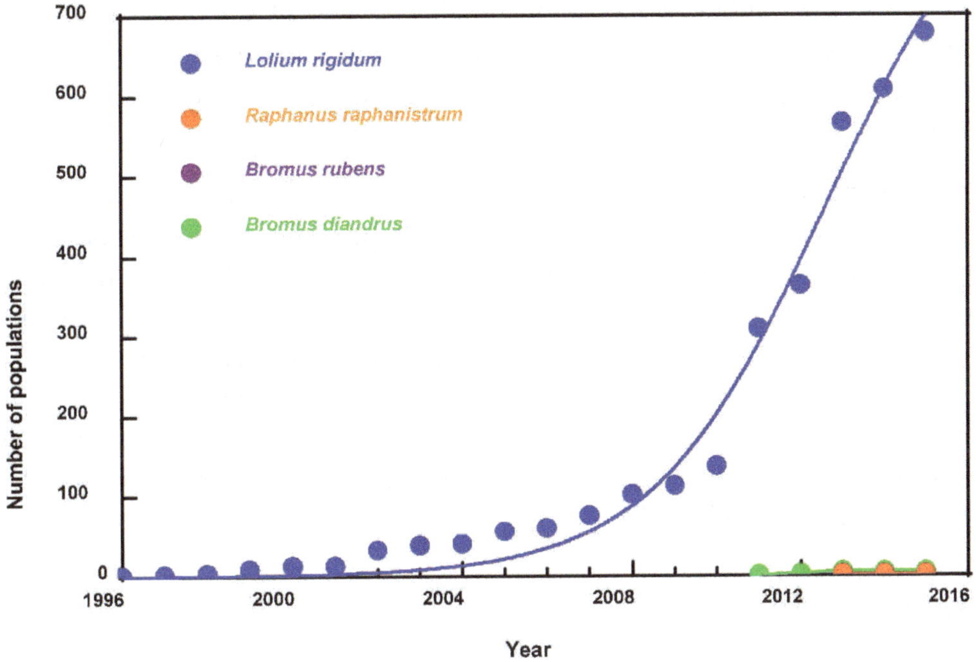

Figure 3. Confirmed glyphosate resistance winter weeds of Australian crop production systems. Source: [15].

Rice production in Australia consumes approximately 2.5–3 Ml ha^{-1} with an average grain harvest of about 1.4 t Ml^{-1}; however, microwave soil treatment increases the production to about 1.92 t Ml^{-1} [16], which is around 37% higher water use efficiency compared to untreated control plots. Globally, an average of 0.003–0.005 Ml of water is required to produce 1 kg of rice (i.e. , a yield of between 0.2 and 0.3 t Ml^{-1}), which is about 2–3 times higher than the water consumption of other cereal crops [17, 18]. Considering this, numerous studies have reported the effects of irrigation water volume on crop yield [19–21]. Interestingly, with a single crop management strategy, it is hard to harvest multiple benefits, including water use efficiency. In the following field studies, we achieved about 37% greater irrigation water use efficiency, where we treated the soil with microwave energy for weed seed bank depletion under field conditions. Therefore, microwave soil heating may also promote effective and efficient water use in Australian rice production systems, in addition to weed management. This assumption needs further research for validation under field conditions.

1.3. Organic rice production

Soil health is the key element in organic farming and as per worldwide agreement; soil fertility in organic farming system should be maintained on a long-term basis. Intensified rice farming has been deteriorating the soil quality [22] in Asian rice growing regions. However, in Australia, limited studies have reported that intensive organic farming enhanced soil fertility as compared to conventional agriculture practices [23]. It has been reported that microwave treatment of soil enhances the humification of soil organic matter [24] and has some positive effects on soil nitrogen availability for crop plants. In a pot trial, Khan et al. [25] reported a persistent effect of microwave soil heating on the second season wheat crop with better grain production benefits than in the first season after a once off microwave treatment, which suggests that there is a persistent effect of this technology on soil health. In addition, the population of ammonia oxidizer bacteria and archaea, studied during this pot trial, showed no significant response to the heating effect of microwave treatment time. The abundance of beneficial microbes is a direct indicator of soil fertility, suggesting that microwave technology can sustain soil health over a long period.

In Australia, weed management in organically produced rice offers extra challenges for the farmer. Sod seeding, where rice is directly sown into a pasture or legume stand, is a common establishment practices in southern New South Wales. The preexisting pasture of legume crop somehow suppresses aquatic weeds like dirty dora, starfruit, arrowhead and water plantain, but has little to no effect on barnyard grass. Another nonchemical weed control strategy is propane gas flame weeder; however, it costs about AU$ 12,000–15,000 ha^{-1}, along with careful water management and post-emergent harrowing [26]. Considering this, organic rice producers are still looking for a nonchemical weed control approach, which could control barnyard grass and sustain organic production.

2. Microwave energy

Microwave frequencies occupy the portion of the electromagnetic spectrum (300 MHz to 300 GHz) that lies between VHF radiowaves and thermal infrared. Their application falls into two categories, depending on whether the wave is used to transmit information or energy. The first category includes terrestrial and satellite communication links, radar, radio-astronomy, microwave thermography, material permittivity measurements, and so on [27]. The second category of applications is associated with microwave heating and wireless power transmission. In case of microwave heating, there is usually no signal modulation and the electromagnetic wave interacts directly with solid or liquid materials.

2.1. Essentials of microwave heating

"It has long been known that an insulating material can be heated by applying energy to it in the form of high frequency electromagnetic waves"([28], pp. 5). Industrial microwave heating has been used since the 1940s ([28], pp. 5). The initial experiments with microwave heating were conducted by Dr. Percy Spencer in 1946, following a serendipitous discovery while he was testing a magnetron [29]. Although Spencer was not the first to observe that microwave energy could impart heat to materials, he was the first to systematically study it. Since then, many heating, drying, thawing [30] and medical applications [31] have been developed.

One key benefit of microwave heating, over conventional convective heating, is speed. The origin of this speed is the volumetric interactions between the microwave's electric field and the material. In contrast, convective heat transfer propagates from the surface into the material, with the final temperature profile depending on the material's thermal diffusion properties [32] and the influence of moisture transport, which often hinders the convective heating process [33].

The factors that contribute to microwave heating include: the physical and chemical structure of the heated material; the frequency of the microwaves [34]; in some cases, such as wood, the orientation of the electrical field relative to the structure of the dielectric material is also important ([35], pp. 13-17); reflections from the interfacial surface of the heated material [27]; electric field strength [34]; the geometry of the microwave applicator [28]; the geometry, size, electrical and thermal properties of the dielectric material [36–38]; the exposure time; and the moisture content of the dielectric material [33, 35].

2.2. ISM band applications

Because microwaves are also used in the communication, navigation and defence industries, and their use in thermal heating is restricted to a small subset of the available frequency bands. A small number of frequencies have been set aside for Industrial, Scientific and Medical (ISM) applications [39]. The main frequencies of interest for industrial applications are 915 ± 13 MHz and 2450 ± 50 MHz [39].

3. Microwave weed treatment

In Australian agricultural industries, the total estimated cost of weed management and loss in crop productivity due to weeds is about AU$4 billion annually [40]. Microwave weed management is an alternative method of weed control in modern agriculture systems. The history of microwave-based weed management is given in **Table 1**. The efficacious application of microwave heating in agricultural systems can substitute for the sometimes hazardous, toxic and environmentally unsafe chemicals that are used to kill weeds [60]. Interest in the use of microwave energy as a tool to weeds control is mainly because of herbicide resistance of various weed species [61] and their long-lasting persistence in the environment [54, 62]. Microwave heating is not influenced by wind direction and speed, therefore prolonging the application periods compared to traditional methods of herbicides spraying [51].

Ayappa et al. [63] reported that the most important features of microwave heating are its accurate control, diminutive start-up time and volumetric heating. Microwave energy density is the most important factor in plant mortality rather than exposure time; therefore, two options for weed management, using microwave energy, become evident: long exposure to diffuse microwave energy; or deliberate application of a strongly focused microwave pulse to quickly debilitate the plants [58].

Microwave radiation, which triggers dielectric heating in plant tissues, is induced by the microwave's electric field. This internal heating ultimately kills or debilitates the plant [54]. Bigu-Del-Blanco et al. [49] treated 2-day-old seedling of maize with microwave energy at a frequency of 9 GHz for 22–24 h. The authors revealed that more exposure time to microwaves even at very low energy densities significantly dehydrated the maize plants and retarded their growth.

In contrast, the recent research on fleabane and paddy melon [58] has concluded that a short exposure (≤ 5 s) of high-intensity microwave heating was enough to hinder plant growth. The plant tissues, which were subjected to microwaves, rapidly dehydrated. Whatley et al. [64] stated that low moisture levels in soil attenuated the microwave transmission less than high moisture content. The authors suggested that pre-emergence microwave treatment for weed control should be worked out when the top soil layer (1–2 cm) contains relatively low moisture.

Microwave frequency	Energy level	Irradiation duration (s)	Treatment scenario	Target species	Percent weed-seed destruction	Reference
39 MHz	–	4–37 s	Pre-emergence	Hard red winter wheat	50% seed mortality	Nelson and Walker [41]
2.45 GHz	600 W	60 s	Pre-emergence (Dry, 4 h soaked and 46 h germinated seeds)	*Zea mays, Arachis hypogaea, Prosopis juliflora, Cucumis sativus, Brassica sp., Rumex crispus, Echinochloa colonum, Amaranthus sp., Gossypium hirsutum, Glycine max, Sorghum vulgare and Triticum vulgare*	17% reduction in germination in dry seeds but 100% in case of moist seeds at 10 s of exposure	Davis *et al.* [42]
2.45 GHz	600 W	8 s	Post emergence (Aquatic weed)	Duckweed (*Wolffia punctata*)	50%	Champ *et al.* [43]
2.45 GHz	2000 & 4000 W	Varying exposure time (not mention properly)	Pre and post emergences	Johnsongrass Morningglory Redroot Pigweed Texas panicum Barnyardgrass Sunflower London rocket Rigseed euphorbia	For post-emergence MW treatment 309 J/cm^2 energy was required for 100% control (field conditions) while for pre-emergence MW weed control 73 J/cm^2 gave 85–100% control (glass house conditions)	Wayland *et al.* [44]
2.45 GHz	45–720 J cm^{-2}	No information	Pre-emergence	London rocket (13 cm deep in soil profile) and Sunflower (2.5 cm seeded depth)	87% for London rocket and 93% for sunflower	Menges and Wayland, [45]
2.45 GHz	100–750 W	120–1200 s	Pre-emergence	Clover and Turnip	60–78% reduction in seeds germination	Hightower *et al.* [46]
2.45 GHz	0.1–1.5 kW	Varying exposure time	Pre-emergence of seeds in soil	Black medic, Barnyard grass, Foxtail purslane, redroot pigweed, large crabgrass,	50%	Rice and Putnam [47]
2.45 GHz	–	360 s	Pre-emergence	*Brassica napus, Linum usitatissimum, Avena fatua*	85–95%	Bhartia *et al.* [48]

Microwave frequency	Energy level	Irradiation duration (s)	Treatment scenario	Target species	Percent weed-seed destruction	Reference
9 GHz	10–30 mW/cm^2	22–24 h	Post emergence	*Zea mays*	100% growth inhibitions	Bigu-del-Blanco *et al.* [49]
2.45 GHz	1.2 kW	5–45 s	Pre-emergence	*Trifolium and Medicago*	85% reduction in germination	Crawford [50]
2.45 GHz	500 W	30 s	Pre-emergence	*Avena fatua*	60% (based on seed moisture)	Diprose *et al.* [51]
2.45 GHz	1.5 kW	0, 10, 20 and 30	Pre-emergence	Wild oat & wheat	90–100%	Lal and Reed, [52]
2.45 GHz		120 s	Pre-emergence	Avena sativa and native weed seeds	Reduced weed seeds emergence	Barker and Craker, [53]
2.45 GHz	900 W	4, 8, 16, 32, 64, 128 and 256 s	Post emergence	*Abutilon theophrasti*, Panicum miliaceum, Lucerne and Rapeseed	Complete dehydrating of plants	Sartorato *et al.* [54]
2.45 GHz	800 W	120, 240, 420 and 960 s	Pre-emergence	Rubber vine, Parthenium and Bellyache bush	88% (Rubber vine), 67% (Parthenium) and 94% (Bellyache bush) mortality at 960 s irradiation	Bebawi *et al.* [55]
2.45 GHz	0.10–1.24 kWh m^{-2}	30–300 s	Pre and post emergence	*Malva parviflora and Triticum astevium*	100% destruction of tested specie at 0.65 kWh m^{-2}	Brodie *et al.* [56]
2.45 GHz	700 W	120, 240, 320 and 720 s	Pre-emergence treatment of soil	*Lolium perenne and Lolium rigidum*	100% seed mortality was achieved at 240 s of MW irradiation	Brodie *et al.* [57]
2.45 GHz	750 W	5, 15, 30 and 60 s	Pre and post emergence	Prickly paddy melon	100% debilitation of plants	Brodie *et al.* [58]
2.45 GHz	2 kW	5, 10, 15, 30, 60 s	Post emergence	Ryegrass and wild radish	100% mortality	Brodie and Hollins [59]

Table 1. History of microwave weed management in different scenarios.

Van Wambeke et al. [65] and Benz et al. [66] reported that seeds, fungi and nematodes could be effectively controlled with a short exposure to microwave treatment; however, the efficacy of this short exposure was highly influenced by soil texture, exposure time (sec), soil depth and soil moisture content. Davis et al. [42] conducted an experiment to evaluate the effects of microwave on the seedling survival percentage of twelve species. They described that the seedling (48 h germination) exhibited no survival after short exposure of microwave energy and concluded that susceptibility of young seedlings to microwave heating was highly correlated with moisture content and absorption of energy. Davis et al. [67] proposed that the specific mass and volume of crop seeds were positively correlated to seed mortality during microwave heating. This might be due to the "radar cross-section" [68] attainable by seeds to transmitting microwave. More radar cross-section enables the seed to interrupt, and thus absorbs more microwave energy. This seems to be the cause of death [69].

The use of electromagnetic radiations for post-emergence control of broad leaves and grasses is the least energy-consuming process available for microwave weed control [70]. Brodie et al. [57] stated that, based on microwave energy calculation for seeds and plants on the sandy soil surface, far more energy was required to kill dry seeds as compared to the previously emerged plants. The actual energy requirement on a large scale would depend on plant density and three-dimensional microwave distribution. Hence, the total energy required for weed management might be significantly reduced if weed seeker systems [71] are employed to control the activation of the microwave unit.

Thermal runway, due to the resonance of electromagnetic field inside the structure of dielectric material, is common in dielectric heating [72, 73]. Total energy and time exposure could be dramatically reduced if thermal runway can be induced in weed plants throughout microwave irradiation treatment; therefore, analogous energy requirements to those related with traditional chemical weed control method could be achieved. This temperature-time exposure scenario can only be discovered and understood through more research into the microwave heating of biological materials.

Based on previous findings and the results of recent studies reported by Khan et al. and Brodie et al. [16, 25]; pre-sowing microwave irradiation of soil for 120 s in first field trial and 60 s in two other field trials, in rice crops, gave significant reduction in weed emergence (**Table 2**). It is possible to reduce weed pressure in direct-seeded rice systems through microwave irradiation of soil in Australia; however, more consolidated research efforts are needed to understand the long-term effects of microwave irradiation and weed control in rice.

3.1. Killing emerged plants

It has been confirmed that microwave energy can debilitate emerged weed plants with a very short exposure time [25, 56, 58, 59]. Some specific microwave energy dose responses are shown in **Figure 4**.

3.2. Soil treatment

Soil is a complex three-dimensional living substance. The propagation of microwave energy through soil depends upon the gravimetric (θ_g) and volumetric (θ_v) moisture content [74], bulk density, organic matter content [75], soil texture [57] and specific heat of soil. Among them, the soil moisture content has three major impacts on microwave heating: (1) moisture increases the soil surface reflectivity [76], which ultimately reduced the microwave penetration into the soil [28]; (2) moist soil readily absorbs the microwave energy to generate heat [28] thus less total microwave energy propagated into the soil; and (3) moisture is also responsible for heat-diffusing phenomena in the soil profile [77].

It has been reported that the dielectric constant (ε') of known soil at known θ_g is proportional to the bulk density of soil. The dependence of soil dielectric constant on bulk density is described by the direct dependence of bulk density on fraction of soil moisture volume [78]. The textural composition of soil (particles sizes distribution) affects the dielectric constant (ε'). The higher percentage of clay particles (with bulk density range of 1.0–1.6 g cm^{-3}) increases the dielectric constant of soil [79]. This might be due to higher water holding capacity of clay particles. Therefore, this will increase the absorption of microwave energy by soil for its synchronized functions.

The temperature profile is dependent on the microwave electric field strength (E) within the soil. Brodie [56] has extensively studied the temperature distribution in soil due to microwave energy application through a horn antenna. The temperature profile can be described by Eq. (1). The Nomenclature of Eq. (1) is presented in **Table 3**.

$$T = \frac{n\omega\varepsilon_o\kappa''(e^{4\gamma\alpha^2 t}-1)}{4k\alpha^2}\left[e^{-2\alpha z}+\left(\frac{h}{k}+2\alpha\right)ze^{-z^2/4\gamma t}\right]\left[\tau\int_{-\frac{B}{2}}^{\frac{B}{2}}\int_{-\frac{A}{2}}^{\frac{A}{2}}\cos\left(\frac{\pi}{A}x'\right)\frac{e^{-j\beta_o\left(\sqrt{(x-x')^2+(y-y')^2+z^2}+\sqrt{R_0^2+(x')^2}+\sqrt{R_0^2+(y')^2}\right)}}{\sqrt{(x-x')^2+(y-y')^2+z^2}}.dx'.dy'\right]^2 \tag{1}$$

Figure 5 compares measured temperature distributions with those predicted by Eq. (1). The highest temperature in the microwave-treated soil was along the centre line of horn antenna and between the 0.02 and 0.05 m below the soil surface. **Figure 6** illustrates the effects of microwave soil treatment using a different system configuration and treatment scenario.

3.3. Effect on soil

3.3.1. *Effect of microwave energy on nutritional dynamic of soil*

The dynamic of soil key nutrients (Carbon, Nitrogen, Phosphorus, Potassium and Sulphur) is explained by the knowledge of size and turnover rate of plant biopolymers such as C-N compounds, cellulose and hemicellulose and lignin [81]. The soil-microwave interaction is the function of various soil properties such as texture, moisture, salinity, bulk density and temperature [58, 78, 79]. Cooper and Brodie [82] investigated the effect of different durations

Weed parameters	**Treatments**		**LSD**	**Percentage change from control**
	microwave treatment	**Untreated control**	**(p = 0.05)**	
Weed density (plants $plot^{-1}$)	17.6^a	94.8^b	37.7	−80%
Weed fresh weight (g $plot^{-1}$)	156.4^a	612.8^b	426.6	−74.6%
Weed dr. Weight (g $plot^{-1}$)	21.6^a	122.6^b	69.6	−82%

Table 2. Effect of microwave energy application for weed seedbank depletion in direct seeded-rice crop under filed conditions in Australia (adapted from [16]).

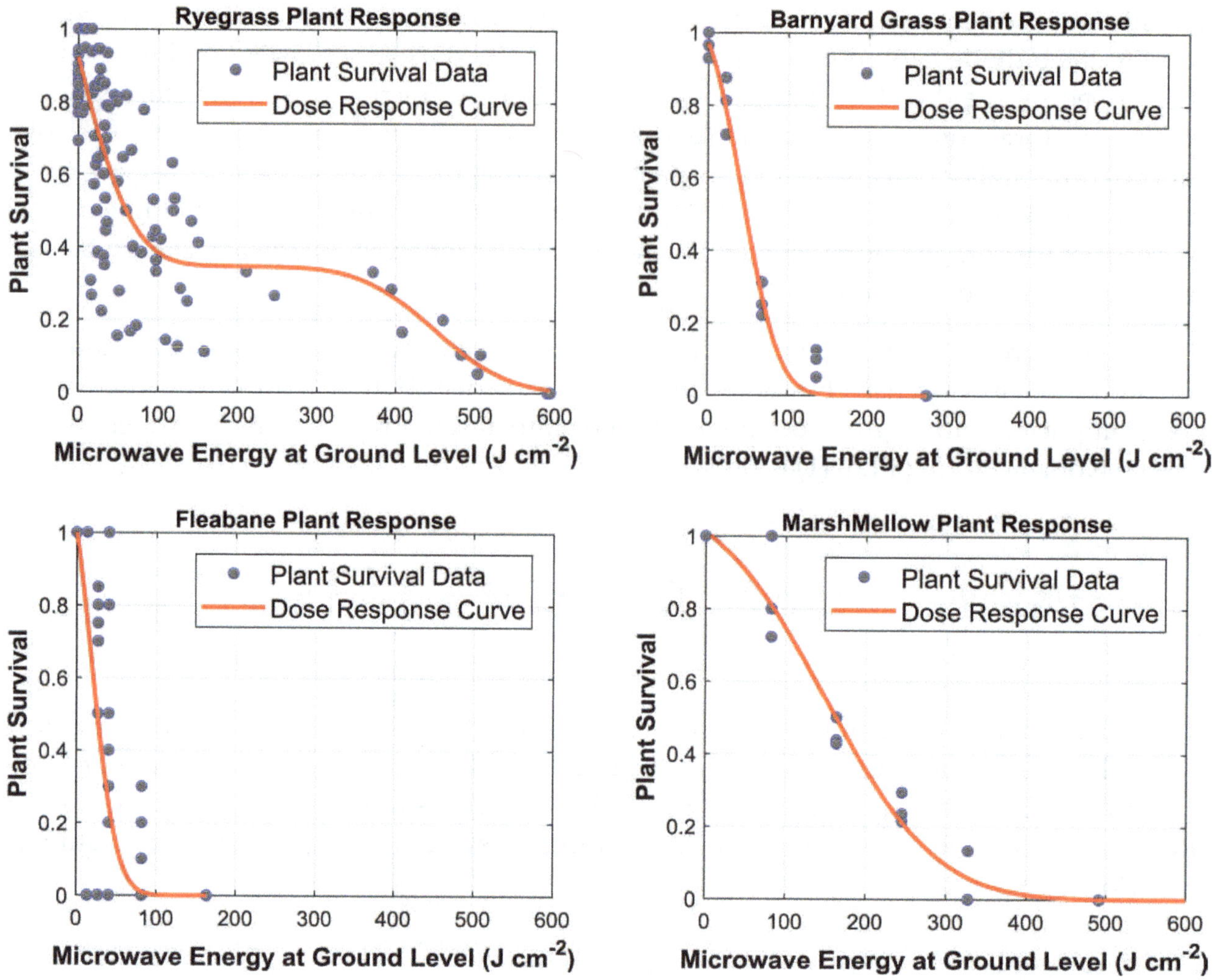

Figure 4. Dose-response curves for microwave treatment of four herbicides resistance weed species using a horn antenna. Source: Khan et al. [25].

of microwave treatment and soil depth on soil nutrient status and pH. They found that microwave treatment of soil had no significant effect on nitrogen, phosphorus, potassium and sulphate concentrations in all the treatment combination, but they reported an increase in nitrite concentration after 120 s of microwave treatment of soil. The nitrate reduction in the irradiated soil could be the principle cause of nitrite formation [83], in their study.

Speir et al. [84] examined the effect of microwave energy on low fertility soil (100 randomly selected cores at depth of 50 mm), microbial biomass, N, phosphorus and phosphatase activity. They reported that an increase in microwave treatment duration (90 s) dramatically increased the N level (106 μg N g^{-1} soils) but the phosphorus concentration declined as treatment time increased. The higher flush in soil N is of microbial origin as microwave has a biocidal effect [85, 86]. The fixation of NH_4^+ or K^+ in soil by inorganic colloids has been well documented [87]. Kittrick [88] hypothesized that the ion fixation in the clay lattice could be described by the expanding and contracting forces in the interlayer position. The contraction is due to electrostatic force of attraction between negatively charged clay mineral and positively charged ions and ion hydration causes the expansion. Fixation occurred when the force of attraction

Parameter	Meaning
n	Scaling factor for simultaneous heat and moisture movement [80]
ω	Angular frequency of electromagnetic wave (rad s^{-1})
ε_o	Permittivity of free space
κ''	Dielectric loss factor
τ	Transmission coefficient of the soil surface
E	Electric field strength (V m^{-1})
γ	Combined diffusivity for simultaneous heat and moisture transfer
α	Field attenuation factor in the soil (m^{-1})
t	Time (s)
A	Width of antenna aperture (m)
B	Height of antenna aperture (m)
Ro	Length of antenna (m)
k	Thermal conductivity of the composite material (W m^{-1} $°C^{-1}$)
x, y, z	Cartesian coordinates of a point in front of the horn antenna (m)
x′, y′	Cartesian coordinates of a point in the aperture of the antenna (m)

Table 3. Nomenclature of mathematical terms.

dominated the cations' hydration energy. Zagal [89] pointed out that the mechanical effect induced by microwave irradiation can stimulate the dispersion of inorganic colloids. This stimulation can increase the decomposition of non-biomass organic matter in soil and release the fixed $NH4^+$. Yang et al. [90] tested the nutrient extractability effect of microwave energy on soil. When fresh soil was exposed to microwave energy, a dramatic increase in the NH_4^+-N concentration was observed for an extended treatment of 120 s. They concluded that this effect was partially from nonmicrobial processes, either from site exchange or from fixed position in inorganic collides (clay minerals).

Alphei and Scheu [91] evaluated the effects of various biocidal treatments on mull-structured soil biota and nutritional dynamics. They reported the survival of soil microorganisms; in particular, higher concentration of ammonium nitrogen and phosphorus was observed when soil was subjected to microwave treatment at high power. The increase in soil C and N mineralization [89] and NH_4^+-N and sulphur oxidation was reported by Wainwright et al. [92]. In contrast, numerous studies documented that the effect of ionizing irradiation (γ-rays) on soil effectively increased the mineralization of NH_4^+-N [93, 94]. They proposed three possible pathways which may be responsible for the release of NH_4^+-N from soil through irradiation: (1) ammonia could be produced by the chemical action of ionizing radiation through a variety of biochemical processes from nitrogenous organic compounds, particularly deamination of amino acid [95] and proteins, (2) several enzymes were functional in irradiated soil including urease, which is active during decomposition and produces ammonia and (3) release of N from dead organisms due to subsequent cell lysis by irradiation [96].

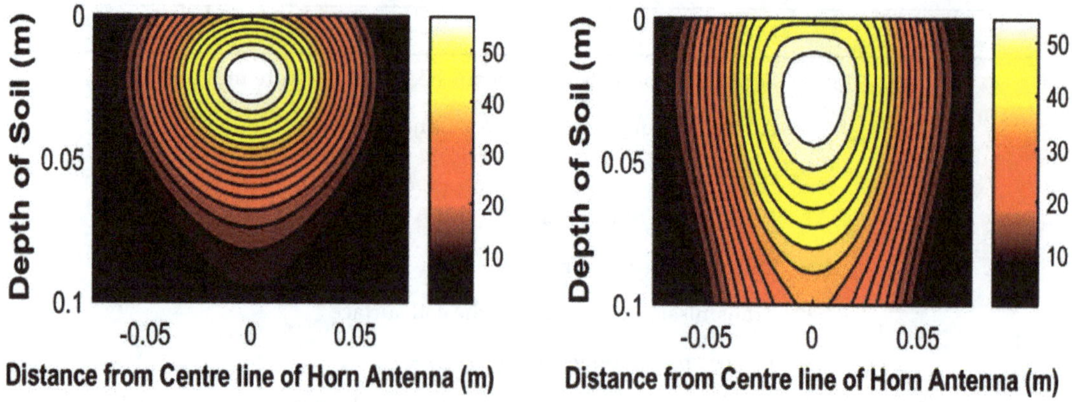

Figure 5. Comparison of expected soil temperature profile with measured soil temperature profile (left) and for the 750 W prototype microwave unit after 150 s of heating. Adapted from [56].

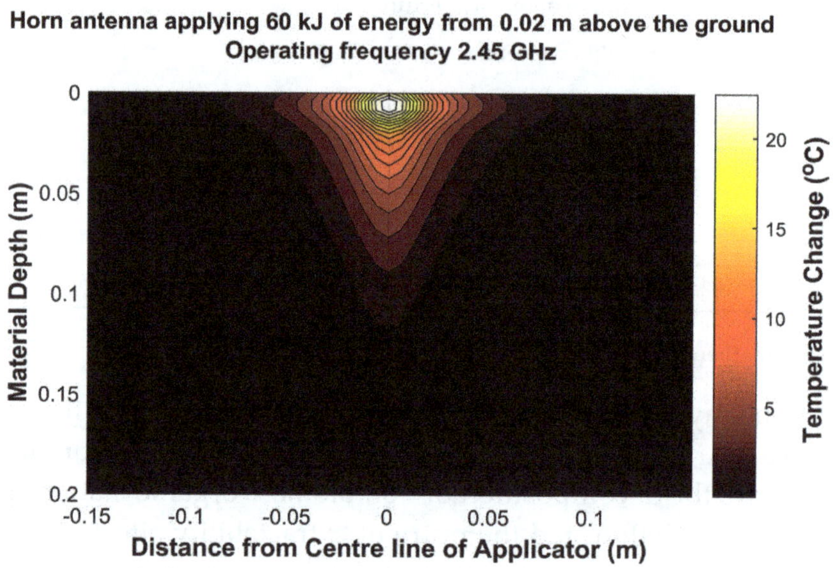

Figure 6. Estimated change in soil temperature treated from the 2 kW microwave system after 30 s with horn antenna at 2 cm above the soil. Adapted from [56].

Soil organic matter (SOM) is an aggregate of organic residues in soil at different degrees of humification [97]. Various biopolymers are serially transferred to humus (fulvic acid, humic acid and humin) in soil through geological SOM development processes such as humification [98]. Protein is the basic structural component of cell and cellular enzymes [99]. Approximately 5–25% of organic inputs are expected to accumulate in soil as proteins, peptides and free amino acids [100]. Amino acids typically incorporate about 10–20% of soil organic carbon and 30–40% of soil org-N [100]. Thermal denaturation of biopolymers induced by microwave irradiation could increase the concentrations of free amino acids for succeeding turnover to CO_2 and ammonia pool NH_4^+. Hur et al. [101] demonstrated that microwave irradiation of soil can enhance the binding efficiency of hydrophobic organic containments with SOM. They irradiated 5 g samples of soil in plastic tubes in aerobic and anaerobic conditions with activated C for 600 s in a lab-scale microwave oven (2.45 GHz) operated at 700 W. They pointed out that MW irradiation significantly alters the physical and chemical properties of SOM and increased its humification. Kim and Kim [24] studied the influences of microwave irradiation on the SOM properties. They reported that thermal cracking induced by irradiation scenario potentially alters the molecular composition (C, H, O and N), chemical structure and humification of SOM. The results of these studies suggest that microwave soil heating has potential to maximize the crop yield.

3.3.2. *Enzyme activity as a function of microwave soil heating*

Enzymes are essential to ecosystem processes because they arbitrate innumerable reactions that have biogeochemical importance in soil [102]. It has been demonstrated that *in vitro* exposure of microbial cells to microwave energy increased cell membrane permeability [103], released DNA and protein [104], soluble carbohydrate concentration [105] and inhibited growth of cells [106]. The enzymatic activity, selectivity and stability could be improved through high-frequency electromagnetic energy in an aqueous medium [107]. d'Ambrosio et al. [108] found that acid phosphatase was highly stable to the microwave deactivation energy of 280 mW g^{-1}. The hydration state and polarity of the reaction medium directly influenced the enzyme functionality under microwave irradiation. Notably, Carrillo-Munoz et al. [109] performed two lipases esterification reactions in a mono-mode microwave system at temperature of 100°C. They found that a 2–9% higher yield and 2.1–2.5-fold increment increase in protease activity [110] were obtained in microwave conditions compared to conventional heating in the hydration state. Furthermore, Yadav and Lahi [111] investigated the influence of microwave on lipase activity in a highly polar solvent and concluded that microwave noticeably accelerated the enzymatic reaction with an increase in hydrophobicity. Pirogova et al. [112] tested the effect of low frequency microwave energy, in the range of 500–900 MHz and at various power levels (1, 0.1 and 0.01 μW) on the activity of l-lactate dehydrogenase in solution for 300 s. They found a 73% increase in the bioactivity of the studied enzyme in microwave-irradiated samples compared to nonirradiated samples. Asadi et al. [113] tested the physiology of cyanobacterium (*Schizothrix mexicana*) against low power microwave modulation of various frequencies; they found that 9.685 GHz significantly increased growth metabolisms. Dreyfuss and Chipley [114] documented that metabolic enzyme activity of *S. aureus* increased after microwave irradiation. The cell biopolymer excitation induced by MW exposure was suggested to alter the enzymes' functionality.

Kothari et al. [115] studied the effect of low-power microwave on protease and urease activity of nine microorganisms (Bacteria, yeast and fungi). They treated enzyme cultures for different durations (0, 120, 240 and 360 s) in a microwave oven and concluded that the significant increase in the enzymes' activity was an athermal effect of microwave energy on the metabolism of the organisms. Numerous previous studies have shown higher enzymatic activity of industrial importance as a function of microwave in various reaction media at a temperature range of 70–110°C [107] and soil enzymes that are resistant to denaturation stress by heat [116, 117]. In contrast, Yeargers et al. [118] investigated the effect of microwave and conventional heating methods on the sensitivity of two enzyme (lysozyme and trypsin) solutions and found no discernible difference in enzyme activity, but the lysozyme was slightly more heat resistant than trypsin. Elzobair et al. [119] reported that microwave energy of 800 J g^{-1} of soil decreased (<10 nmolg^{-1} h^{-1}) dehydrogenase enzyme activity but 3200 J g^{-1} increased (>20 nmolg^{-1} h^{-1}) its functionality.

3.3.3. *Influence of microwave soil heating on soil microbes*

Soil biota is known to survive under severe physio-chemical environmental changes [120–122]. Microwave heating of soil can eradicate soil-borne fungi with minimal reduction of prokaryotic organisms [123]. Microbial cell response to microwave irradiation depends on the location, power density, time, frequency, pulses and physiology of cells. The nitrogen fixing bacteria persist, even after relatively high energy dosages. Vela and Wyss conducted a microwave heating experiment on soil *Azotobacter* and found that they survived microwave exposure of 480 s in very moist soil while, they were inactivated after only 20 s of treatment in laboratory culture conditions. Vela et al. [124] found that soil-nitrifying bacteria were highly resistant to microwave energy applied at the rate 40,000 J cm^{-2} to the soil surface. However, nitrifiers (mesophilic) are much more sensitive to high temperature than ammonifying (thermophilic) bacteria. This implies that native habitat and intrinsic environment are the most important factors in resistance of soil organisms to microwave irradiation [125]. Soil bacterial communities are resistant to microwave energy; some scientists concluded that the soil shelters microflora, while others discovered that the rate of proliferation causes resistance. This rate of proliferation is determined by nutrient concentrations. The heat-shock activation of the soil bacterial community was reported by Vela et al. [124]. Bacteria can form various thermal-resistance structure (i.e., spore and cysts), which keep them resistant against harmful effects of physical environments [126]. Based on work done by Hollins [127], she reported that a sharp reduction in colony forming unit of *E. colie* with 10 s of treatment of 2.3 kg soil (**Figure 7**), treated through 2 kW microwave system under horn antenna and complete soil sterilization was achieved through 120 s of irradiation.

3.4. Effect on crop growth

Rice productivity is strongly influenced through weed management strategies. Recently, a field experiment was conducted to evaluate the effect of pre-cropping microwave soil heating for weed seedbank depletion in direct-seeded rice crop based on the above soil heating methodology [16]. In addition to weed suppression (**Table 1**), the application of microwave energy (2.45 GHz; 120 s; 560 J cm^{-2}) into soil significantly (P = 0.05; **Table 4**) increased the tiller density (419 m^{-2}), dry biomass yield (27.8 t ha^{-1}) and grain yield (9.0 t ha^{-1}) of rice, compared to the untreated control scenario 292 m^{-2}, 22.8 t ha^{-1} and 6.7 t ha^{-1}, respectively. These results are strongly supported with findings of Brodie [128], who found that in pot trial maximum rice grain yield was attained with energy application of 600 J cm^{-2} to soil before crop sowing. The higher crop productivity could be attributed to 70–80% reduction in weed establishment achieved through microwave irradiation of soil, ultimately leaving more room for crop growth. Thermal devitalisation of weed seedbanks in the vertical soil profile may be the possible cause of minimum weed interference with the rice crop. This was evidenced by Vidotto et al. [129] who explored the effectiveness of high temperature on seed viability of six weed species including *Echinochloa crus-galli*: the problematic weed of rice growing regions globally. They stated that 80–100% germination reduction was achieved through raising the soil temperature to 79.6°C. The same temperature regime (70–80°C) that was acquired by microwave irradiation of the soil in the present study. This effectively induced an inhibitory effect on the weed population and therefore increased the rice crop yield.

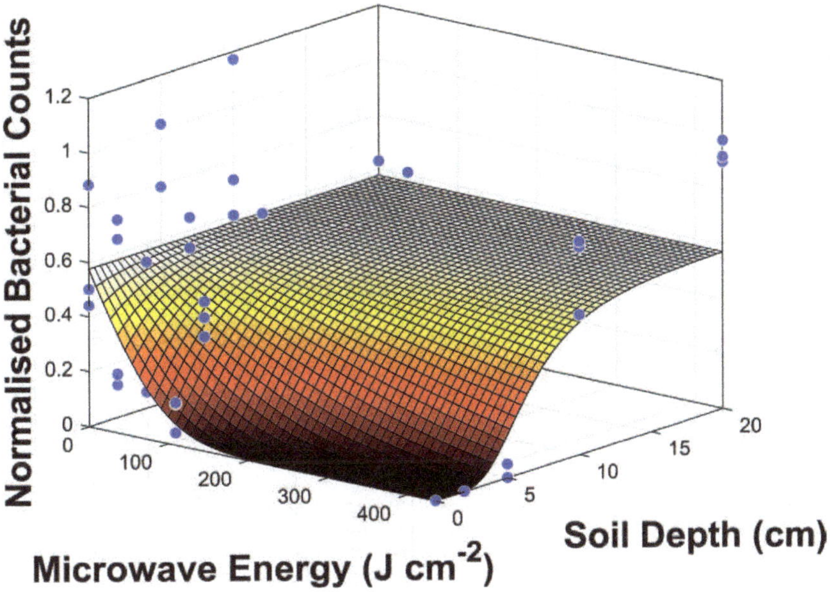

Figure 7. Assessment of E. coli survival in top 2 cm of soil as a function of applied microwave energy (Source: [127]).

For further validation of this yield changing effect with microwave soil heating, two field trials were conducted during October, 2016 to April, 2017 in a randomized complete block design with five replications at two different locations. The first location Dookie Campus of the University of Melbourne (36.395°S, 145.703°E) is a central grain growing region of the Goulburn Valley, which is in north of the state of Victoria, Australia; part of this region grows temperate rice. This region has a temperate climate with an average annual rainfall of 575 mm and an average monthly temperature range of 9.4–20.9°C (Australian Bureau of Meteorology). Soil at this experimental site is medium clay and classified as an Upotipotpon Clay [130] or an Orthic Basic Rudosol [131]. Historically, the same paddock has since been used for sheep grazing and highly invaded with a numerous grass species. The second location Old Coree, Jerilderie, New South Vales (35.210 °S, 145.440 °E) is the rice research farm a totally owned property of the Rice Research Australia Pty. Ltd. – SunRice™. Soil was treated using a prototype 2 kW microwave system, it has four independently controlled, 2 kW microwave generators operating at 2.45 GHz. The trailer is powered from two on-board 7 kVA, three-phase electrical generators [25]. Treatment was applied for 60 s and the temperature achieved through microwave energy application into soil was about 70–75°C in top soil layer (0–5 cm) at both study locations. Brodie reported that the microwave energy application to soil of about 400–500 J cm^{-2} gave 1.2–1.5 t extra grain yield compared to untreated control soil (**Figure 8**). The same range of microwave energy has used to treat the soil in the above field experiments. Therefore, the microwave soil treatment for pre-emergence weed suppression gave substantial increase in rice crop yield at both study location (**Table 4**; **Figures 9** and **10**). This is an additional benefit of soil heating through MW energy; we assumed that temperature has influenced on the soil nutrient profile particularly nitrogen.

3.5. Evaluation of rice crop production potential

Sustainable production of rice crop is the present need of the agriculture sector to fulfil increasing demand. In general, herbicide resistance, lower water use efficiency and

Rice parameters	Dookie location 1					Jerilderie location 2				
	Treatments		LSD $_{(p = 0.05)}$	P-value	Percentage change	Treatments		LSD $_{(p = 0.05)}$	P-Value	Percentage change
	Microwave treated	Untreated control				Microwave treated	Untreated control			
Number of tillers (m^{-2})	387 a	268 b	62.0		44.4%	480 a	418 a	145.2	0.29	44.4%
Dry biomass weight (t ha^{-1})	16.90 a	14.0 a	4.2	0.14	20.7%	19.80 a	17.05 b	0.57	<0.001	16.1%
Grain yield (t ha^{-1})	3.88 a	2.56 a	1.76	0.12	51.5%	9.21 a	7.63 b	0.65	<0.001	20.7%
Harvest Index	22.3 a	17.4 a	8.08	0.19	6.1%	46.77 a	44.59 a	4.29	0.26	6.1%
¥Water use efficiency (t Ml^{-1})	1.3	0.85	—	—	—	3.07	2.54	—	—	—
ᵞPartial factor productivity of nitrogen (kg rice grain per kg application of N)	31.04	20.42	—	—	—	73.68	61.04	—	—	—

Note: different letters in a row reflecting a significant difference at 5% probability level.Note: different letters in a row reflecting a significant difference at 5% probability level.

ᵞPartial factor productivity of nitrogen (PFP) $= \frac{Y_0}{N_r} + \frac{\Delta Y}{N_r} = \frac{Y_0}{N_r} + \frac{\Delta Y}{N_r}$, change in crop yield with nitrogen application was calculated based on work done by [132]. Note: Applied nitrogen during cropping period was 125 kg N ha^{-1} at both study locations.

¥Water use efficiency was calculated based on the change in grain yield per unit application of water. Note: Irrigation water volume was about 3 Ml ha^{-1} as per recommendation of Ricegrowers Association of Australia.

Table 4. Influence of pre-sowing microwave soil heating for weed seedbank depletion on rice productivity at two different agro-ecological zones of the Australia.

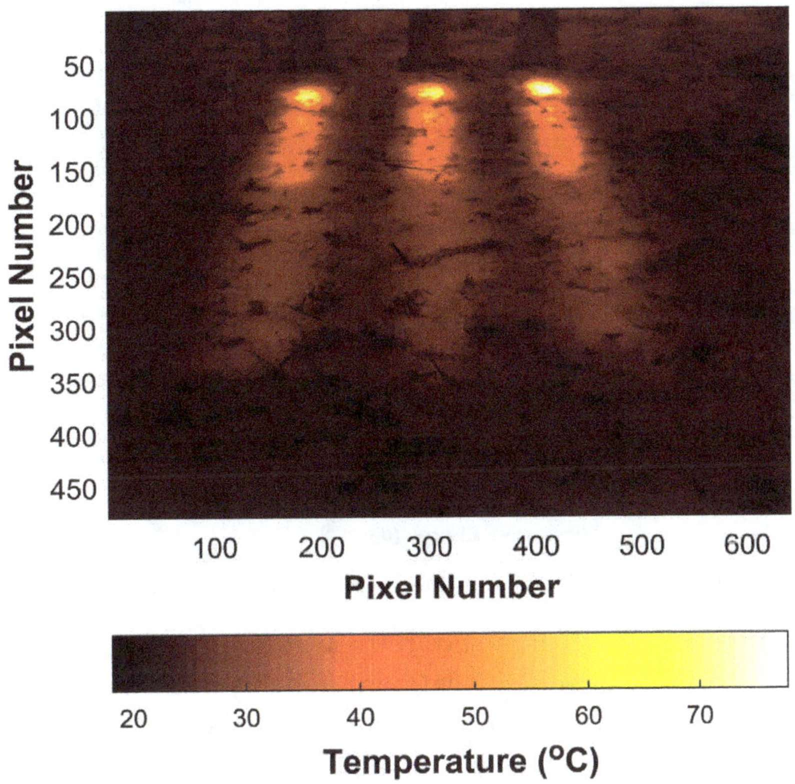

Figure 8. Infrared thermal images of microwave treated plot for weed seedbank depletion in rice crop under field conditions.

Figure 9. Comparison of early growth establishment of rice crop. Plants on left collected from microwave treated plot and plants on right collected from untreated control plot. (Left image taken from Dookie Trial Site and right image taken from Old Coree, Jerilderie site).

nitrogen use efficiency are key sustainability limiting factors, globally. For herbicide resistant weed suppression, Khan et al. [16] compared a microwave energy cost in rice crop with pre-sowing soil fumigation [133, 134] and reported that in terms of fuel

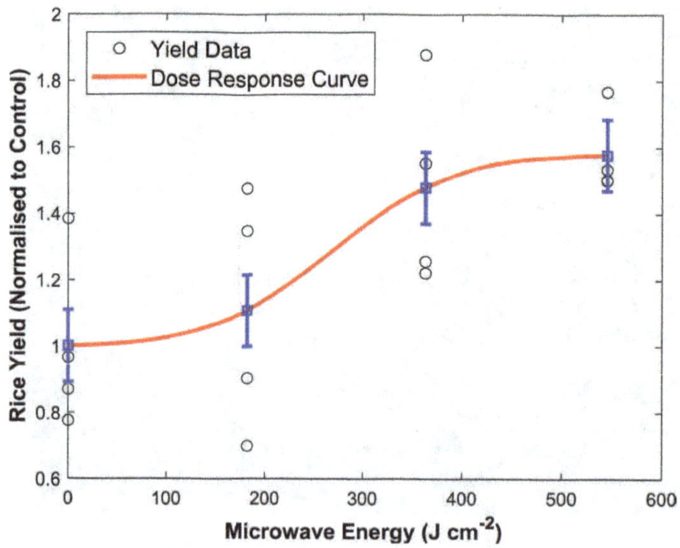

Figure 10. Relative increase in rice grain yield as a function of applied microwave energy. Source: [128].

consumption, the microwave system used in their study was quite comparable or better than soil fumigation and soil steaming treatment done by Samtani et al. [135]. Higher rice crop productivity without soil nutrient depletion has been confirmed with microwave soil heating methodology with an average of 20–50% increase under field conditions (**Table 4**). The microwave soil heating did not significantly alter the grain mineral concentration of rice (**Figure 11**), which suggests that higher yield producing crops effectively utilize the yield-changing nutrients from the soil. Based on this estimate, the profitability of rice production through this technology is better than conventional weed control technology. In other domain, however, soil health and persistence effects of the treatment for up to two growing seasons give an additional productivity advantage to rice farming community.

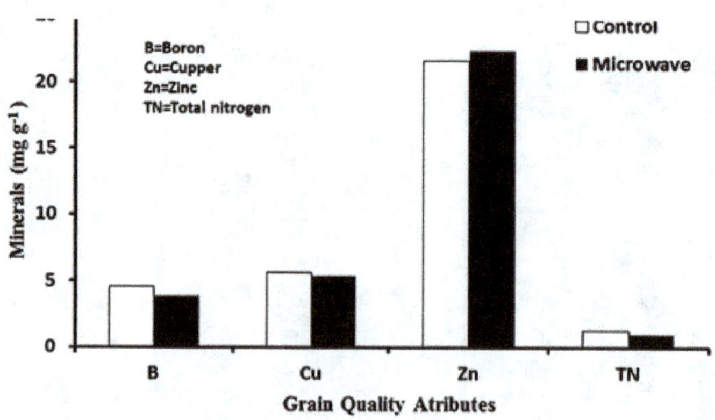

Figure 11. Effect of microwave soil heating on quality related parameter of rice grain. Adapted from [16].

4. Discussion

Weed seedbank is a resting place of dormant seeds in the top soil horizon. Various biotic and abiotic factors have a tremendous effect on seed viability. Among the many abiotic factors, however, soil temperature has an ability to debilitate weed seeds *in situ* [136, 137]. Therefore, it was hypothesized that the projection of non-ionizing energy into a top soil layer through a horn antenna may cause thermal devitalisation of weed seeds. The results of field studies have strongly supported our hypothesis and achieved about 70–80% reduction in rice weeds establishment. Overall, energy of a microwave system for weed management program has a direct relationship with application duration. Therefore, for a pre-emergence weed control under field conditions, Wayland et al. [44] reported an energy level of $80 < J\ cm^{-2} < 160$, which is quite low compared to the present investigation. In contrast, for a post-emergence weed control Sartorato et al. [54] tested the efficacy of microwave energy on seedling of *Abutilon theophrasti* and *Panicum miliaceum*. They reported that an energy range of $101.5 < J\ cm^{-2} < 343.3$ gave significant reduction in dry weight of about 90%. However, it is highly unlikely that certain set of MW energy may give a same control spectrum, because soil moisture [74] and seed geometry [38, 67] have a considerable influence on microwave absorption. These vary according to cropping system.

Independent of soil heating methodology for weed control; various studies also reported the profound effect of high temperature on weed establishment. Gay et al. [133] reported on a soil steaming experiment with various duration (0, 6, 8 and 10 min) in a soil, to depths of about 1.5–16.5 cm, giving a temperature gradient of 100–37°C (decreasing with depth), in a lettuce crop for weed control. They found an average weed density of less than 50 plants m^{-2} in the case of soil steam treated plots compared to untreated control plots (400 plants m^{-2}). Vidotto et al. [129] found that exposure of a soil-seed mixture to high temperature gradually decreased seed germination. Almost all the tested weed species seeds were completely devitalized through soil thermal treatment at a temperature between 70°C and 80°C.

The same temperature distribution was achieved through microwave application in the present study, which might have a degrading effect on the weed seedbank and ultimately led to a significant weed reduction. Therefore, based on previous findings and the results of this study, it may be possible to minimize the weed pressure through microwave irradiation of soil in no-till wheat production systems of Australia. However, a further research effort is needed to understand the long-term effects of microwave soil irradiation for weed control in crops. Furthermore, the fuel cost associated with a pre-sowing microwave weed management has been previously estimated by Khan et al. [16], therefore, about 0.98 L diesel m^{-2} were consumed in their experiment. Samtani et al. [135] calculated the fuel cost for pre-sowing steam treatment for weed control and reported a diesel consumption of between 0.81 and 2.16 L m^{-2}. Considering the fuel consumption, the MW system used in the present investigation for soil heating was comparable or even better than soil steaming used by Gay et al. [133] and Samtani et al. [135].

In addition to weed suppression, a few previous studies have reported the supplementary effect of microwave energy on soil nutrient dynamics; Yang et al. [90] tested the nutrient extractability effect of microwave on soil. When fresh soil was exposed to microwave energy a dramatic increase in the NH_4^+-N concentration was observed for an extended treatment of 120 s. They concluded that this effect was partially from nonmicrobial processes, either from site exchange or from fixed position in inorganic collides (clay minerals). Hur et al. [101] demonstrated that microwave irradiation of soil can enhance the binding efficiency of hydrophobic organic containments with soil organic matter. They irradiated 5 g samples of soil in plastic tubes in aerobic and anaerobic conditions with activated C for 600 s in a lab-scale microwave oven (2.45 GHz) operated at 700 W. They pointed out that MW irradiation significantly alters the physical and chemical properties of soil organic matter and increased its humification. In another study, Kim and Kim [24] studied the influences of microwave irradiation on the soil organic matter properties. They reported that thermal cracking induced by irradiation potentially alters the molecular composition (C, H, O and N), chemical structure and humification of soil organic matter. Based on these previous findings, we assumed that thermal denaturation of recalcitrant humic substance induced by microwave irradiation may increase the concentrations of free amino acids for succeeding turnover to CO_2 and ammonia pool NH_4^+, which might have substantially increased wheat productivity in the present investigation. Moreover, microwave soil heating gave 10 times higher nitrogen use efficiency and about 20–50% higher irrigation water use efficiency in those field experiment conducted to manage the herbicides resistance weeds.

5. Conclusion

Based on these experiments, we conclude that microwave weed and soil treatment can be implemented as an alternative method of weed control in direct-seeded rice crop. Additional benefit of this technology has prompted a motivation for further research in this area to enhance sustainability in agricultural industry.

Author details

Muhammad Jamal Khan and Graham Ian Brodie*

*Address all correspondence to: grahamb@unimelb.edu.au

Faculty of Veterinary and Agricultural Sciences, Dookie Campus, The University of Melbourne, Melbourne, Australia

References

[1] Mahajan G, Chauhan BS, Kumar V. Integrated weed management in rice. In: Recent Advances in Weed Management. New York: Springer; 2014. pp. 125-153

[2] ABS. Agricultural commodities, Australia. 2016. Accessed from: http://www.abs.gov.au

[3] Oerke EC. Crop losses to pests. The Journal of Agricultural Science. 2006;**144**:31-43

[4] Gianessi LP. The increasing importance of herbicides in worldwide crop production. Pest Management Science. 2013;**69**:1099-1105

[5] Rao AN, Johnson DE, Sivaprasad B, Ladha JK, Mortimer AM. Weed management in direct-seeded rice. Advances in Agronomy. 2007;**93**:153-255

[6] Norsworthy JK, Burgos NR, Scott RC, Smith KL. Consultant perspectives on weed management needs in Arkansas rice. Weed Technology. 2007;**21**:832-839

[7] Holm LG, Plucknett DL, Pancho JV, Herberger JP. The world's worst weeds: Distribution and biology. Honolulu, HI: University Press of Hawaii; 1991

[8] Maun MA, Barrett SCH. The biology of Canadian weeds. 77. *Echinochloa crus-galli* (L.) Beauv. Canadian Journal Plant Science. 1986;**66**:739-759

[9] Davis AS, Dixon PM, Liebman M. Cropping system effects on giant foxtail (Setaria faberi) demography: II. Retrospective perturbation analysis. Weed Science. 2003;**51**:930-939

[10] Bagavathiannan MV, Norsworthy JK, Smith KL, Neve P. Modeling the evolution of barnyardgrass (*Echinochloa crus-galli* L.) resistance to glyphosate in cotton and its management implications In proceedings of Beltwide Cotton Conference. 2011

[11] Burnside OC, Moomaw RS, Roeth FW, Wicks GA, Wilson RG. Weed seed demise in soil in weed-free corn (Zea mays) production across Nebraska. Weed Science. 1986;**34**:248-251

[12] Heap IM. International survey of herbicide resistant weeds. 2016. Accessed from: http://www.weedscience.org/

[13] Beckie HJ, Tardif FJ. Herbicide cross resistance in weeds. Crop Protection. 2012;**35**:15-28

[14] Beckie HJ, Warwick SI, Sauder CA. Basis for herbicide resistance in Canadian populations of wild oat (Avena fatua). Weed Science. 2012;**60**:10-18

[15] Preston C. The Australian Glyphosate Sustainability Working Group Accessed Tuesday 17 October. 2017. Available from: http://www.glyphosateresistance.org.au/

[16] Khan JM, Brodie G, Gupta D. The Effect of Microwave Soil Treatment on Rice Production under Field Conditions. Transactions of ASABE. 2017;**60**:517-525

[17] Bouman BA, Hengsdijk H, Hardy B, Bindraban PS, Tuong TP, Ladha JK. Water-wise Rice Production. Vol. 1. Los Banos: Philippians: International Rice Research Institute; 2002

[18] Yao Z, Zheng X, Liu C, Lin S, Zuo Q, Butterbach-Bahl K. Improving rice production sustainability by reducing water demand and greenhouse gas emissions with biodegradable films. Scientific Reports. 2017;**7**:39855

[19] Li YS. Evaluation of field soil moisture condition and the ways to improve crop water use efficiency in Weibei region. Journal of Agronomy Shananxi. 1982;**2**:1-8

[20] Hamblin AP, Tennant D. Root length density and water uptake in cereals and grain legumes: How well are they correlated. Australian Journal of Agricultural Research. 1987;**38**:513-527

[21] Zhang J, Sui X, Li B, Su B, Li J, Zhou D. An improved water-use efficiency for winter wheat grown under reduced irrigation. Field Crops Research. 1998;**59**:91-98

[22] Cassman KG, Pingali PL. Intensification of irrigated rice systems: Learning from the past to meet future challenges. GeoJournal. 1995;**35**:299-305

[23] Stevenson G, Tabart T. Tasmanian Organic Farm Monitoring Project (1995-1998). National Landcare Program Hobart: Tasmanian Organic-Dynamic Producers Cooperative Ltd; 1998

[24] Kim MC, Kim HS. Artificial and enhanced humification of soil organic matter using microwave irradiation. Environmental Science and Pollution Research. 2013;**20**:2362-2371

[25] Khan MJ, Brodie G, Gupta D, He J. Residue effect of microwave soil treatment on growth and development of wheat. In: Proceedings of International Microwave Power Institute Symposium. Miami, Florida: International Microwave Power Institute. 2017;**2017**:32-34

[26] Neeson R. Organic rice and soybean production: A guide to convert to organic production. Canberra, Australia: Rural Industries Research and Development Corporation; 2008

[27] Adamski W, Kitlinski M. On measurements applied in scientific researches of microwave heating processes. Measurement Science Review. 2001;**1**:199-203

[28] Metaxas AC, Meredith RJ. Industrial Microwave Heating. London: Peter Peregrinus; 1983

[29] Gallawa JC. The Complete Microwave Oven Service Handbook: Operation, Maintenance, Troubleshooting and Repair. Englewood Cliffs: Prentice Hall; 1998

[30] Liu B, Marchant TR, Turner IW, Vegh V. A comparison of semi-analytical and numerical solutions for the microwave heating of lossy material in a three-dimensional waveguide In proceedings of Third World Congress on Microwave and Radio Frequency Applications. 2003. 331-339

[31] Bond EJ, Li X, Hagness SC, Van Veen BD. Microwave imaging via space-time beamforming for early detection of breast cancer. IEEE Transaction on Antennas and Propagation. 2003;**51**:1690-1705

[32] Holman JP. Heat Transfer. New York: McGraw-Hill; 1997

[33] Crank J. The Mathematics of Diffusion. Bristol: J. W. Arrowsmith Ltd.; 1979

[34] Van Remmen HHJ, Ponne CT, Nijhuis HH, Bartels PV, Herkhof PJAM. Microwave heating distribution in slabs, spheres and cylinders with relation to food processing. Journal of Food Science. 1996;**61**:1105-1113

[35] Torgovnikov GI. Dielectric Properties of Wood and Wood-Based Materials. Berlin: Springer-Verlag; 1993

[36] Perre P, Turner IW. The use of numerical simulation as a cognitive tool for studying the microwave drying of softwood in an over-sized wave-guide. Wood Science and Technology. 1999;**33**:445-464

[37] Zhao H, Turner IW, Torgovnikov G. An experimental and numerical investigation of the microwave heating of wood. The Journal of Microwave Power and Electromagnetic Energy. 1998;**33**:121-133

[38] Brodie G. The influence of load geometry on temperature distribution during microwave heating. Transactions of the American Society of Agricultural and Biological Engineers. 2008;**51**:1401-1413

[39] International Telecommunication Union. Spectrum Management for a Converging World: Case Study on Australia, Geneva, Switzerland: International Telecommunication Union; 2004

[40] DAFF. Weeds. Australian Department of Agriculture. Canberra: Fisheries and Forestry; 2006

[41] Nelson SO, Walker ER. Effects of radio-frequency electrical seed treatment. Agricultural Engineering. 1961;**42**:688-691

[42] Davis FS, Wayland JR, Merkle MG. Ultrahigh-frequency electromagnetic fields for weed control: phytotoxicity and selectivity. Science. 1971;**173**:535-537

[43] Champ MA, Davis FS, Wayland JR. Ultrahigh-frequency electromagnetic radiation utilized for aquatic vegetation control. In: Proceedings of Annual Conference. 1973

[44] Wayland JR, Davis FS, Merkle MG. Toxicity of an UHF device to plant seeds in soil. Weed Science. 1973;**21**:161-162

[45] Menges RM, Wayland JR. UHF electromagnetic energy for weed control in vegetables. Weed Science. 1974;**22**:584-590

[46] Hightower NC, Burdette EC, Burns CP. Investigation of the Use of Microwave Energy for Weed Seed and Wood Products Insect Control. Final Technical report. Project E-230-901. Atlanta, Gerogia, U.S.A.: Georgia Institute of Technology; 1974

[47] Rice RP, Putnam AR. Some factors which influence the toxicity of UHF energy to weed seeds. Weed Science. 1977;**25**:179-183

[48] Bhartia P, Lebell M, Baron J. Weed control with electromagnetic fields. Journal of Microwave Power. 1977;**12**:52-54

[49] Bigu-Del-Blanco J, Bristow JM, Romero-Sierra C. Effects of low-level microwave radiation on germination and growth rate in corn seeds. Proceedings of the IEEE. 1977;**65**:1086-1088

[50] Crawford AE. Phytotoxicity Threshold Levels of Microwave Radiation for Trifolium and Medicago Seeds. Netherlands: Seed Science and Technology; 1977

[51] Diprose MF, Benson FA, Willis AJ. The effect of externally applied electrostatic fields, microwave radiation and electric currents on plants and other organisms, with special reference to weed control. The Botanical Review. 1984;**50**:171-223

[52] Lal R, Reed WB. The effect of microwave energy on germination and dormancy of wild oat seeds. Canadian Agricultural Engineering. 1980;**22**:85-88

[53] Barker AV, Craker LE. Inhibition of weed seed germination by microwaves. Agronomy Journal. 1991;**83**:302-305

[54] Sartorato I, Zanin G, Baldoin C, De Zanche C. Observations on the potential of microwaves for weed control. Weed Research. 2006;**46**:1-9

[55] Bebawi FF, Cooper AP, Brodie GI, Madigan BA, Vitelli JS, Worsley KJ, Davis KM. Effect of microwave radiation on seed mortality of rubber vine (*Cryptostegia grandiflora* R.Br.), parthenium (*Parthenium hysterophorous* L.) and bellyache bush (*Jatropha gossypiifolia* L.). Plant Protection Quarterly. 2007;**22**:136-142

[56] Brodie G, Hamilton S, Woodworth J. An assessment of microwave soil pasteurization for killing seeds and weeds. Plant Protection Quarterly. 2007;**22**:143-149

[57] Brodie G, Harris G, Pasma L, Travers A, Leyson D, Lancaster C, Woodworth J. Microwave soil heating for controlling ryegrass seed germination. Transactions of ASABE. 2009;**52**:295-302

[58] Brodie G, Ryan C, Lancaster C. The effect of microwave radiation on prickly paddy melon (*Cucumis myriocarpus*). International Journal of Agronomy. 2012:article 287608

[59] Brodie G, Hollins E. The effect of microwave treatment on ryegrass and wild radish plants and seeds. Global Journal of Agricultural Innovation Research & Development. 2015;**2**:16-24

[60] Wayland JR, Davis FS, Merkle MG. United States Patent No.: C. S. Phytox Corporation, Texas; 1978

[61] Heap IM. The occurrence of herbicide-resistant weeds worldwide. Pest Management Science. 1997;**51**:235-243

[62] Cox C. Glyphosate. Journal of Pesticide Reform. 2004;**24**:10-15

[63] Ayappa KG, Davis HT, Crapiste G, Davis EA, Gordon J. Microwave heating: an evaluation of power formulations. Chemical Engineering Science. 1991;**46**:1005-1016

[64] Whatley TL, Davis FS, Merkle MG. Effects of soil moisture on phytotoxicity of microwave fields. In: proceedings of Southern Weed Science Society. 1973

[65] Van Wambeke E, Wijsmans J, d'Hertefelt P. Possibilities in microwave application for growing substrate disinfestation In proceedings of II International Symposium on Soil Disinfestation. 1983. 209-218

[66] Benz W, Moosmann A, Walter H, Koch W. Wirkung einer Bodenbehandlung mit Mikrowellen auf Unkrautsamen, Nematoden und bodenburtige Microorganismen. Mitteil. Biologish. Bunoesanst. Land Forstvirtsch. Berlin-Dalhem. 1984;**223**:128

[67] Davis FS, Wayland JR, Merkle MG. Phytotoxicity of a UHF electromagnetic field. Nature. 1973;**241**:291-292

[68] Wolf WW, Vaughn CR, Harris R, Loper GM. Insect radar cross-section for aerial density measurement and target classification. Transactions of the American Society of Agricultural and Biological Engineers. 1993;**36**:949-954

[69] Nelson SO. Frequency, moisture, and density dependence of the dielectric properties of small grains and soybeans. Transactions of the American Society of Agricultural and Biological Engineers. 1987;**30**:1538-1541

[70] Wayland J, Merkle M, Davis F, Menges RM, Robinson R. Control of weeds with UHF electromagnetic fields. Weed Research. 1975;**15**:1-5

[71] Langner HR, Böttger H, Schmidt H. A special vegetation index for the weed detection in sensor based precision agriculture. Environmental Monitoring and Assessment. 2006;**117**:505-518

[72] Debye P. Polar Molecules. New York: Chemical Catalog; 1929

[73] Heddleson RA, Doores S. Factors affecting microwave heating of foods and microwave induced destruction of foodborne pathogens-a review. Journal of Food Protection. 1994;**57**:1025-1037

[74] Nelson SO. A review and assessment of microwave energy for soil treatment to control pests. Transactions of the ASAE. 1996;**39**:281-289

[75] O'Neill PE, Jackson TJ. Observed effects of soil organic matter content on the microwave emissivity of soils. Remote Sensing of Environment. 1990;**31**:175-182

[76] Von Hippel AR. Dielectric Materials and Applications. Cambridge: M.I.T. Press; 1954

[77] Brodie G. Simultaneous heat and moisture diffusion during microwave heating of moist wood. Applied Engineering in Agriculture. 2007;**23**:179-187

[78] Dobson MC, Ulaby FT, Hallikainen MT, El-Rayes MA. Microwave dielectric behavior of wet soil-Part II: Dielectric mixing models. IEEE Transactions on Geoscience and Remote Sensing. 1985;**GE-23**(1):35-46

[79] Hallikainen MT, Ulaby FT, Dobson MC, El-Rayes MA. Microwave dielectric behaviour of wet soil—Part II: Dielectric-mixing models. IEEE Transactions on Geoscience and Remote Sensing. 1985;**GE-23**:35-46

[80] Henry PSH. The diffusion of moisture and heat through textiles. Discussions of the Faraday Society. 1948;**3**:243-257

[81] Paul EA. Dynamics of organic matter in soils. Plant and Soil. 1984;**76**:275-285

[82] Cooper AP, Brodie G. The effect of microwave radiation and soil depth on soil pH, N, P, K, SO4 and bacterial colonies. Plant Protection Quarterly. 2009;**24**:67-70

[83] Cawse PA, Crawford DV. Accumulation of nitrite in fresh soils after gamma irradiation. Nature. 1967;**216**:1142-1143

[84] Speir TW, Cowling JC, Sparling GP, West AW, Corderoy DM. Effects of microwave radiation on the microbial biomass, phosphatase activity and levels of extracN and P in a low fertility soil under pasture. Soil Biology and Biochemistry. 1986;**18**:377-382

[85] Shamis Y, Patel S, Taube A, Morsi Y, Sbarski I, Shramkov Y, Croft RJ, Crawford RJ, Ivanova EP. A new sterilization technique of bovine pericardial biomaterial using micro- wave radiation. Tissue Engineering Part C: Methods. 2009;**15**:445-454

[86] Shamis Y, Taube A, Shramkov Y, Mitik-Dineva N, Vu B, Ivanova EP. Development of a microwave treatment technique for bacterial decontamination of raw meat. International Journal of Food Engineering. 2008;**4**:1556-3758

[87] Kissel DE, Cabrera ML, Paramasivam S. Ammonium, ammonia, and urea reactions in soils. Agronomy. 2008;**49**:101

[88] Kittrick JA. Forces involved in ion fixation by vermiculite. Soil Science Society of America Journal. 1966;**30**:801-803

[89] Zagal E. Effects of microwave radiation on carbon and nitrogen mineralization in soil. Soil Biology and Biochemistry. 1989;**21**:603-605

[90] Yang JE, Skogley EO, Schaff BE. Microwave radiation and incubation effects on resin-extractable nutrients: I. Nitrate, ammonium, and sulfur. Soil Science Society of America Journal. 1990;**54**:1639-1645

[91] Alphei J, Scheu S. Effects of biocidal treatments on biological and nutritional properties of a mull-structured woodland soil. Geoderma. 1993;**56**:435-448

[92] Wainwright M, Killham K, Diprose MF. Effects of 2450 MHz microwave radiation on nitrification, respiration and S-oxidation in soil. Soil Biology and Biochemistry. 1980;**12**:489-493

[93] Bowen HJM, Cawse PA. Some effects of gamma radiation on the composition of the soil solution and soil organic matter. Soil Science. 1964;**98**:358-361

[94] Cawse PA, White T. Accumulation of nitrate in fresh soils after gamma irradiation. The Journal of Agricultural Science. 1969;**72**:331-333

[95] Maxwell CR, Peterson DC, Sharpless NE. The effect of ionizing radiation on amino acids: I. The effect of X-rays on aqueous solutions of glycine. Radiation Research. 1954;**1**:530-545

[96] Eno CF, Popenoe H. The effect of gamma radiation on the availability of nitrogen and phosphorus in soil. Soil Science Society of America Journal. 1963;**27**:299-301

[97] Stevenson FJ. Humus Chemistry: Genesis, Composition. Reactions: Wiley; 1994

[98] Huang W, Weber WJ. A distributed reactivity model for sorption by soils and sediments: Relationships between desorption, hysteresis, and the chemical characteristics of organic domains. Environmental Science & Technology. 1997;**31**:2562-2569

[99] Madigan MT, Martinko JM, Parker J. Biology of Microorganisms. New Jersey: Prentice Hall, Inc.; 2003

[100] Stevenson FJ. Nitrogen in Agricultural Soils. Soil Science Society of America Inc., Crop Sciences Society of America Inc., and American Society of Agronomy Inc; 1982

[101] Hur J, Park S-W, Kim MC, Kim HS. Enhanced binding of hydrophobic organic contaminants by microwave assisted humification of soil organic matter. Chemosphere. 2013;**93**:2704-2710

[102] Dick RP, Burns RG. A brief history of soil enzymology research. In: Methods of soil enzymology. Dick RP, Editor. Madison, WI: Soil Science Society of America; 2011. pp. 1-34

[103] Barański S, Czerski P. Biological Effects of Microwaves. Dowden: Hutchinson & Ross; 1976

[104] Woo IS, Rhee IK, Park HD. Differential damage in bacterial cells by microwave radiation on the basis of cell wall structure. Applied and Environmental Microbiology. 2000;**66**:2243-2247

[105] Zhou BW, Shin SG, Hwang K, Ahn JH, Hwang S. Effect of microwave irradiation on cellular disintegration of Gram positive and negative cells. Applied Microbiology and Biotechnology. 2010;**87**:765-770

[106] Peters WJ, Jackson RW, Iwano K. Effect of controlled electromagnetic radiation on the growth of cells in tissue culture. Journal of Surgical Research. 1979;**27**:8-13

[107] Rejasse B, Lamare S, Legoy MD, Besson T. Influence of microwave irradiation on enzymatic properties: Applications in enzyme chemistry. Journal of Enzyme Inhibition and Medicinal Chemistry. 2007;**22**:519-527

[108] d'Ambrosio G, Massa R, Gianfreda L, Greco G Jr, Scaglione A. Influence of microwave radiation on acid phosphatase deactivation process. Alta Frequenza. 1989;**58**:361-364

[109] Carrillo-Munoz JR, Bouvet D, Guibé-Jampel E, Loupy A, Petit A. Microwave-promoted lipase-catalyzed reactions. Resolution of (±)-1-phenylethanol. The Journal of Organic Chemistry. 1996;**61**:7746-7749

[110] Roy I, Gupta MN. Non-thermal effects of microwaves on protease-catalyzed esterification and transesterification. Tetrahedron. 2003;**59**:5431-5436

[111] Yadav GD, Lathi PS. Synergism between microwave and enzyme catalysis in intensification of reactions and selectivities: transesterification of methyl acetoacetate with alcohols. Journal of Molecular Catalysis A: Chemical. 2004;**223**:51-56

[112] Pirogova E, Vojisavljevic V, Cosic I. Non-thermal effects of 500MHz-900MHz microwave radiation on enzyme kinetics In proceedings of Engineering in Medicine and Biology Society, 2008. 30th Annual International Conference of the IEEE 1052-1055. 2008

[113] Asadi A, Soltani N, Asadi A. Effect of various microwave frequencies on the physiology of a cyanobacterium, Schizothrix mexicana. Acta Physiologiae Plantarum. 2013; **35**:1367-1372

[114] Dreyfuss MS, Chipley JR. Comparison of effects of sublethal microwave radiation and conventional heating on the metabolic activity of Staphylococcus aureus. Applied and Environmental Microbiology. 1980;**39**:13-16

[115] Kothari V, Dholiya K, Patel D. Effect of Low Power Microwave on Microbial Growth and Metabolism. Grin Verlag OHG; 2013

[116] Miller M, Dick RP. Thermal stability and activities of soil enzymes as influenced by crop rotations. Soil Biology and Biochemistry. 1995;**27**:1161-1166

[117] Deng SP, Tabatabai MA. Effect of tillage and residue management on enzyme activities in soils: III. Phosphatases and arylsulfatase. Biology and Fertility of Soils. 1997;**24**:141-146

[118] Yeargers EK, Langley JB, Sheppard AP, Huddleston GK. Effects of microwave radiation on enzymes. Annals of the New York Academy of Sciences. 1975;**247**:301-304

[119] Elzobair KA, Stromberger ME, Ippolito JA. Stabilizing effect of biochar on soil extracellular enzymes after a denaturing stress. Chemosphere. 2016;**142**:114-119

[120] Jenkinson DS. Studies on the decomposition of plant material in soil. European Journal of Soil Science. 1966;**17**:280-302

[121] Powlson DS, Jenkinson DS. The effects of biocidal treatments on metabolism in soil—II. Gamma irradiation, autoclaving, air-drying and fumigation. Soil Biology and Biochemistry. 1976;**8**:179-188

[122] Huhta V, Wright DH, Coleman DC. Characteristics of defaunated soil. I: A comparison of three techniques applied to two different forest soils. Pedobiologia. 1989;**33**:417-426

[123] Ferriss RS. Effects of microwave oven treatment on microorganisms in soil. Phytopathology. 1984;**74**:121-126

[124] Vela GR, Wu JF, Smith D. Effect of 2450 Mhz microwave radiation on some soil microorganisms in situ. Soil Science. 1976;**121**:44-51

[125] Proctor BE, Goldblith SA. Food processing with ionizing radiations. Food Technology. 1951;**5**:376-380

[126] Socolofsky MD, Wyss O. Resistance of the Azotobacter cyst. Journal of Bacteriology. 1962;**84**:119-124

[127] Hollins E. Microwave Weed Control and the Soil Biota. Unpublished Honours thesis Melbourne School of Land and Environment University of Melbourne; 2013

[128] Brodie G. Derivation of a cropping system transfer function for weed management: Part 2. Microwave weed management. Global Journal of Agricultural Innovation Research & Development. 2016;**3**:1-9

[129] Vidotto F, De Palo F, Ferrero A. Effect of short-duration high temperatures on weed seed germination. Annals of Applied Biology. 2013;**163**:454-465

[130] Downes RG. A Soil, Land Use, and Erosion Survey of Parts of the Counties of Moira and Delatite, Victoria. Melbourne, Australia: CSIRO; 1949

[131] Isbell RF. The Australian Soil Classification. Collingwood: CSIRO Publishing; 1993

[132] Cassman KG, Peng S, Olk DC, Ladha JK, Reichardt W, Dobermann A, Singh U. Opportunities for increased nitrogen-use efficiency from improved resource management in irrigated rice systems. Field Crops Research.1998;**56**:7-39

[133] Gay P, Piccarolo P, Aimonino DR, Tortia C. A high efficacy steam soil disinfestation system, part II: Design and testing. Biosystems Engineering. 2010;**107**:194-201

[134] Gay P, Piccarolo P, Aimonino DR, Tortia C. A high efficiency steam soil disinfestation system, part I: Physical background and steam supply optimisation. Biosystems Engineering. 2010;**107**:74-85

[135] Samtani JB, Gilbert C, Weber JB, Subbarao KV, Goodhue RE, Fennimore SA. Effect of steam and solarization treatments on pest control, strawberry yield, and economic returns relative to methyl bromide fumigation. Hortscience. 2012;**47**:64-70

[136] Ascard J. Dose-response models for flame weeding in relation to plant size and density. Weed Research. 1994;**34**:377-385

[137] Kolberg RL, Wiles LJ. Effect of steam application on cropland weeds. Weed Technology. 2002;**16**:43-49

Permissions

All chapters in this book were first published in RCCD, by InTech Open; hereby published with permission under the Creative Commons Attribution License or equivalent. Every chapter published in this book has been scrutinized by our experts. Their significance has been extensively debated. The topics covered herein carry significant findings which will fuel the growth of the discipline. They may even be implemented as practical applications or may be referred to as a beginning point for another development.

The contributors of this book come from diverse backgrounds, making this book a truly international effort. This book will bring forth new frontiers with its revolutionizing research information and detailed analysis of the nascent developments around the world.

We would like to thank all the contributing authors for lending their expertise to make the book truly unique. They have played a crucial role in the development of this book. Without their invaluable contributions this book wouldn't have been possible. They have made vital efforts to compile up to date information on the varied aspects of this subject to make this book a valuable addition to the collection of many professionals and students.

This book was conceptualized with the vision of imparting up-to-date information and advanced data in this field. To ensure the same, a matchless editorial board was set up. Every individual on the board went through rigorous rounds of assessment to prove their worth. After which they invested a large part of their time researching and compiling the most relevant data for our readers.

The editorial board has been involved in producing this book since its inception. They have spent rigorous hours researching and exploring the diverse topics which have resulted in the successful publishing of this book. They have passed on their knowledge of decades through this book. To expedite this challenging task, the publisher supported the team at every step. A small team of assistant editors was also appointed to further simplify the editing procedure and attain best results for the readers.

Apart from the editorial board, the designing team has also invested a significant amount of their time in understanding the subject and creating the most relevant covers. They scrutinized every image to scout for the most suitable representation of the subject and create an appropriate cover for the book.

The publishing team has been an ardent support to the editorial, designing and production team. Their endless efforts to recruit the best for this project, has resulted in the accomplishment of this book. They are a veteran in the field of academics and their pool of knowledge is as vast as their experience in printing. Their expertise and guidance has proved useful at every step. Their uncompromising quality standards have made this book an exceptional effort. Their encouragement from time to time has been an inspiration for everyone.

The publisher and the editorial board hope that this book will prove to be a valuable piece of knowledge for researchers, students, practitioners and scholars across the globe.

List of Contributors

Savitha Palanisamy
Plant Breeding and Genetics, PGP College of Agricultural Sciences, Namakkal, Tamil Nadu, India

Muhammad Saeed
Department of Botany, Faculty of Life Sciences, Government College University, Faisalabad, Pakistan

D.M. Ruwani G. Mayakaduwa and Tara D. Silva
University of Colombo, Sri Lanka

Andrew De-Xian Kok, Low Lee Yoon and Lai Kok Song
Department of Cell and Molecular Biology, Faculty of Biotechnology and Biomolecular Sciences, Universiti Putra Malaysia, Selangor, Malaysia

Rogayah Sekeli and Wee Chien Yeong
Malaysian Agricultural Research and Development Institute (MARDI), Kuala Lumpur, Malaysia

Zetty Norhana Balia Yusof
Department of Biochemistry, Faculty of Biotechnology and Biomolecular Sciences, Universiti Putra Malaysia, Selangor, Malaysia

Prabhat Pramanik
Soils Department, Tocklai Tea Research Institute, Tea Research Association, Jorhat, Assam, India

Pil Joo Kim
Division of Applied Life Science, Gyeongsang National University, Jinju, South Korea

Ginigaddara Appuhamilage Sanjeewanie Ginigaddara and Sampath Priya Disanayake
Faculty of Agriculture, Rajarata University of Sri Lanka, Anuradhapura, Sri Lanka

Josefine Kant and Michelle Watt
Institute for Bio- and Geosciences, Plant Sciences (IBG-2), Forschungszentrum Juelich, Juelich, Germany

Takuma Ishizaki
Tropical Agricultural Research Front, Japan International Research Center for Agricultural Sciences (JIRCAS), Tsukuba, Japan

Juan Pariasca-Tanaka and Matthias Wissuwa
Crop, Livestock and Environment Division, Japan International Research Center for Agricultural Sciences (JIRCAS), Tsukuba, Japan

Terry Rose
Southern Cross Plant Science, Southern Cross University, Lismore, Australia

Anitha Mani and Kavitha Sankaranarayanan
Ion Channel Biology Laboratory, AU-KBC Research Centre, MIT Campus of Anna University, Chennai, Tamil Nadu, India

Lei Gao and Jie Xiong
College of Life Sciences, Zhejiang Sci-Tech University, Hangzhou, People's Republic of China

Shyama R. Weerakoon, Seneviratnage Somaratne, E. M. Sachini I. Ekanayaka and Sachithri Munasighe
Department of Botany, The Open University of Sri Lanka, Nawala, Nugegoda, Sri Lanka

Ananda Scherner, Fábio Schreiber, Matheus Bastos Martins and Andressa Pitol
Federal University of Pelotas, Pelotas, Brazil

André Andres and Germani Concenço
Embrapa Temperate Agriculture, Pelotas, Brazil

Muhammad Jamal Khan and Graham Ian Brodie
Faculty of Veterinary and Agricultural Sciences, Dookie Campus, The University of Melbourne, Melbourne, Australia

Index

A

Abiotic Stress, 9, 16, 46, 65, 105-106 108, 120, 124

Aerenchyma, 85-86, 88, 91-92, 100-101 103, 108, 117, 126

Androgenesis, 21-24, 27-30, 32-35

Association Mapping, 9, 11-12, 14 17-18, 20

B

Bioavailability, 36-37, 40, 51, 55, 99 103, 107, 142, 145, 149, 153, 155 157-158

Biofortification, 1-4, 8, 36-37, 40-44 46-55, 136, 142, 145, 148-149, 154-155 157-158

Biotechnology, 19-22, 32-34, 36-37, 48 50-52, 55, 70, 100, 102, 127, 150, 155 158, 220

C

Callus Induction, 21-22, 24, 26-29, 32-35

Chelation, 38-39, 110-111, 118-119

Coenzyme M, 56, 61-65, 70, 72

Cultivar, 12, 15, 33, 35, 41-43, 46, 51, 81 130, 146, 183-184

Culture Media, 21-22, 26-28, 32-34, 70

F

Ferritin, 41-45, 50, 52-54, 145-146, 155 158-159

G

Genetic Engineering, 3, 36-37, 41, 47 150, 163

Genotype, 4, 6, 15, 21-24, 26-27, 33-34 41, 89-90, 92-97, 146

Germination, 12-13, 15, 41, 93, 96, 98 106, 110, 123, 127, 154, 180, 184, 187 200-201, 208, 213, 217-218, 222

Germplasm, 3, 10-11, 14-15, 20, 96

Glufosinate, 160, 162-174, 176-177, 188

Glyphosate, 160, 162-172, 174, 176-177 181, 185-186, 188, 193, 195-197, 215 218

H

Heavy Metal Stress, 120-123, 125, 134

Herbicide, 160, 162-167, 170, 172, 174 176-186, 188-192, 194-195, 199, 209 211, 215, 218

Herbicide Resistance, 160, 163 165-167, 177, 179-181, 191, 194-195 199, 209, 215

Heritability, 2-6, 8, 12

Homeostasis, 20, 37, 44, 52-53, 108-110 116-117, 122-125, 127, 138, 149, 152 155

I

Inbred Line, 21-22, 31

Indica Rice, 14, 17, 21-25, 27-30, 32-35 53, 146, 159, 176

Iron Biofortification, 36-37, 40-44 47-50, 52-54, 145, 148, 155, 158

Iron Content, 37, 40-41, 44-46

Iron Deficiency, 2-4, 36-38, 40, 44-47 52, 134, 152, 154

Iron Uptake, 36-42, 44, 49-50, 52 109-110, 126, 128, 149, 151

J

Japonica Rice, 14, 20-24, 28, 30, 33, 146

M

Methanogenesis, 56, 59, 62-63, 65-66 70-71

Micronutrients, 1-4, 36, 38, 46, 49, 51 91, 110, 157, 159

Microspore, 23, 25-28, 30, 32, 34-35

Microwave Heating, 198-199, 202, 208 216, 218-219

Microwave Radiation, 199, 217-222

Microwave Soil Heating, 194, 197 207-209, 212, 214, 218

Molecular Mapping, 9-12, 15-16, 18-19

Molecular Mechanism, 110, 117

N

Nicotianamine, 44, 49-50, 52-54 110-111, 118, 127-128, 131-132, 138 146, 149-151, 153-155, 157-158

O

Oryza Sativa, 1, 4, 7-9, 17-20, 32-36, 42 83, 85, 102-104, 106, 125-127, 129-131 133, 149, 152, 154-157, 160, 162, 165 167, 177, 179, 185, 192, 194

P

Phenotype, 41, 85-86, 89

Phloem, 44, 53, 109, 111-112, 114, 118 128, 132, 139, 141-142, 151, 154

Phytosiderophore, 50, 52, 111-112 126-128, 151

Plant Regeneration, 21-22, 24, 27-30 32-35

Polished Grain, 36, 42-44, 46-47

R

Rhizosphere, 38-40, 44-45, 56, 87-88 90-92, 99-100, 102, 111, 137-141, 144 153, 156

Rice Anther Culture, 22-24, 26-32, 35

Rice Breeding, 8, 21-24, 33, 51, 84, 99 160, 168, 177
Rice Cultivar, 15, 41-43, 46, 81, 130, 146
Rice Genotype, 15, 89-90, 95-96
Rice Paddy Soil, 56-59, 62, 65-67
Rice Production, 1-2, 9, 15, 21, 58, 69 73, 80, 103, 143, 148, 161-162, 164 175-176, 179-180, 194-197, 212, 215
Root Hair, 85, 88, 90-98, 101-102, 104
Root System, 85-87, 89-90, 92, 102, 108 123, 143-144

S
Salinity Stress, 10-11, 15, 17
Salinity Tolerance, 9-10, 12, 14-15, 17 19-20
Seed Bank, 178-180, 183-187, 191 194-195, 197
Stress Tolerance, 9, 15-17, 21-22, 102 123, 134, 190

T
Traditional Rice Varieties, 73-82, 84 166-167
Transgene, 43-44, 46-47, 55, 97, 104, 176
Transgenic Rice, 36, 43-46, 50, 52 54-55, 86, 177
Translocation, 40, 42, 44-45, 53, 70 106-108, 111, 114-119, 122, 130, 132 136, 138-139, 141-142, 144-145, 147 149-152, 154, 157, 195

V
Vacuole, 115-116, 118-119, 122-123
Vascular Bundle, 110, 142

W
Water Use Efficiency, 194-195, 197 209, 214-215
Weed Control, 161-164, 178-179, 182 186, 192, 194-195, 198-200, 202 212-214, 217, 222
Weed Management, 164, 176-180, 184 189-191, 194-195, 197-199, 201-202 208, 213-215, 222
Weed Seedbank, 203, 208, 211-213
Weedy Rice, 164-165, 178-186, 191

X
Xylem, 115-118, 130, 134, 139-142, 151